Mending the Ozone Hole

Mending the Ozone Hole

Science, Technology, and Policy

Arjun Makhijani
Kevin R. Gurney

The MIT Press
Cambridge, Massachusetts
London, England

Set in Sabon by Asco Trade Typesetting Ltd., Hong Kong.
Printed and bound in the United States of America.

Library of Congress Cataloging-in-Publication Data

Makhijani, Arjun.
Mending the ozone hole : science, technology, and policy / Arjun Makhijani, Kevin R. Gurney.
p. cm.
Includes bibliographical references and index.
ISBN 0-262-13308-3
1. Ozone layer depletion. I. Gurney, Kevin R. II. Title.
QC879.7.M34 1995 95-11630
363.73'84—dc20 CIP

dedicated to Natasha, Shakuntala, and Julie

Contents

Preface

This book is the result of a six-year project on the causes and consequences of ozone depletion, the technologies available to replace ozone-depleting chemicals, and policies that would arrest and reverse ozone-layer damage.

Persuasive arguments, models, and evidence have indicated since 1974 that, at least as a protective measure, corporations and governments should have taken steps to limit the use of ozone-depleting compounds and prepared contingency plans for their quick elimination. Instead, a post-mortem approach to environmental problems prevailed—let the damage be done and the corpses appear before serious action is taken. As a result, action was scattered and insufficient compared to the need. Findings of a huge and unexpected ozone hole over the Antarctic, which had been appearing each southern-hemisphere (austral) spring since the late 1970s, were published in 1985. Since that time there has been a continual stream of negative news about the status of the ozone layer. The assessments have generally been far worse than those predicted by atmospheric models.

In 1987 the first attempts to end this reckless global experiment were taken by the international community through the adoption of a treaty known as the Montreal Protocol. The Montreal Protocol has been through two major revisions, known as the London Amendments (agreed to in June 1990) and the Copenhagen Amendments (agreed to in November 1992). Each revision, however, has been followed by news of worsening ozone-layer depletion and by suggestions that greater action would be necessary in order to reverse ozone-layer depletion.

Today, we again face the danger that action may not be strong enough.

Ozone levels have fallen dramatically since 1992 over populated mid-latitude locations. Ozone-hole-type deplction is suspected over the North Pole. Increases in UV-B radiation have been measured. Direct evidence of biological damage has been observed. Under current regulatory control, atmospheric chlorine and bromine will not even begin to decrease in the stratosphere until about 2005. Until then, we face further declines in stratospheric ozone, and we risk the possibility of nonlinear ozone loss, similar to that found over the poles, occurring over more populous locations.

A number of shortcomings exist in the current regulatory regime. Recovery of ozone-depleting chemicals from existing systems is far lower than it could be. The phaseout schedule for ozone-depleting compounds in Third World countries is far too lax, and the fund to assist those countries is inadequate to equitably meet their needs. Some large chemical manufacturers and some corporations with strong financial interests in the traditional refrigeration and air conditioning technologies have succeeded in extending the production of the partially halogenated compounds called HCFCs for another 30–40 years. These compounds, while less ozone-depleting than CFCs, contribute far more ozone-depleting chlorine in the short term and in the medium term than conventional calculations would indicate. Finally, the control of methyl bromide, a powerful ozone-depleting compound, is insufficient and has not fully addressed all the industrial sources. Methyl bromide and a similar ozone-depleting compound, methyl chloride, are also emitted in large quantities via anthropogenic burning of biomass in smoldering low-temperature fires. This issue is connected to the daily economic existence of hundreds of millions of people and to a variety of health and environmental problems.

No single step will stop the buildup of ozone-depleting chlorine and bromine in the atmosphere during the next few years. Many steps, each one important, are needed. The primary aim of our analysis is to provide a solid reference for those who want to be fully informed about the potential for eliminating ozone-depleting compounds and the possible consequences of failing to do so.

The book is divided into three parts. Part I deals with the mechanisms of ozone depletion and with its health, epidemiological, and ecological consequences. Part II discusses the sources of emissions of halogenated compounds, their uses, and alternative technologies. Part III discusses the policy background, assesses the impacts of various measures for protecting the ozone layer by developing three scenarios, and addresses the context in which policy is made. Part III also contains a summary of researchers' findings and our recommendations for protecting the ozone layer.

All data are in metric units unless otherwise specified. One tonne (metric ton) equals about 1.1 U.S. (short) tons of 2000 pounds. The term "ozone-depleting compounds" refers to the human-derived component of substances emitted into the atmosphere which are known to deplete stratospheric ozone. Unless specifically stated otherwise, this does not include the relatively small amounts of naturally emitted ozone-depleting compounds. Except where we specify otherwise, our statistics are for worldwide production and emissions.

This work builds on four previous Institute for Energy and Environmental Research reports: *Saving Our Skins: Technical Potential and Policies for the Elimination of Ozone-Depleting Chlorine Compounds* (1988), *Reducing Ozone-Depleting Chlorine and Bromine Accumulations in the Stratosphere: A Critique of the U.S. Environmental Protection Agency's Analysis and Recommendations* (1989), *Biomass Burning and Ozone Depletion: An Assessment of the Problem and Its Implications for the Protection of the Ozone Layer* (1990), and *Saving Our Skins: The Causes and Consequences of Ozone Layer Protection and Policies for Its Restoration and Protection* (1992). Annie Makhijani was a co-author of all of these reports. Amanda Bickel was a co-author of the studies dated 1988 and 1989. We are indebted to them for their contributions to these earlier works.

We owe special thanks to Friends of the Earth (FoE) for permitting us to use portions of the first two reports mentioned above, for which FoE and IEER hold joint copyrights. We would also like to thank Dr. Abha Sur for her assistance in research regarding ozone-depletion potentials and Dr. Sunder Mehta for his assistance with the sections on the immune system and the function of the skin. Helpful comments on specific chapters were received from Neil Donahue, Jim Jenal, Julia May, Dr. Pushpa Mehta, and Mary O'Brien. Our librarian Lois Chalmers was very helpful with literature searches, Diana Kohn in preparing the reference list, and Freda Hur in providing office support. Finally, thanks to Dave Kershner whose tireless efforts made possible an earlier version of this work in time to have some impact on the Copenhagen negotiations.

Funding for this work was provided by the C. S. Mott Foundation, the C.S. Fund, the H.K.H. Foundation, the Levinson Foundation, the Amelia Peabody Charitable Trust, the Peace Development Fund, a private contribution from Susan Clark, and general support grants from the Public Welfare Foundation and Rockefeller Financial Services (formerly Rockefeller Family Associates). In addition, funds for outreach have been provided by the Schumann Foundation.

The Term "Ozone Hole"

When depicting the concentration of a chemical compound in the atmosphere, scientists often generate maps onto which lines of equal concentration are traced. Low values of concentration appear as valleys, high values of concentration as hills or ridges. When such mapping was performed for a few years over the Antarctic, a deepening hole in ozone concentration was readily apparent. This has come to be referred to as the "ozone hole." Technically speaking, the seasonal depletion over the southern pole does not result in a complete lack of ozone over this portion of the planet; however, it does represent a severe thinning, and "ozone hole" is an apt description of this phenomenon when it is depicted visually.

I

Causes and Consequences of Stratospheric Ozone Depletion

1

The Scientific Understanding of Stratospheric Ozone Loss

Dramatic progress has been made over the last two decades toward understanding the scientific aspects of ozone depletion. Not only has the research aimed at uncovering some of the underlying physical principles associated with ozone depletion led to remarkable advances in our understanding of this problem; it has also brought significant advances in the general understanding of the global environment. As with many other global environmental problems, scientific progress has been achieved through multidisciplinary and interdisciplinary research drawing on a wide variety of scientific disciplines. Understanding the science of ozone depletion therefore requires an integration of various aspects of biology, chemistry, physics, and mathematics.

In this chapter we survey the currently accepted scientific understanding of ozone depletion and its implications. A much deeper understanding of specific aspects of ozone depletion can be attained from the many references cited throughout. An excellent place to start is with the World Meteorological Organization's reports, which collect scientific work on this problem performed by the leading researchers around the world.

1.1 Introduction to the Atmosphere

The Earth's early atmosphere was created by the volcanic expulsion of gases from the interior. These original compounds have been transformed by the presence of sunlight and the evolution of life. The oxygen in the atmosphere began to accumulate after the first living organisms capable of photosynthesis appeared in the oceans, where water screened out ultraviolet radiation. The creation of ozone molecules in the atmosphere was one result of the initial accumulation of oxygen. Gradually a stratospheric ozone layer formed. That ozone layer screened out lethal ultraviolet radiation, thus (many believe) allowing life to migrate from the oceans onto the land. As Levine (1991, p. xxvi) has noted: "Once on land, life flourished both in numbers and biological diversity. Plant and animal life soon became significant controllers and regulators of the composition of the atmosphere through the biogeochemical cycling of elements and the production and emission of atmospheric gases through respiration, nitrification, denitrification, and methanogenesis."

The present atmosphere is composed of a variety of substances. Table 1.1 lists most of the important constituents and their present concentrations. As the table shows, all but four of the important constituents exist in extremely small amounts and are generally referred to as **trace gases.** Ozone is one of these trace gases. The word "ozone" is derived from the Greek *ozein* ("to smell").

Table 1.1
Average composition of Earth's atmosphere below 100 km (based on Wallace and Hobbs 1977, p. 5, table 1.1, and Wayne 1992, p. 410). Note that local concentrations of water vapor vary considerably; the range is 0–4 percent.

Constituent	Molecular weight	Content[a]
Nitrogen (N_2)	28.0160	00.7808[a] (75.51% by mass)
Oxygen (O_2)	32.00	0.2095[a] (23.14% by mass)
Argon (Ar)	39.94	0.0093[a] (1.28% by mass)
Water vapor (H_2O)	18.02	0.01[a]
Carbon dioxide (CO_2)	44.01	355 ppmv[b]
Neon (Ne)	20.18	18 ppmv
Helium (He)	4.00	5 ppmv
Methane (CH_4)	16.00	1.66 ppmv
Krypton (Kr)	83.7	1 ppmv
Hydrogen (H_2)	2.02	0.5 ppmv
Ozone (O_3)	48.00	0–12 ppmv

a. Fraction of total molecules.
b. Parts per million by volume: number of molecules of the listed constituent per million molecules of background air.

The atmosphere is divided into a number of layers defined by the variation of temperature with altitude, as shown in figure 1.1. The lowest layer is the **troposphere**—derived from *tropo,* meaning "change." It extends from the surface to an altitude of approximately 12 kilometers. Like a pot of boiling water heated from below, the troposphere exhibits strong vertical mixing, of which weather systems Are a direct outcome. The level at which temperature ceases to decline with increasing altitude, marking the upper boundary of the troposphere, is called the **tropopause.** This level varies somewhat with latitude and season.

The layer directly above the troposphere, the **stratosphere** (derived from *stratum,* meaning "layer") extends from the top of the troposphere to approximately 50 km in altitude. The stratosphere, which comprises only 10 percent of the mass of the atmosphere, is characterized by increasing temperature with increasing altitude and little vertical mixing. Nevertheless, the stratosphere plays a crucial role in the energy balance of the atmospheric system.

When sunlight first enters the atmosphere, it is composed of radiation of various energy levels, including potentially harmful ultraviolet radiation. Most ultraviolet radiation, however, never reaches the surface. Its energy is absorbed in the upper atmosphere, mainly by oxygen and ozone. Figure 1.1 shows the distribution of ozone concentration with altitude. Approximately 90 percent of atmospheric ozone lies within the stratosphere. Although ozone is a toxic oxidizing agent and a constituent of smog near the surface, its presence in the stratosphere is essential for the protection of life. The toxicity of the ozone in the air we breathe derives from the fact that its oxygen atoms are more weakly bound than those of ordinary molecular oxygen. Ozone is therefore a far more powerful oxidizing agent, and it can cause lung irritation.

The maximum ozone concentration occurs in the middle of the stratosphere, in a region commonly referred to as the **ozone layer.** The atmosphere as a whole contains very little ozone. For example, if all the ozone above a particular spot on Earth were compressed to standard atmospheric pressure (the average pressure at sea level), it would be only 3 millimeters (a little over 0.1 inch) thick. For this reason, the amount of ozone is generally stated in **atmosphere-millimeters** (atm-mm) or **atmosphere-centimeters** (atm-cm). Since gas pressure in the atmosphere

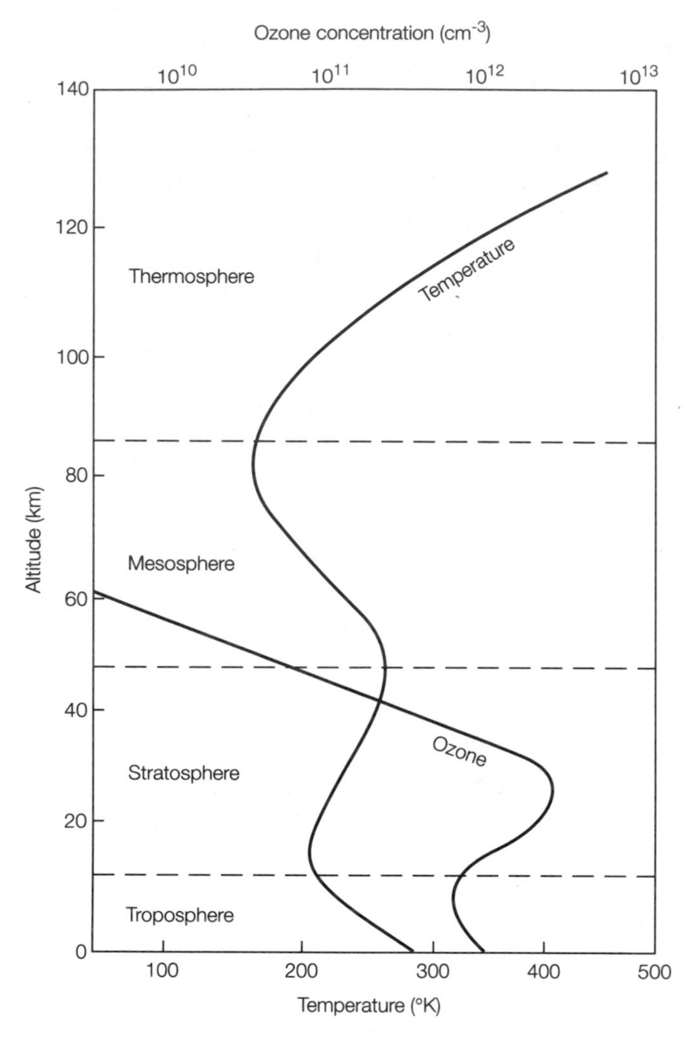

Figure 1.1
Temperature and ozone distribution in the atmosphere (NASA 1986, p. 20, figure 1).

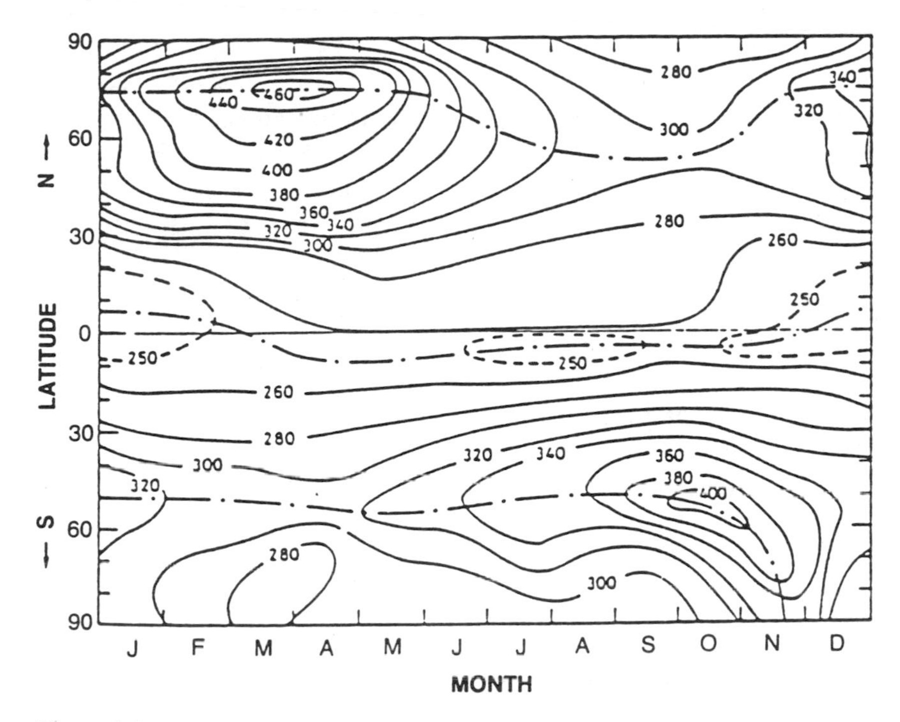

Figure 1.2
Distribution of column ozone, based on total column observations (Dobson units) prior to 1980s (WMO 1989, p. 5, figures 1.1.2 and 1.1.3).

decreases with increasing altitude, this quantity of ozone, known as **column ozone,** is spread thinly above the surface. The unit of measure used to represent the amount of ozone above a particular position on the surface is the **Dobson unit,** and one such unit is equal to 0.01 atm-mm.[1] Thus, a typical column of ozone is equal to approximately 300 Dobson units. The observed global distribution of total column ozone with latitude and season prior to the 1980s is shown in figure 1.2.

The lowest amounts of column ozone (excluding the seasonal depletion occurring at the poles) are found near the equator, independent of season. At higher latitudes, seasonal variations become much more noticeable, with a distinct springtime maximum in both hemispheres. This maximum is greater in the northern hemisphere and greatest near the North Pole. These regional and seasonal variations are due to varying rates of ozone formation and destruction and to transport of ozone from one region of the stratosphere to another. Transport between the stratosphere and the troposphere is small relative to the rates of formation and destruction of ozone in the stratosphere; however, north-south transport within the stratosphere is quite significant, and it is partially responsible for the greater amount of column ozone over mid- to high-latitude locations than near the equator.

Nonseasonal variations in transport and temperature also influence stratospheric ozone distribution. A phenomenon known as the **quasi-biennial oscillation** (QBO), which affects temperature and transport throughout the stratosphere, is thought to give rise to regular fluctuations in column ozone with a period of just over 2 years (Angell and Korshover 1964, 1967). Variations in solar output associated with the 11-year solar cycle also cause nonseasonal variations in stratospheric ozone, although the variation is small—about 2–3 percent over the entire cycle (Wayne 1991, p. 149).

1.2 The Natural Ozone Balance

Ozone is naturally produced and destroyed in the stratosphere. The resulting amount is a balance between these two processes. We refer to the concentration of ozone prior to the depletion observed within the last two decades as the **natural ozone balance**.

The sun emits a broad spectrum of radiation, only a portion of which is visible light. Low-energy solar radiation with longer wavelengths (and lower frequencies) than visible light is called **infrared radiation**; more energetic radiation with shorter wavelengths (and higher frequencies) than visible light is called **ultraviolet radiation**. Energy reaches Earth almost wholly as infrared and visible radiation. Radiant energy is carried in discrete units known as **photons**. Photons of shorter wavelength (higher frequency) carry more energy than photons of longer wavelength.[2]

If radiation possesses enough energy per photon, it can split molecules of gas—a process referred to as **photodissociation**. The radiant energy of the photon is absorbed in the atmosphere and converted into other forms of energy, such as chemical or thermal energy of gas molecules. High-energy photons in the solar radiation spectrum are absorbed in this way.

The rest of the sun's energy either is reflected back into space or passes through the atmosphere to the surface.

Visible light has wavelengths in the range 400–700 nanometers. (A nanometer, abbreviated nm, is a billionth of a meter.) This light either is reflected, is scattered, or passes through the stratosphere. The weak ultraviolet radiation known as UV-A, with wavelengths in the range 320–400 nm, also passes through the stratosphere and reaches the surface.

In contrast, strong ultraviolet radiation with wavelengths shorter than 320 nm has enough energy to dissociate, or break apart, certain gas molecules. Energy is absorbed in the upper atmosphere by this process. More specifically, ultraviolet radiation with wavelengths in the range 40–240 nm can photodissociate oxygen molecules. An oxygen molecule (O_2) consists of two relatively tightly bound atoms of oxygen. High energy photons photodissociate an oxygen molecule into two separate atoms of oxygen (O). The reaction is written

$$O_2 + hv(40\text{–}240\ \text{nm}) \rightarrow O + O, \tag{1.1}$$

where hv is the energy of a photon of frequency v. Some of these single atoms of oxygen then combine with oxygen molecules to form a molecule of ozone (O_3):

$$O_2 + O + M \rightarrow O_3 + M, \tag{1.2}$$

where M represents any molecule in proximity to the reaction that acts as a stabilizing force—typically either molecular nitrogen (N_2) or molecular oxygen. This is the primary process by which ozone is produced in the atmosphere.

Since the three atoms of oxygen in ozone are weakly bound relative to molecular oxygen, weaker ultraviolet radiation (with wavelengths in the range 240–320 nm) has enough energy to photodissociate ozone, but not molecular oxygen. The reaction can be depicted as

$$O_3 + hv(240\text{–}320\ \text{nm}) \rightarrow O_2 + O. \tag{1.3}$$

Although UV radiation is the principal radiation involved in the photodissociation of ozone molecules, a small but non-negligible amount of ozone is also photodissociated by visible light.

Thus, through a combination of absorption by oxygen and ozone molecules, essentially all the energy of ultraviolet radiation between 40 and 290 nm (called **UV-C**) is prevented from reaching the troposphere. Because more energy is absorbed in the higher reaches of the stratosphere than in the lower, we get the characteristic temperature profile in the stratosphere.

Ultraviolet radiation with wavelengths between 290 and 320 nm, known as **UV-B**, is progressively less effective at breaking up ozone molecules as the wavelength increases. Therefore, some UV-B radiation penetrates the stratosphere and reaches the surface. Ultraviolet radiation with wavelengths in the range 320–400 nm (UV-A) is transmitted through the stratosphere. Figure 1.3 illustrates this phenomenon. Because of the relatively low energy of the photons in this range, UV-A is not associated with as severe health effects as UV-B radiation. Indeed, humans require some UV-A—it helps form vitamin D.

Stratospheric ozone is constantly produced and destroyed in daylight as a result of the photochemical reactions 1.1–1.3. It is convenient to consider the sum of atomic oxygen and ozone, which are together referred to as **odd oxygen** (because they have an odd number of oxygen atoms in each molecule). By this description, reaction 1.1 produces odd oxygen; reactions 1.2 and 1.3 do not alter the total amount of odd oxygen but change the ratio of ozone to atomic oxygen.

In the absence of catalytic processes leading to ozone destruction, the reaction of two odd oxygen compounds places an upper limit on the overall amount of odd oxygen ($O + O_3$) in the atmosphere. The reaction produces molecular oxygen:

$$O_3 + O \rightarrow O_2 + O_2. \tag{1.4}$$

This reaction is quite slow, however. Chemical reaction sequences called **catalytic cycles** (involving molecules containing nitrogen, hydrogen, chlorine, and bromine) exert much greater control over stratospheric odd oxygen concentrations. The most important of these cycles and their net results (specified in the third line in each reaction sequence) can be represented as follows:

$$\begin{array}{l} OH + O_3 \rightarrow HO_2 + O_2 \\ \underline{HO_2 + O \rightarrow OH + O_2} \\ O_3 + O + OH \rightarrow O_2 + O_2 + OH \end{array} \tag{1.5}$$

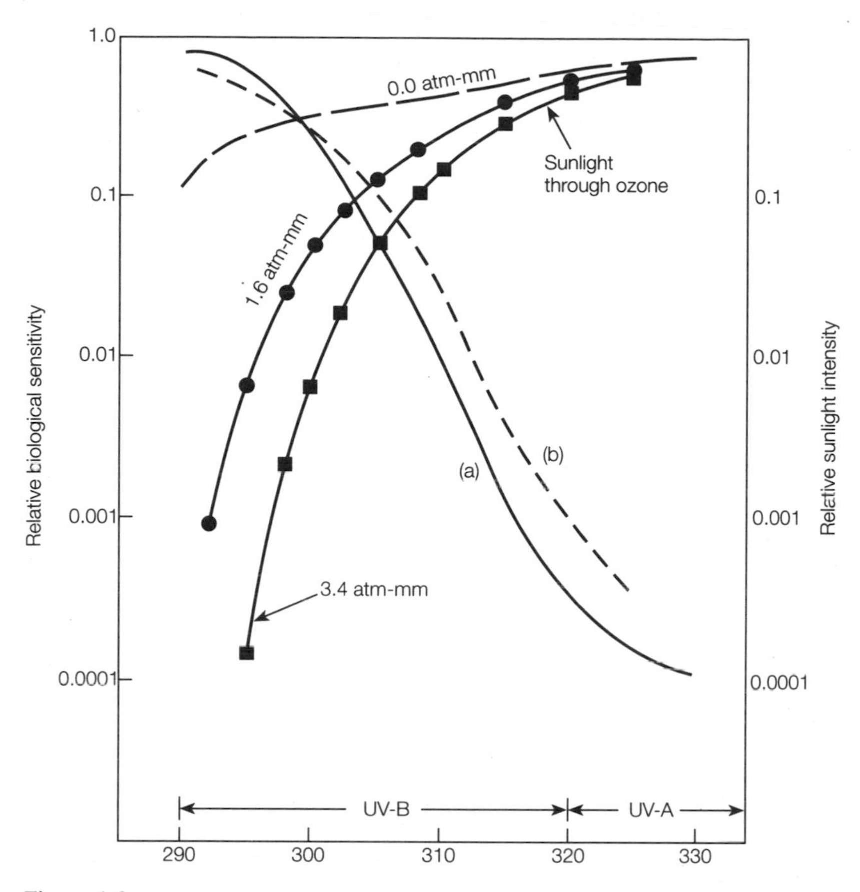

Figure 1.3
Screening of UV by various amounts of column ozone (NRC 1982, p. 40, figure 2.2).

$$OH + O_3 \rightarrow HO_2 + O_2$$
$$HO_2 + O \rightarrow OH + O_2$$

$$O_3 + O + NO \rightarrow O_2 + O_2 + OH \quad (1.6)$$

$$NO + O_3 \rightarrow NO_2 + O_2$$
$$NO_2 + O \rightarrow NO + O_2$$

$$O_3 + O + NO \rightarrow O_2 + O_2 + NO \quad (1.7)$$

$$Cl + O_3 \rightarrow ClO + O_2$$
$$ClO + O \rightarrow Cl + O_2$$

$$O_3 + O + Cl \rightarrow O_2 + O_2 + Cl \quad (1.8)$$

$$Br + O_3 \rightarrow BrO + O_2$$
$$BrO + O \rightarrow Br + O_2$$

$$O_3 + O + Br \rightarrow O_2 + O_2 + Cl \quad (1.9)$$

(Warneck 1988, pp. 125–130; WMO 1985, p. 30).

Hydroxyl (OH), nitric oxide (NO), chlorine (Cl), and bromine (Br) are not used up in reaction sequences 1.5–1.9. They are catalysts which activate and speed up these ozone-removal processes. These compounds are released at the end of the respective reaction sequences to repeat this process with a new group of ozone molecules many times until they are chemically altered or removed from the stratosphere. These catalysts can chemically interact in a variety of complex ways, depending on their relative concentrations; thus, their contributions to ozone removal can vary.

The sources for these catalysts are thought to have been relatively constant over decadal time scales until the 1950s, and the production and removal of ozone described by reactions 1.1–1.9 are thought to have been nearly in balance. For example, nitric oxide (NO) is formed by the decomposition of nitrous oxide (N_2O) in the stratosphere, and most of that N_2O came from natural sources associated with the nitrogen cycle (Warneck 1988, pp. 441–467). Similarly, before the 1950s most of the stratospheric chlorine was derived from volcanic emissions of hydrogen chloride (HCl) and methyl chloride (CH_3Cl)—the former a by-product of oceanic biological activity and the latter an effluent of biomass combustion (ibid., pp. 268–272).

1.3 Homogeneous Ozone Depletion

Since the mid 1950s the concentration of human-derived ozone-destroying catalysts in the atmosphere has grown, disturbing the balance between natural production and destruction of stratospheric ozone. For example, emissions of nitrous oxide derived from the use of fertilizers, the disturbance of tropical land, and the combustion of fossil fuels have increased levels in the atmosphere beyond that expected from natural processes alone (Pearman et al. 1986). The concentration of nitrous oxide in the atmosphere is now growing at approximately 0.28 percent per year (Prinn et al. 1990).

Of particular importance is the increasing amount of total chlorine and bromine in the stratosphere. This increase is due primarily to the production and release of certain industrial chemicals, the most commonly known of which are the **chlorofluorocarbons** (CFCs). These compounds were suspected of contributing to ozone depletion as early as 1974, when Mario Molina and Sherwood Rowland of the University of California at Irvine outlined the magnitude and the mechanisms of ozone loss (Molina and Rowland 1974).

Ozone-depleting compounds (ODCs) are used as solvents, in refrigeration, and in the production of insulation and foam. The principal ODCs are stable compounds, generally nontoxic to the immediate user. They are stable in the troposphere precisely because oxygen and ozone in the stratosphere screen out high-energy UV-B and UV-C radiation. Normal troposphere dynamics (weather) gradually mix these compounds throughout the troposphere. They slowly rise across the tropopause to reach the stratosphere. Once in the stratosphere, they are broken apart by ultraviolet radiation in the same way that such radiation photodissociates oxygen molecules. Upon photodissociation, a chlorine atom is released, contributing to the amount of "reactive" chlorines (Cl and ClO) in the stratosphere.[3] This additional reactive chlorine contributes to the removal of stratospheric ozone as discussed above.

Figure 1.4 shows measurements at remote locations of two of the more commonly used ODCs, indicating increases in their concentration in the atmosphere over time. The average rate of increase during the 1980s has

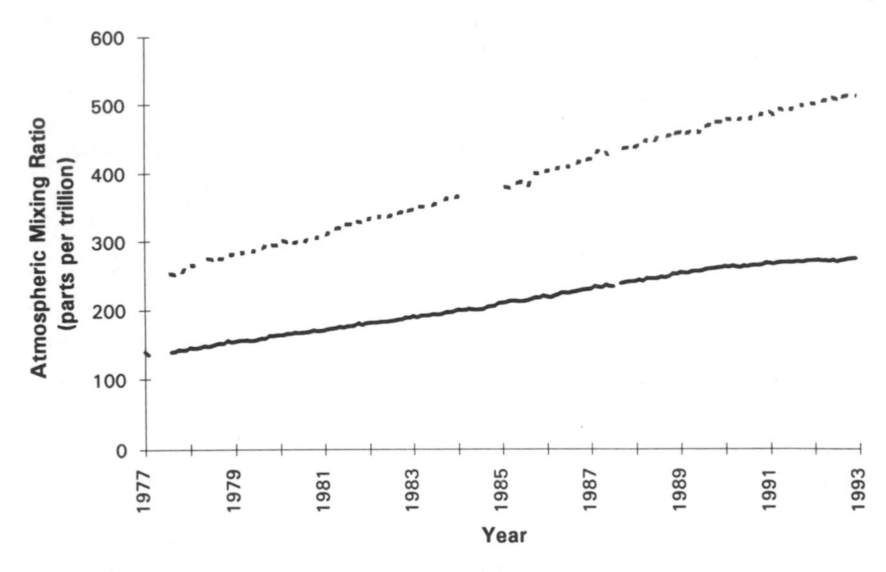

Figure 1.4
Globally averaged measured concentrations of CFC-11 (solid line) and CFC-12 (dotted line) (derived from data on pp. 426–427 of Elkins et al. 1994).

been approximately 4 percent per year (WMO 1989, p. 246; 1991, p. 1.4). In late 1989 this rate of increase began to lessen as production of these compounds diminished as a result of regulatory controls (Elkins et al. 1993; Cunnold et al. 1994).

Some ODCs, known as **halons**, contain bromine. Although halons are emitted in relatively smaller amounts than nonbrominated ODCs, the reactive bromine (Br and BrO) derived from their photodissociation of halons tends to be much more efficient at depleting stratospheric ozone on an atom-for-atom basis than reactive chlorine. Even though the remainder of our discussion of ozone depletion will focus on the role of chlorine in the stratosphere, it must be kept in mind that the arguments also apply to bromine.

It takes considerable time for a given amount of an ODC emitted at the surface to travel to the stratosphere and undergo destruction. Thus, for many of the ODCs an amount released today will result in a release of chlorine decades from now. Slow destruction is also the reason why, despite drastically decreased emissions, the cumulative quantities of most ODCs in the atmosphere continue to increase.

Aside from the large upward trend in total stratospheric chlorine, chlorine-catalyzed stratospheric ozone depletion is important because chlorine is able to remain in reactive chemical forms for a relatively long time and because chlorine participates in a special type of chemistry involving particles in the stratosphere, particularly in polar regions.

In the catalytic cycle described by reaction sequence 1.8, one reactive chlorine atom can destroy about 1000 stratospheric ozone molecules before being chemically transformed into an "inactive" form of chlorine (Rowland 1990, p. 284). One such inactive form is hydrogen chloride (HCl), formed by the reaction of chlorine with methane (CH_4). Some of the methane enters the atmosphere through natural processes (such as decomposition of marsh materials); considerable amounts arise from human activities (such as raising cattle, mining coal, and drilling for oil and natural gas). The reaction between chlorine and methane is

$$CH_4 + Cl \rightarrow CH_3 + HCl. \quad (1.10)$$

Stratospheric chlorine can be incorporated into other inactive forms in addition to HCl. The most important of these is chlorine nitrate ($ClONO_2$), which can be formed by the reaction

$$ClO + NO_2 + M \rightarrow ClONO_2 + M. \quad (1.11)$$

At most latitudes, well over half the total stratospheric chlorine is stored in these two inactive forms (HCl and $ClONO_2$).

Permanent removal of stratospheric chlorine occurs when hydrogen chloride eventually diffuses across the tropopause into the troposphere and, owing to its solubility in water, is rained out. Unfortunately, transport to the troposphere from the stratosphere takes approximately 2–5 years (WMO 1991, p. 8.41). Before permanent removal, hydrogen chloride and chlorine nitrate can release reactive chlorine approximately 50–200 times through photodissociation or reaction with OH, making

for further ozone destruction (Rowland 1990, p. 284). This cycling of chlorine in the stratosphere between its inactive and reactive forms means that one chlorine atom can destroy about 100,000 ozone molecules before being permanently removed from the atmosphere.

Reaction 1.11 points to one of the ways in which the chlorine and nitrogen catalytic cycles are coupled. In the lower stratosphere, the

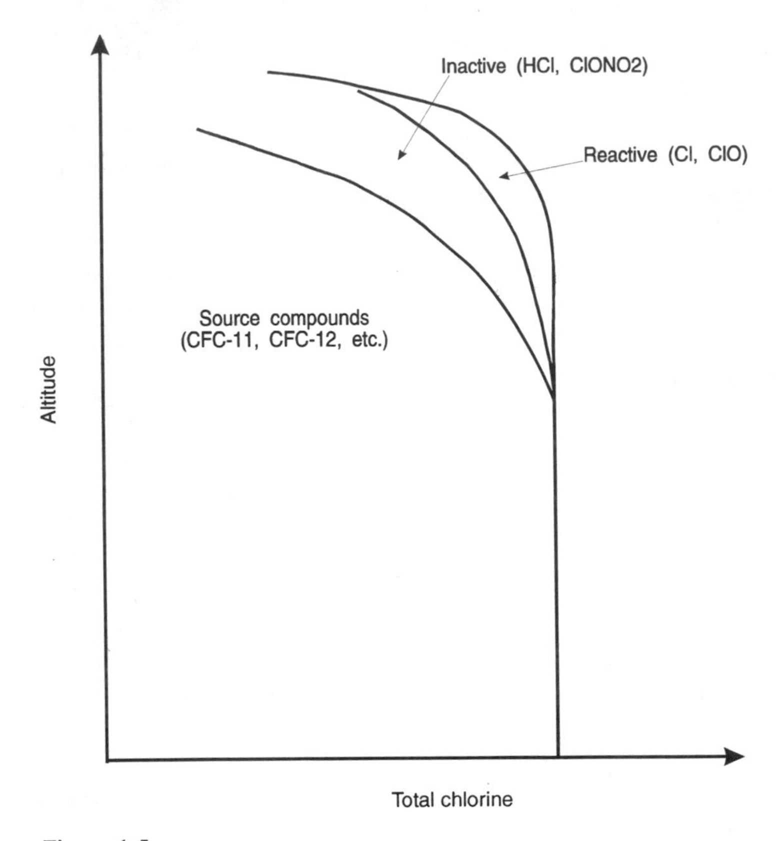

Figure 1.5
Schematic representation of atmospheric chlorine reservoirs.

availability of chlorine to destroy ozone catalytically is limited by this reaction and depends on the amount of both NO_2 and NO (which readily oxidizes to NO_2). Together, NO_2 and NO are referred to as NO_X.

The relationship between the inactive and reactive forms of stratospheric chlorine is represented schematically in figure 1.5. At any given time, the total amount of chlorine in the stratosphere is contained within chlorine-containing compounds (ODCs among others), within reactive forms of chlorine, and within inactive forms of chlorine. The increase in the total amount of reactive stratospheric chlorine is not due entirely to emissions of ODCs; some of it is due to a shift in the partition between inactive and reactive forms of chlorine. This shift is now thought to play a significant role in stratospheric ozone depletion at the poles and at middle latitudes.

1.4 Heterogeneous Ozone Depletion

The stratospheric ozone-depleting process described above is usually referred to as "homogeneous" or "gas-phase" depletion, because all the reacting chemical compounds are in the gaseous phase. A newly recognized series of reaction sequences, involving chemical compounds in various phases (gas + ice and/or gas + liquid), has been hypothesized to deplete ozone, particularly that occurring in the polar regions. Both laboratory experiments and observations in the stratosphere have strengthened the validity of this hypothesis, which was first proposed to explain the sudden, unexpected formation of the Antarctic ozone hole in the late 1970s.

The seasonal thinning of ozone in the Antarctic polar stratosphere requires an ozone-depletion process considerably more efficient than the homogeneous mechanism described above. The leading theories incorporating such a process show that the same total concentration of all chlorine-containing compounds in the stratosphere can produce considerably different amounts of ozone depletion under certain conditions, such as those that prevail over the Antarctic during winter and spring. Under these conditions, the inactive forms of chlorine (HCl and $ClONO_2$) are rapidly transformed into reactive forms (Cl and ClO) by reactions occurring on aerosol surfaces, greatly increasing ozone depletion.[4] There is now evidence suggesting that this mechanism (generally referred to as "heterogeneous" ozone depletion, because there is interaction between chemicals in the gaseous, liquid, and/or solid phases) is likely occurring over the Arctic and over middle latitudes. This mechanism is particularly prominent over the Antarctic, so we will first describe it as it occurs there.

Air over the Antarctic becomes extremely cold during the southern hemisphere's winter because of the lack of sunlight and because there is little mixing of the stratospheric air over the South Pole with air from

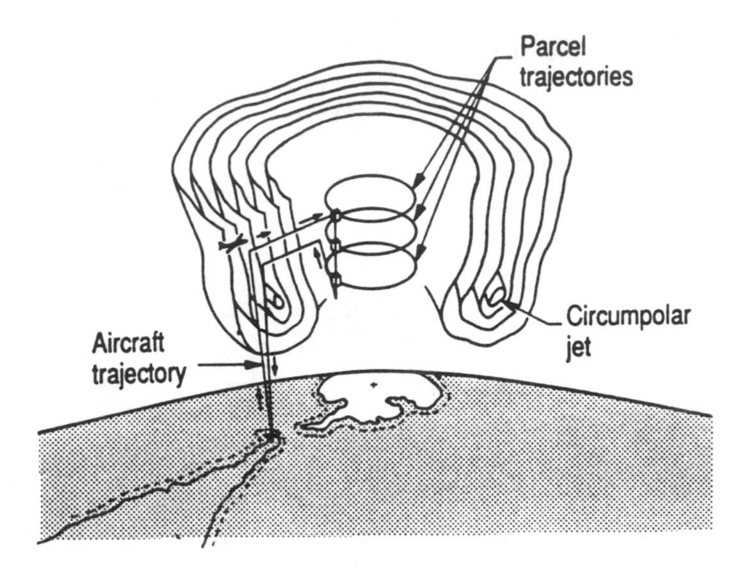

Figure 1.6
Schematic of the Antarctic vortex (Anderson et al. 1991).

other latitudes. The isolated air is contained by the circumpolar vortex, a stratospheric jet of wind circulating around the pole at speeds of approximately 85–170 miles per hour between approximately 50° and 65° south latitude (Schoeberl and Hartmann 1991). Cold temperatures, as low as −78° Celsius, allow for the formation of polar stratospheric clouds (PSCs) at altitudes of approximately 15–25 km within the vortex (McCormick et al. 1982). PSCs formed at this temperature are composed of nitric acid that attaches to water as it condenses on background sulfate aerosol particles (these are known as Type I PSCs) (Toon et al. 1986; Crutzen and Arnold 1986). The initial diameter of these particles is about 1 micrometers. (A micrometer is a millionth of a meter.) If temperatures drop to approximately −83°C, ice can accumulate on the outside of the PSC particles, allowing them to attain diameters up to tens of micrometers (Type II PSCs) (Toon et al. 1989).

It was originally suggested in the mid 1980s that PSCs might lead to an alteration of the chlorine chemistry in the Antarctic stratosphere (Solomon et al. 1986; McElroy et al. 1986). Laboratory evidence suggests that the PSC particles interact rapidly (within hours) with the two inactive chlorine compounds, chlorine nitrate and hydrogen chloride (Molina et al. 1987; Tolbert et al. 1987). The heterogeneous reactions are

$$ClONO_2 + HCl(s) \rightarrow Cl_2 + HNO_3(s) \quad (1.12)$$

and

$$ClONO_2 + H_2O(s) \rightarrow HOCl + HNO_3(s), \quad (1.13)$$

where s denotes species in the solid phase (other compounds are in the gas phase). Further calculations and laboratory experiments have shown that an additional heterogeneous reaction involving inactive chlorine may be

$$HOCl + HCl(s) \rightarrow Cl_2 + H_2O \quad (1.14)$$

(Crutzen et al. 1992; Prather 1992a; Abbatt and Molina 1992; Hanson and Ravishankara 1992). Two other important heterogeneous reactions are

$$N_2O_5 + HCl(s) \rightarrow HNO_3(s) + ClNO_2 \quad (1.15)$$

and

$$N_2O_5 + H_2O(s) \rightarrow 2HNO_3(s) \quad (1.16)$$

(Tolbert et al. 1988; Wofsy 1988; Evans 1985).

Because NO_2 is transformed into N_2O_5 in the absence of light, it is an important reservoir of NO_X ($NO + NO_2$) during polar night (Warneck 1988, pp. 109–110).

The result of heterogeneous reactions 1.12–1.15 is the transfer of inactive stratospheric chlorine to the potentially reactive forms Cl_2, $ClNO_2$, and HOCl. In addition, the ability of reaction 1.11 to transfer chlorine back to inactive reservoirs is diminished by the conversion of NO_X to solid nitric acid (reactions 1.15 and 1.16), a process referred to as **denoxification** (Hamill and Toon 1991). Growth and gravitational settling of the larger PSC particles into the troposphere can further reduce the availability of NO_X. This is referred to as **denitrification** (WMO 1989, p. 17; Toon et al. 1986; Crutzen and Arnold 1986; Wofsy et al. 1990; Hübler et al. 1990).

When sunlight first penetrates the Antarctic stratosphere, in late August, a series of photodissociation reactions occur, transforming the potentially reactive chlorine compounds into reactive forms. These reactions are

$$HOCl + h\nu \rightarrow Cl + OH, \tag{1.17}$$

$$Cl_2 + h\nu \rightarrow Cl + Cl, \tag{1.18}$$

and

$$ClNO_2 + h\nu \rightarrow ClO + NO \tag{1.19}$$

(WMO 1989, p. 54).

The net result of reactions 1.12–1.19 is to increase the proportion of reactive chlorine atoms. This process can be visualized as enlarging the area designated as "reactive" in figure 1.5.

The chlorine atoms resulting from reactions 1.17 and 1.18 quickly combine with ozone, yielding ClO and molecular oxygen. The result of these reaction mechanisms at this point is the production of ClO. But rather than continuing to the second step of the homogeneous process described in reaction sequence (1.8), the high concentration of ClO, in the absence of abundant atomic oxygen (which is needed for reaction sequence 1.8), initiates the following catalytic cycle:

$$\begin{array}{l} ClO + ClO + M \rightarrow ClOOCl + M \\ ClOOCl + h\nu + Cl + ClOOO \\ ClOO + M \rightarrow Cl + O_2 + M \\ 2(Cl + O_3 \rightarrow ClO + O_2) \\ \hline 2O_3 \rightarrow 3O_2 \end{array} \tag{1.20}$$

(Molina and Molina 1987).

It is important to note that the rate at which the reaction sequence 1.20 proceeds is determined by its first step, which leads to the creation of the chlorine monoxide dimer (ClOOCl).[5] As a result, the rate of ozone loss in this reaction sequence goes as the square of the concentration of chlorine monoxide (since two molecules of chlorine monoxide are needed to make one chlorine monoxide dimer). A doubling of ClO means a quadrupling of the rate at which ozone is lost. It has been estimated that this nonlinear mechanism is responsible for approximately 75 percent of the severe stratospheric ozone loss observed each southern spring over Antarctica in the last decade (Anderson et al. 1991).

Dramatic confirmation of this mechanism was established during a 1987 scientific expedition to the Antarctic (Anderson et al. 1989). Simultaneous aircraft measurements of both ClO and O_3 were made before and during the period of severe ozone depletion. The results of these measurements, shown in figure 1.7, confirm the anti-correlation between ClO and O_3 suggested in reactions 1.17–1.20. These results were so convincing that they are commonly referred to as the "smoking gun" of Antarctic ozone depletion.

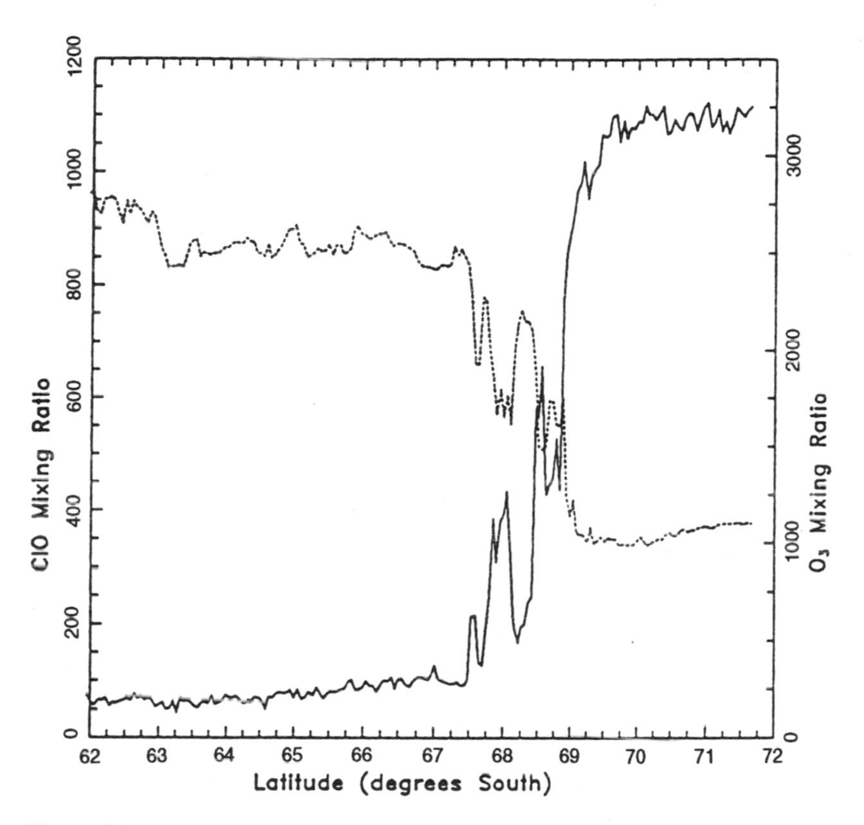

Figure 1.7
Anticorrelation of ClO (solid line, representing parts per thousand) and ozone (dotted line, representing parts per billion) over the Antarctic: observations obtained September 16, 1987 (Anderson et al. 1989).

In addition, a coupling of reactive chlorine and bromine occurs in the fast reaction sequence

$$\begin{array}{l} Br + O_3 \rightarrow BrO + O_2 \\ Cl + O_3 \rightarrow ClO + O_2 \\ \underline{BrO + ClO \rightarrow Br + Cl + O_2} \\ 2O_3 \rightarrow 3O_2 \end{array} \qquad (1.21)$$

(McElroy et al. 1986). After reaction 1.20, this reaction is thought to account for nearly all of the remaining ozone loss over the Antarctic.

Stratospheric loss of ozone due to bromine is particularly efficient primarily because bromine is present in its reactive form much longer than chlorine. Reaction 1.21 indicates that ozone loss due to bromine is also dependent on the amount of chlorine which is larger and rising considerably faster (WMO 1985, p. 38).

The rapid loss of stratospheric ozone continues until higher sun angles of the advancing spring warm the stratosphere sufficiently to slowly dissipate the PSCs. In the process, they release the bound nitrogen and initiate the transfer of reactive chlorine back to inactive forms. The circumpolar vortex also dissipates, further warming the polar air and allowing for mixing with relatively ozone-rich air from the middle latitudes. The result is a near recovery of stratospheric ozone amounts over the polar region. Figure 1.8 is a simple schematic of the evolution of polar ozone depletion.

Ozone loss over the Arctic has been considerable, though not as dramatic as that observed over the Antarctic. Heterogeneous phenomena are also strongly suspected to play a role in Arctic ozone depletion (Hofmann et al. 1989; Proffitt et al. 1990; Brune et al. 1990; Hofmann and Deshler 1991; Brune et al. 1991; NASA 1992). The differences in ozone loss are due primarily to the dissimilar meteorological conditions found at the poles during their respective winter and spring seasons. For example, a circumpolar vortex forms over the Arctic, but it generally is smaller and less symmetric, breaks up sooner after polar sunrise, and is less isolated than its counterpart over the Antarctic (Hirota et al. 1983; Schoeberl and Hartmann 1991). This allows for a greater amount of mixing with air from lower latitudes—air that typically is warmer and richer in ozone.

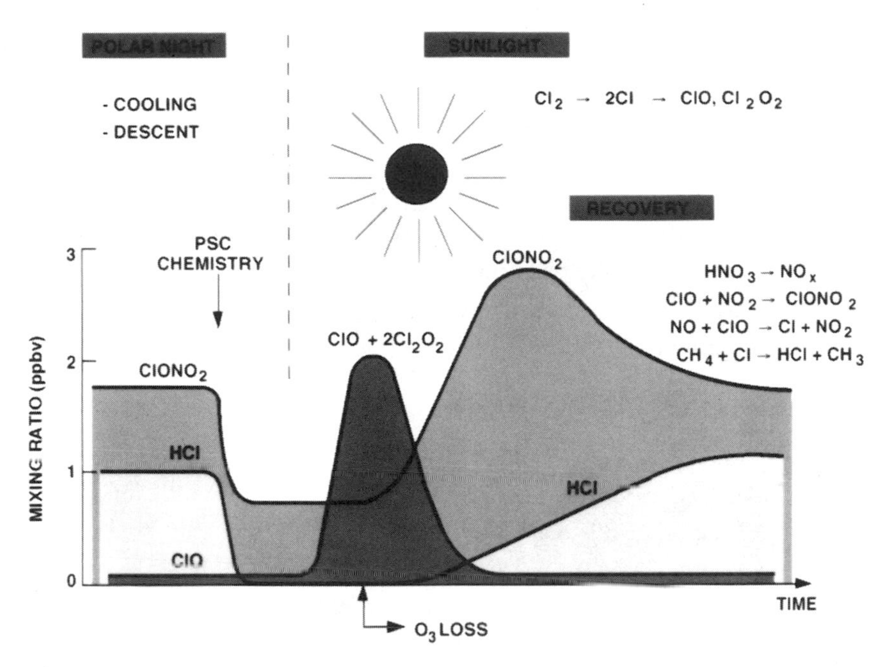

Figure 1.8
Schematic of springtime polar ozone depletion (Webster et al. 1993).

Frequent sudden warming of the Arctic stratosphere can cause the northern vortex to break up in the middle of winter, far earlier than its southern counterpart. As a result, PSC formation is less extensive in the north and tends not to persist into winter to the extent found over the Antarctic (McCormick et al. 1982; Fahey et al. 1990; SORG 1993). In fact, the dissipation of PSCs and a subsequent rise in NO_X often occurs before the onset of polar sunrise with little apparent denitrification, though significant amounts of inactive chlorine have been transferred to potentially active forms by reactions 1.12–1.15 (Rodriguez 1993; Newman et al. 1993; Toohey et al. 1993). The potentially active chlorine is converted back into reservoir species before sunlight returns and catalytic ozone destruction can begin. As a result of these differences, ozone depletion has so far been lower over the Arctic, in terms of both spatial extent and total column depletion, than over Antarctica.

Stratospheric ozone loss outside the poles is also suspected to be due to heterogeneous chemical processes, independently and in conjunction with loss at the poles. "Ozone-poor" and/or "chemically processed" air (low amounts of NO_X, large amounts of reactive chlorine) from the Arctic region may migrate to mid-latitude locations, contributing to the observed ozone losses there. Some estimates have apportioned between one-quarter and one-half of the 6–8-percent-per-decade loss of ozone over middle latitudes to this mechanism (Proffitt et al. 1993). Since models have yet to fully account for the ozone loss measured over the middle latitudes, the connection to heterogeneous loss at the poles is compelling. However, limited observational data on the precise chemical and dynamical nature of the polar vortices have fueled a debate concerning the mechanism responsible for this proposed connection.

One school of thought maintains that both polar vortices are isolated from outside air prior to their breakdown in the late winter or spring (Hartmann et al. 1989; Schoeberl et al. 1992). When this breakdown occurs, ozone-depleted air is transported to locations outside the poles. The subsequent mixing of this ozone-depleted air with air over mid-latitude locations results in a dilution of the column ozone outside the poles (Atkinson et al. 1989; Sze et al. 1989). The competing theory contends that considerable exchange of mid-latitude and polar air occurs throughout the winter period (Tuck et al. 1989). The vortices behave like "flowing processors," converting inactive chlorine to active forms as air passes through their PSC-laden regions (Proffitt et al. 1989; Jones et al. 1990). This chemically perturbed air then flows over the sunlit middle latitudes and undergoes rapid ozone depletion. Measurements in the early 1990s over the Arctic suggest that the vortex may be analogous to a "leaky bucket," allowing mid-latitude air in at the top and flushing it out the bottom rather than the sides (Proffitt et al. 1993).

Independent of polar effects, increasing amounts of sulfate aerosols in the lower stratosphere around the globe may be suitable sites for the heterogeneous conversion of NO_X to HNO_3 in a reaction similar to 1.16:

$$N_2O_5 + H_2O(\text{sulfate aerosol}) \rightarrow 2HNO_3(s) \qquad (1.22)$$

(Rodriguez et al. 1988; Hofmann and Solomon 1989; Mather and Brune 1990; Rodriguez et al. 1991).[6] This reaction has been shown to proceed rapidly in laboratory experiments for sulfuric acid solutions similar in composition to those found in the stratosphere (Hanson and Ravishankara 1991; Fried et al. 1994).

The increase in stratospheric sulfate has been tied to sulfur emissions from the exhaust of jet aircraft flying in the upper troposphere and the lower stratosphere (Hofmann 1991). Superimposed upon these increasing anthropogenic emissions is the occasional injection of sulfur compounds from large volcanic eruptions.

The heterogeneous conversion of nitrogen compounds shown in reaction 1.22 effectively suppresses the ability of NO_X to convert active chlorine to inactive forms by reaction 1.11. Model calculations have shown that removal of NO_X due to heterogeneous reactions on volcanic aerosols allows both chlorine-related and hydrogen-related mechanisms (reactions 1.8, 1.5, and 1.6) to control ozone removal to a much greater extent (Rodriguez et al. 1994).

Because the sulfate aerosol layer extends the expanse of the stratosphere, reaction 1.22 may be contributing to the existing heterogeneous loss over the poles. Furthermore, observations over the Antarctic and the Arctic during their respective winters suggest that dilute, supercooled sulfate aerosol with a containing dissolved HCl may be reacting directly with chlorine nitrate as in reaction 1.15 (Wolff and Mulvaney 1991; Toon et al. 1993; Molina 1993). There are still considerable uncertainties regarding the extent and relative importance of these phenomena in mid-latitude ozone depletion.

1.5 Observations of Ozone Loss

The evidence of stratospheric ozone depletion has rarely coincided with advances in theoretical understanding. Catalytic destruction of ozone from the breakdown of ODCs in the stratosphere was originally proposed as a theory in 1974, roughly a decade before there was observational evidence (Molina and Rowland 1974). On the other hand, the Antarctic ozone hole, first detected in 1982 and brought to the attention of the general scientific community in 1985, was a complete surprise. It was a couple of years before the primary mechanism responsible for the large seasonal losses over the Antarctic was understood. Wintertime losses in the middle latitudes in the early 1990s have considerably exceeded model predictions and, once again, have preceded theoretical understanding.

Continuous measurement of stratospheric ozone has been made by various instruments since 1960. The earliest continuous measurements of total column ozone (total amount above a point on the surface) were made from the ground using the Dobson spectrophotometer and M-83 Filter Ozonemeters starting in the mid 1950s (WMO 1989, pp. 164–167). This network of ground-based instruments has been generally limited to land-based positions in the northern hemisphere. Starting in late 1978, measurements from NASA's Nimbus 7 satellite using both the TOMS (Total Ozone Monitoring Spectrometer) and the SBUV (Solar Backscatter Ultraviolet Spectrometer) allowed a great refinement of the coverage of ozone measurements because it enabled measurements to be made with much greater spatial resolution (Heath et al. 1975).[7] Other satellite-based measurements include those from the NOAA 11 SBUV/2 (launched January 1, 1989), the Meteor 3 TOMS (launched August 15, 1991), and the UARS-MLS (Upper Atmosphere Research Satellite–Microwave Limb Sounder, launched September 12, 1991). In addition, instruments that measure the vertical profile of ozone in the stratosphere from the ground and by balloon or rocket, have been in operation since the mid 1950s (WMO 1989, p. 164).

Since the late 1970s, the World Meteorological Organization, in conjunction with other organizations and scientists from around the world, has been assessing the data collected by these instruments. The 1991 update of this scientific assessment summarizes the ongoing loss of stratospheric ozone, placing the responsibility for this loss on the continuing emission of human-derived ozone-depleting compounds into the atmosphere. The conclusion reveal a worsening problem whose extent, once again, exceed expectations. Reacting to the release of the 1991 assessment, William K. Reilly, Administrator of the U.S. Environmental Protection Agency, said: "The announcement today, suggesting a worse situation than we thought, affirms the warning I issued last Spring: Upper atmosphere ozone depletion remains one of the world's most pressing environmental threats." (Reilly 1991) Measurements in 1992 and 1993, exhibiting an increase over ozone loss rates in the 1980s, underscore this warning.

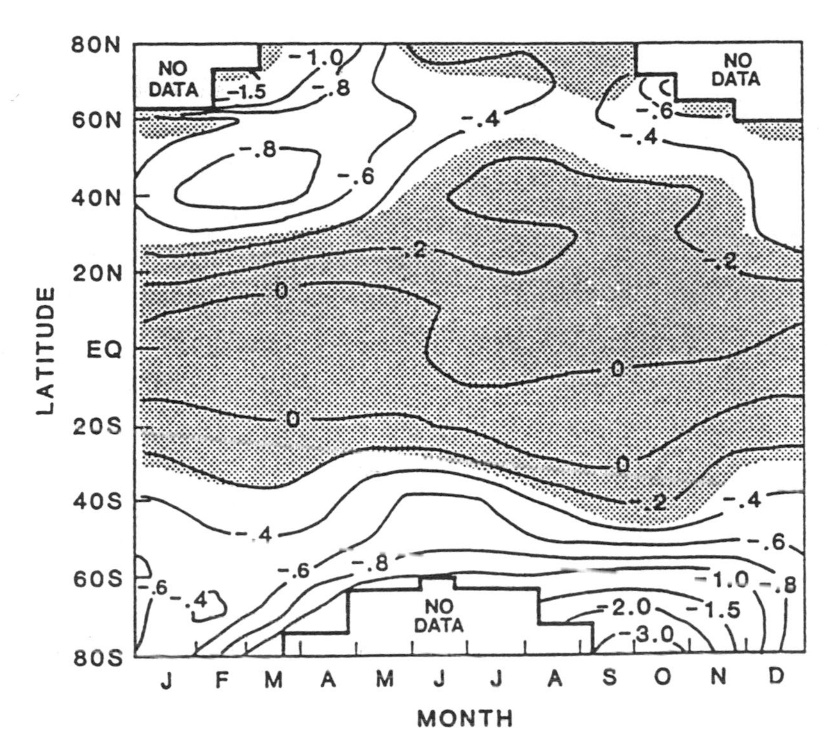

Figure 1.9
TOMS column ozone trends (percent per year; statistical fit to TOMS column-ozone measurements; "no data" refers to regions of polar night where observations are not possible with TOMS) (Stolarski et al. 1991).

Middle Latitudes

The extent of observed ozone depletion varies by season and by latitude. In the northern hemisphere ozone depletion has a strong seasonal dependence, the greatest depletion occurring during the winter months. In the southern hemisphere the greatest depletion occurs in spring and summer. The greatest depletion occurs close to the poles; losses closer to the equator are negligible. Figure 1.9 shows total column ozone trends in percent change per year derived from satellite measurements according to latitude and season.

Because of the extensive spatial coverage allowed by satellite-based measurements of the atmosphere, hemispheric or global averages of ozone loss are best derived from satellite data. Table 1.2 lists the trends in total column ozone depletion up to 1991 measured by both the TOMS

Table 1.2
Total ozone trends (percent per decade, with 95 percent confidence limits) (WMO 1991, p. xii).

	TOMS: 1979–1991			Ground-based: 26–64°N	
Season	45°S	Equator	45°N	1979–1991	1970–1991
Dec–Mar	-5.2 ± 1.5	$+0.3 \pm 4.5$	-5.6 ± 3.5	-4.7 ± 0.9	-2.7 ± 0.7
May–Aug	-6.2 ± 3.0	$+0.1 \pm 5.2$	-2.9 ± 2.1	-3.3 ± 1.2	-1.3 ± 0.4
Sep–Nov	-4.4 ± 3.2	$+0.3 \pm 5.0$	-1.7 ± 1.9	-1.2 ± 1.6	-1.2 ± 0.6

(1979–1991) and ground-based instruments (1970–1991 and 1979–1991).

The year 1992 and much of 1993 exhibited ozone losses that exceeded extrapolations of previous measured trends. According to NASA scientist J. F. Gleason and colleagues[8]: "A statistical model including the effects of the seasonal cycle, QBO, 10.7-cm solar flux cycle, and a linear trend accurately reproduces the observed ozone variation from 1979 to 1991, For 12 years of data, 1979 to 1990, this model accurately predicts the 13th year, 1991. However, when we fit the data through 1991 (13 years) and used the model to predict amounts through the end of 1992, the observed values were 1 to 2 percent lower than the predicted values." (Gleason et al. 1993) Extension of this analysis to May 1993 indicates that the observed values in 1992 and 1993 were consistently 2–3 percent below the range of ozone values in the preceding 13 years. In 1993 this deviation was deemed highly significant by the United Kingdom's Stratospheric Ozone Research Group when they stated in their 1993 report (SORG 1993, p. 3): "... the last eighteen months of the record indicate a very significant shift in global average ozone, without precedent in the lifetime of TOMS. Never before has the yearly average difference been outside the range ± 0.5 percent. In 676 weeks (13 years, 1979–1991) only 19 of the weekly average differences were outside the range ± 1.0 percent, and never for more than two weeks consecutively." (SORG 1993, p. 3)

Data from the NOAA 11 satellite covering the northern hemisphere (25–60°N) indicate that ozone losses in early 1993 were approximately 8 percent below the 1989–1992 average (Planet et al. 1994). In the latter half of 1993 and early in 1994, ozone levels appeared to be returning to the 1989–1992 average.

The observed *regional* losses of total column ozone have been greater than the zonal and latitudinal averages. For instance, from January to April of 1993 ozone levels at U.S. measurement stations were approximately 13 percent below normal (defined as prior to 1981) (Komhyr et al. 1994). Canadian measurements were as much as 17 percent below normal (defined as prior to 1980) during the same January–April period (Kerr and McElroy 1993). European ozone levels were equally depressed; ozone levels over Siberia for the month of March were approximately 20 percent below normal (Bojkov et al. 1993). Stratospheric ozone over the equator, previously unaffected, showed declines in both 1991 and 1992 (Gleason et al. 1993).

In the middle latitudes, ozone loss is most prominent in the lower stratosphere at altitudes of approximately 15–25 km. Figure 1.10 shows the trend in the vertical profile of ozone between 1970 and 1990. Measurements over Canadian locations reconfirm the vertical distribution of ozone loss in the middle latitudes, although, in contrast with figure 1.10, no increase in tropospheric concentrations was noted (Kerr and McElroy 1993).

All of the ozone-loss observations in 1992 and 1993 indicate an increase in loss rates over previous observed trends. Some of this unprecedented ozone depletion may be a result of the 1991 eruption of Mount Pinatubo, which injected large amounts of sulfate aerosol into the stratosphere. If this is the case, ozone loss should return to its pre-eruption rate sometime in 1994 or 1995. Furthermore, weather-related events can alter regional ozone values for weeks or months. It has been noted by researchers that persistent wintertime high pressure systems were frequent during 1992 and 1993 over the northern hemisphere. These are known to lower column ozone amounts for weeks or months at a time by altering transport of ozone-poor air from the tropics (Bojkov et al. 1993). Confirmation of these causal linkages must await further data.

Antarctica

The discovery of significant seasonal ozone loss over the continent of Antarctica is the most alarming episode in the history of ozone monitor-

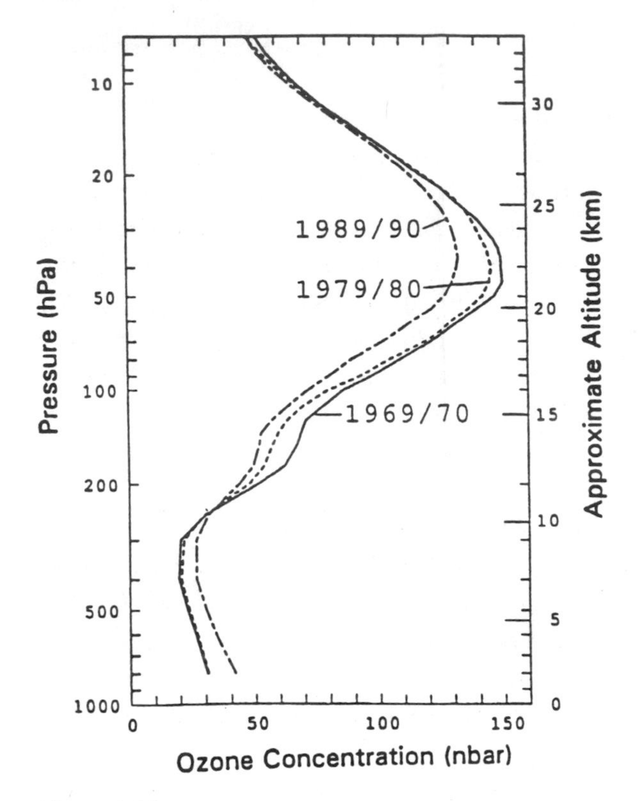

Figure 1.10
Changes in the vertical distribution of ozone (WMO 1991, p. 2.25, figure 2-16).

ing. Large declines in total column ozone during the Antarctic spring were first observed in the 1982 by the British Antarctic Survey team. By 1984, intensive examination of data from ground-based instruments in the Antarctic and correlation of the measurements with NASA's satellite data had confirmed that very serious depletion of ozone had been occurring each Antarctic spring since 1979.

Figure 1.11 shows total column ozone levels over the Antarctic in October from the late 1950s to 1994. By the late 1980s, decreases of as much as 60 percent over locations in the Antarctic had been observed, while depletion at altitudes between 15 and 20 km had been nearly total (WMO 1989, pp. 7, 19). The lowest total column ozone measurement ever—90 Dobson units—was recorded on October 6, 1993 (NOAA

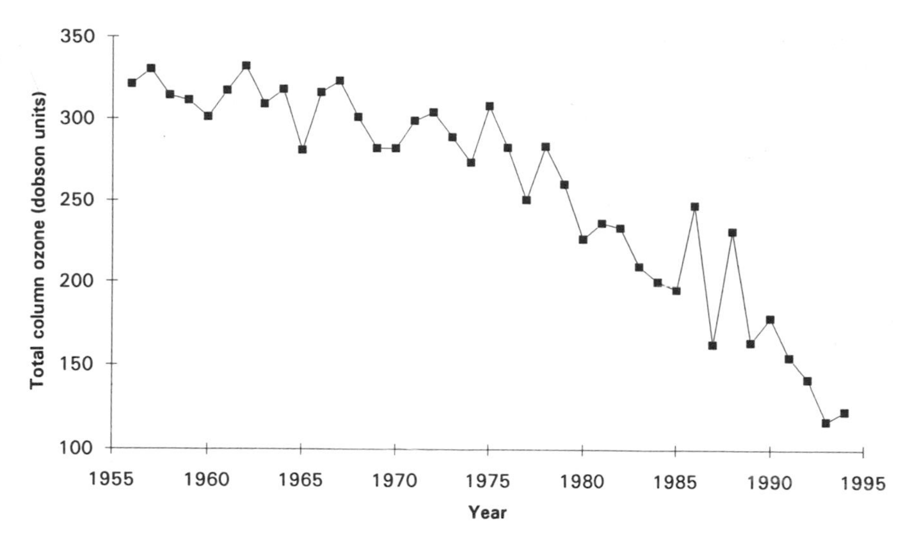

Figure 1.11
October mean of column ozone over the Halley Bay (data from British Antarctic Survey, communicated by J. D. Shanklin in December 1994).

1993). (A reading of approximately 300 Dobson units is considered normal.)

The spatial extent of the Antarctic ozone hole is also increasing. As much as 10 percent of the southern hemisphere lies under the area of seasonally depleted Antarctic ozone for a period of approximately a month (Newman et al. 1991). On September 23, 1992, the Antarctic ozone hole covered a 9.4 million square mile area.[9] In 1993 the maximum extent of the hole covered 9 million square miles (NOAA 1993), and in 1994 the areal extent nearly matched that of 1992 (*Science* 1994). For reference, North America is about 9.1 million square miles and the Antarctic continent is about 5.4 million square miles. Significant loss of total column ozone associated with the Antarctic ozone hole is now occurring over New Zealand, Australia, and southern Latin America.

The breakup of the circumpolar vortex, which signals the end of the seasonal ozone depletion, has been observed to occur later in years of deep ozone holes, extending significant ozone loss into the month of December. Furthermore, the ozone depletion is showing signs of beginning earlier each year. Figure 1.12 shows the temporal evolution of ozone loss

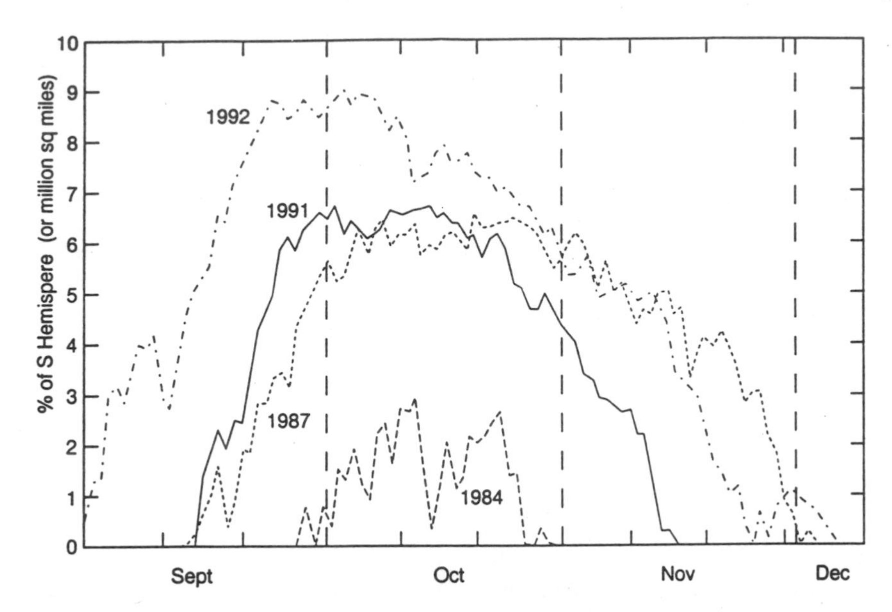

Figure 1.12
Proportion of southern hemisphere under the ozone hole (area under which levels of column ozone were less than 212 Dobson units) (SORG 1993, p. 8, figure 1.7).

for selected years. Vertical profiles of ozone loss over the Antarctic indicate losses are greatest from approximately 12 to 24 km (Hofmann et al. 1987). Data from the 1991 Antarctic ozone hole suggest that losses in the 11–13-km and 25–30-km regions were significant and may be tied to the presence of sulfate aerosols from volcanic eruptions (Hofmann et al. 1992). Both 1992 and 1993 exhibited near-total ozone loss in the 13–19-km layer (*Science* 1992; SORG 1993, p. 8).

Measurements of ozone and of chlorine monoxide (a chemically reactive form of chlorine) taken by aircraft flying within highly depleted regions provide conclusive evidence that chlorine (alone and in combination with bromine) is primarily responsible for the Antarctic ozone hole (Anderson et al. 1989).

The Arctic

Although Arctic ozone depletion has not been as severe as springtime ozone depletion in the Antarctic, considerable episodic wintertime Arctic depletion associated with the presence of polar stratospheric clouds has been measured. Confirmation of PSC processing in the Arctic vortex derives from measurements of chlorine monoxide (ClO). In 1991 and 1992, concentrations exceeding 1 part per billion by volume (ppbv) were detected in the vortex; these were comparable in magnitude to concentrations observed in the Antarctic ozone hole (Water et al. 1993; Schoeberl et al. 1993). Furthermore, these concentrations were found in air that had traveled through the coldest portions of the northern vortex. This implied activation by PSCs (Newman et al. 1993; Toohey et al. 1993). As temperatures rose in February, implying dissipation of PSCs, concentrations of ClO fell.

Total column ozone (TCO) losses of up to approximately 10 percent were measured over the Arctic during the winter of 1989–90 (Proffitt et al. 1990; Hofmann and Deshler 1991). Measurements made during the winter of 1991–92 imply TCO losses of approximately 10 percent and losses of up to 25 percent at altitudes between 15 and 20 km (NASA 1992; Browell et al. 1993). Calculations indicate that this depletion was due in considerable measure to reactions involving both bromine and chlorine.

Measurements of Ultraviolet Radiation

The intensity of ultraviolet radiation reaching the Earth's surface depends on many factors other than stratospheric ozone, including urban and global tropospheric air pollution, cloud cover, aerosol abundance, and surface reflectivity. Because of these factors, which may increase or decrease UV radiation reaching the surface both locally and regionally, several years' data with adequate geographical coverage are required before a trend in UV-B can be distinguished. Unfortunately, accurate measurement of UV-B radiation only began in the late 1980s.

Two different instruments are commonly used to measure ultraviolet radiation: the Robertson-Berger (RB) meter and the spectrometer. RB meters are referred to as "broad-band" instruments because they essentially sum the incident radiant energy over a broad spectral interval (typically 290–330 nm). Spectrometers, on the other hand, can measure radiation at a series of discrete wavelengths.

In the United States, RB meters have been in operation since 1974 at eight locations. The data from the years 1974–1985 indicate a small decline (about 1 percent or less) in 290–330-nm-wavelength irradiance. This conflicts with the UV increase expected in this spectral range due stratospheric ozone loss during this period (Scotto et al. 1988). It has been suggested that this trend is a reflection of the increasing levels of urban air pollution, since these meters are all located in urban areas, many next to airports (Grant 1988). Calculations indicate that in some mid-latitude regions UV-B increases due to ozone loss may be partially or wholly offset by the rise in UV-absorbing pollutants such as sulfur dioxide (SO_2), nitrogen oxides (NO_X), dust, and ground-level ozone (Bruhl and Crutzen 1989; Frederick et al. 1989; Liu et al. 1991). This offset is not expected to continue, because the increases in pollutant emissions, particularly SO_2, are slowing. This suspicion is confirmed somewhat by RB measurements at high altitudes. Between 1981 and 1989, UV-B levels increased an average of about 1 percent per year (Blumthaler and Amback 1990). If this is correct, it means that reductions in urban air pollution may be accompanied by increases in UV-B radiation unless stratospheric ozone depletion is simultaneously ameliorated.

RB-meter measurements have some disadvantages other than those deriving from their urban placement (SORG 1993). For example, the measured UV is dominated by longer wavelengths, whereas shorter-wavelength UV is the ultraviolet radiation most affected by ozone depletion. Furthermore, they are designed to respond to wavelengths that produce sunburn in humans. They cannot, therefore, be used to explore other impacts, such as damage to DNA and plants (which tend to be more sensitive to shorter wavelengths).

UV measurements have also made with spectrometers, though the historical record is limited. In the northern hemisphere, measurements begun in 1989 show sizable increases in UV-B radiation. For example, between 1989 and 1993 the intensity of UV radiation at a wavelength of 300 nm increased at average rates of 35 percent and 7 percent per year during winter and summer, respectively, over Toronto (Kerr and McElroy 1993).

Because of the large spatial extent and depth of the Antarctic ozone hole, UV-B increases are expected to be more dramatic in the southern hemisphere—particularly after the breakup of the ozone hole in November and December. For example, spectrometer measurements made in Australia after the breakup of the Antarctic vortex in 1987 revealed a negative correlation between column ozone amounts and UV-B radiation. UV-B levels were found to have risen approximately 21 percent over a five-day period in December (Roy et al. 1990). Measurements taken over Argentina during December of 1991 show an increase of 45 percent over the climatological norm. In the summers of 1990 and 1991, New Zealand experienced UV-B radiation levels nearly twice the levels at similar latitudes in the northern hemisphere (Frederick et al. 1993; Seckmeyer and McKenzie 1992).

A similar correlation showing the impact of polar ozone depletion has been measured at the south pole (SORG 1993). In 1990 enhanced UV-B levels persisted into December, and during the spring 20 percent of the days had radiation levels approximately twice the climatological norm (Frederick and Alberts 1991).

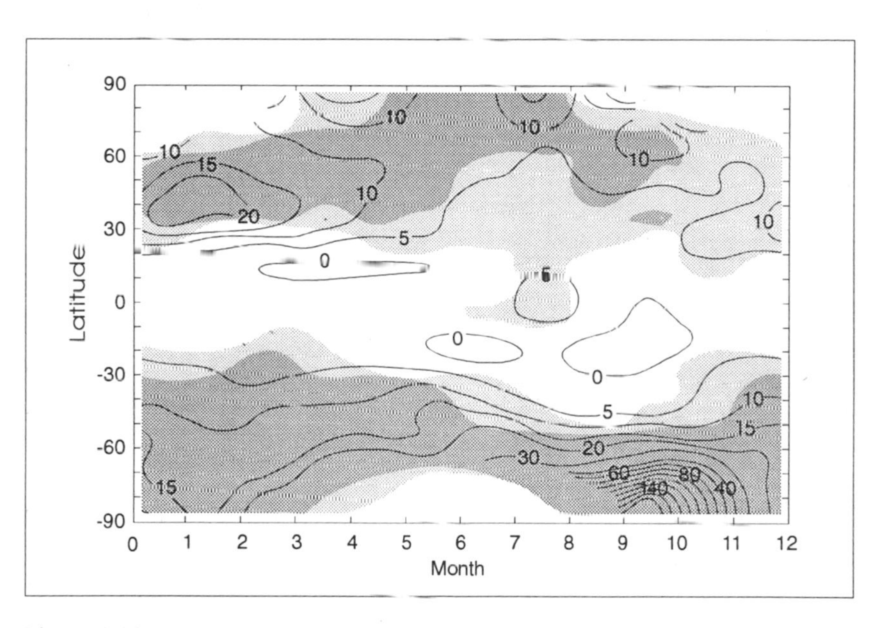

Figure 1.13
Percent change in DNA damage, computed using TOMS column ozone data between 1979 and 1989 (Madronich 1992).

Because of the limited observational data on UV-B radiation, calculations based on ozone declines have been used to characterize the expected changes in UV-B.[10] Figure 1.13 shows the computed percent increase in daily UV dose specific to DNA damage between 1979 and 1989.

1.6 Implications of Continued Ozone Depletion

Predictions versus Observations

The contrast between predictions of ozone depletion and actual observed ozone loss is startling. While scientists have made tremendous advances in understanding the complex processes controlling the amount and distribution of stratospheric ozone, model calculations have consistently underestimated the observations. Figure 1.14 shows this trend superimposed on the total column ozone losses observed in the Antarctic and at northern middle latitudes during the early 1990s. This consistent un-

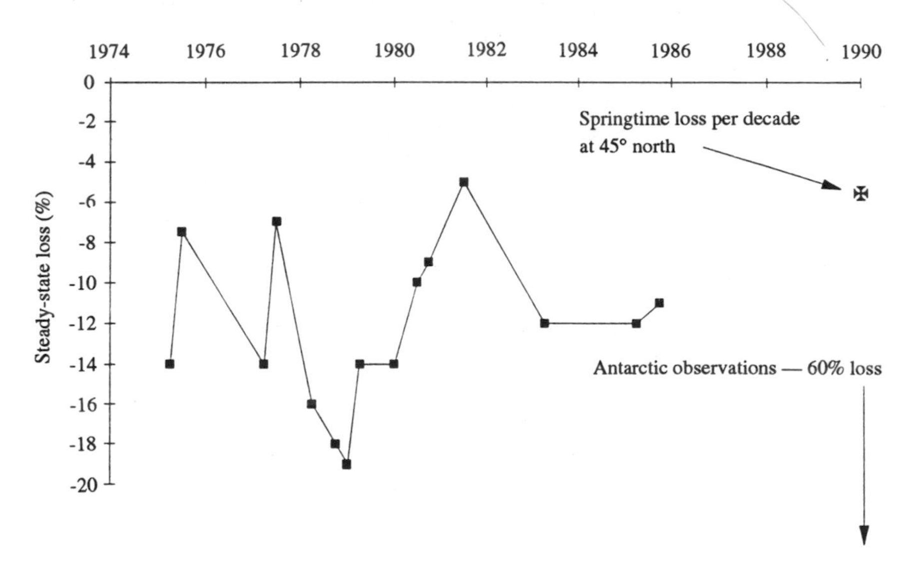

Figure 1.14
Model predictions of ozone depletion made during the period 1975–1986 (EPA-RIA 1987, p. 2-8, exhibit 2-6). Projections are for steady-state concentrations resulting from emissions of CFC-11 and CFC-12 at 1974 levels expected to occur in the latter half of the next century. Recent ozone losses over middle latitudes and over Antarctica are included for comparison.

derestimation poses serious problems for decision makers. Sherwood Rowland, a co-author of the paper that first pointed out the possible role of CFCs in ozone depletion, expressed the following concerns in Congressional testimony several years ago:

> Since September [1987], the chief cause of the Antarctic ozone losses has been unequivocally demonstrated to be traceable to man-made chlorine compounds. Since September, the existence of significant losses of stratospheric ozone in the northern winter has been demonstrated, more than can be accounted for by current theory. The existing models are underestimating current ozone losses both in the northern winter and in the Antarctic spring. Under the circumstances, predictions of ozone change over the next few decades are of little use—how much reliance should be placed on a model for future predictions when it doesn't correctly mimic the present? (Rowland 1988, p. 17)

It is difficult to know whether today's long-term predictions of ozone depletion are more accurate than past predictions. While there is surely greater understanding of the nonlinear process causing the Antarctic ozone hole, there may be other unknown processes with the ability to produce unanticipated ozone loss, especially given the continued increase of chlorine and bromine concentrations. In view of the seriousness of this problem and in view of experience, it is prudent to consider the possibility that even the most dire predictions may turn out to underestimate future ozone depletion. As James Anderson, an ozone researcher at Harvard University, explained: "We cannot predict at what rate the thinning will accelerate. We've learned often enough that every time we've thought things were reasonably predictable, that's when we get our next shock." (Zurer 1993)

Implications of Heterogeneous Ozone Depletion

Mid-latitude stratospheric ozone loss has not been fully accounted for by models incorporating the present understanding of atmospheric chemistry and dynamics. It is suspected that these losses are due to heterogeneous processes similar to those occurring at the poles. This suspicion raises serious implications for stratospheric ozone loss in the middle latitudes. Could nonlinear losses occur there?

Much circumstantial evidence exists for heterogeneous processing outside the polar regions. The material most likely to provide reactive surfaces on which heterogeneous reactions can proceed is the layer of sulfate aerosols found throughout the stratosphere at an altitude of approximately 20 km. The sources of sulfur compounds in the stratosphere are direct injection by volcanic eruptions, jet exhaust, combustion of biomass and fossil fuels, and natural sulfur-emitting processes. It is now confirmed that the concentration of stratospheric sulfate aerosols is increasing; it is likely due to a rise in the anthropogenic component of sulfur emissions (Hofmann 1990).[11]

Heterogeneous reactions on sulfate aerosols tend to reduce the amount of nitrogen compounds available to sequester reactive chlorine, and even directly enhance the availability of reactive chlorine by converting inactive chlorine to reactive form. This process has been likened to suppression of the immune system in human beings in the sense that heterogeneous reactions inhibit the atmosphere's ability to combat the rising concentration of ozone-depleting chlorine and bromine (*C&EN* 1992b, p. 4).

This implies that human influence extends not only to the concentrations of ozone-depleting compounds but also to the partitioning of stratospheric chlorine into reactive and inactive forms. These two influences may, in concert, lead to serious stratospheric ozone loss at middle latitudes.

Whether or not nonlinear ozone loss can occur over middle latitudes is difficult to predict. The current understanding of ozone-depletion chemistry would imply that it is a remote possibility. Unfortunately, there is much that is still not known about the subtle chemistry and physics of the atmosphere. As the concentrations of chlorine and sulfate aerosols continue to rise, the possibility exists that unanticipated nonlinear mechanisms may arise.

Volcanic Eruptions

An important implication of heterogeneous mechanisms occurring on sulfate aerosols is the impact that large volcanic eruptions may have on stratospheric ozone levels. In the past, the injection of sulfate aerosols into the stratosphere from volcanic processes may have caused ozone losses of a few percent, which is within the natural variability of ozone concentrations. As stratospheric ozone loss due to anthropogenic sources may already exceed natural variability, volcanic injection of sulfate aerosols will push losses further. In short, increasing the levels of chlorine and bromine far beyond the bounds of natural variation leaves the Earth without a necessary buffer and vulnerable to severe ozone depletion triggered by natural events.

In order for volcanic eruptions to alter stratospheric chemistry, they must have sufficient eruptive force for their emissions to penetrate into the stratosphere. Very few eruptions qualify as such because of the momentum required to lift material a significant distance against gravity and because the temperature structure of the tropopause inhibits vertical motion. However, four volcanic eruptions within the last decade or so have qualified, and one of these eruptions (that of Mount St. Helens) was horizontal rather than vertical. The other three volcanoes on this list are El Chichon (1982, Mexico, 17°N), Mount Pinatubo (June 1991, Philipines, 15°N), and Mount Hudson (August 1991, Chile, 46°S).

A mid-latitude ozone loss of 5–6 percent was observed in the winter and spring following the eruption of El Chichon, though it has been suggested that half of this loss might be accounted for by periodic transport anomalies temporarily redistributing total ozone latitudinally (Chandra and Stolarski 1991). The vertical distribution of ozone indicated that the greatest loss coincided with the height at which aerosol loading from El Chichon was at its greatest (Hofmann and Solomon 1989).

It is estimated that the eruption of Mt. Pinatubo injected approximately two to three times more sulfate aerosol into the stratosphere (approximately 20 million tonnes) than El Chichon (WMO 1991, p. 2.5). The additional aerosol burden is expected to last 2 or 3 years and affect different regions of the atmosphere at different times due to the uneven transport and distribution of the injected aerosol mass.

Model results suggest that additional total column ozone losses of 3–10 percent at middle latitudes should have followed the eruption of Mt. Pinatubo (Brasseur and Granier 1992; Rodriguez et al. 1994). And losses of this magnitude or greater were indeed observed over mid- and high-latitude regions. Much as in the case of the El Chichon eruption, low ozone levels and enhanced levels of reactive chlorine coincident with high aerosol loading were found in both the northern and southern hemispheres over mid- and high-latitude locations (Grant et al. 1992; Weaver et al. 1993; Avallone et al. 1993; Hofmann et al. 1994).

At the poles, similar evidence of Mt. Pinatubo's influence have been found (Wilson et al. 1993; Webster et al. 1993; Hofmann and Oltmans 1993; Solomon et al. 1993). In addition to direct heterogeneous processing, the data suggest that sulfate aerosols can alter the formation of PSCs within the polar vortices. During the 1992 Antarctic ozone-depletion season, for example, greater amounts of PSC formation and PSCs with larger surface areas occurred at the altitudes with the greatest amounts of Pinatubo aerosol (Deshler et al. 1994).

An examination of ozone vertical profiles during the 1991 Antarctic ozone hole suggest that sulfate aerosols from the eruption of Mt. Hudson may have had a similar effect on ozone levels (Hofmann et al. 1992). Because of the more southerly location of this volcano and the atmospheric motions at the time of its eruption, Mt. Hudson primarily affected the 1991 Antarctic ozone hole, whereas the Pinatubo aerosol cloud was not a contributing factor until the 1992 springtime depletion. Furthermore, these factors contributed to the greater sulfate concentration over the Antarctic from the Mt. Hudson cloud, though the amount of aerosol released was less than in the case of Mt. Pinatubo (SORG 1993).

The precise share of the recent ozone declines (within the polar vortices and elsewhere) that is attributable to these eruptions is still uncertain. As the added sulfate from Mt. Hudson and especially from Mt. Pinatubo diminishes during the mid 1990s, the extent to which ozone loss returns to pre-eruption rates may provide some insight.

Though the effects of these eruptions are now likely diminishing, the implications of future eruptions during times of high stratospheric chlorine and bromine levels continue to warrant concern. Model calculations suggest that sulfate aerosols from volcanic eruptions could initiate local "catastrophic" ozone losses as high as 1 percent per day over mid-latitude locations, comparable to loss rates at the southern pole (Prather 1992b). This amplification depends critically on the concentration of chlorine in the stratosphere. What effect this might have on hemispheric-scale ozone levels in the wake of a large eruption is not certain.

Volcanic eruptions can also inject chlorine directly into the stratosphere in the form of hydrogen chloride (HCl). However, eruptions must have sufficient energy to reach the stratosphere and sufficient quantities of HCl in order to increase chlorine levels and influence stratospheric ozone depletion. The 1982 eruption of El Chichon in Mexico was a rare example of such an event. Approximately 40,000 tonnes of HCl were injected into the stratosphere. Scientists monitoring trace constituents detected an "increase of about 40 percent in the stratospheric column of hydrogen chloride between 20 and 40 degrees north" (WMO 1985, p. 115). This regional increase was calculated to have increased total stratospheric HCl by roughly 9 percent (Mankin and Coffey 1984). After the Mt. Pinatubo eruption, stratospheric chlorine levels increased by less than 1 percent (Tabazadeh and Turco 1993).

Volcanic eruptions that do not send hydrogen chloride into the stratosphere do not materially affect ozone depletion, since HCl is quickly rained out of the troposphere. Model calculations and historical investigations both support this view. Modeling studies suggest that less than 1 percent of the total HCl emitted by a volcanic eruption will make it to the stratosphere (ibid.). Analyses of ice-core data indicate large increases in SO_2 but not coincident increases in HCl after major eruptions (Delmas 1992, p. 12, Figure 13).

Implications of Polar Ozone Loss for Middle and High Latitudes

The seasonal ozone-depletion events that occur over the Arctic and the Antarctic are not completely isolated from the rest of the atmosphere. We have described the mechanisms currently proposed for exchange of either ozone-poor or chemically processed air between the poles (particularly the northern vortex) and the high latitudes. Whether or not there is air exchange during the polar ozone-depletion season, the annual breakup of the vortices in late winter or early spring inevitably results in a transient decrease of ozone amounts over the remainder of the stratosphere as ozone-poor air is mixed in.

In view of this connection to locations other than the poles, it is important to determine to what extent ozone depletion can worsen in the polar regions. The Antarctic ozone hole, for example, is not expected to deepen significantly, because the altitudes at which PSCs form experience near-total ozone removal during the height of the southern depletion season. However, it can continue to widen. The boundary of the ozone hole (defined by ozone levels of less than 200 Dobson units) is still well within the boundary of the polar vortex. If the ozone hole were to extend to the edge of the vortex boundary, the hole would roughly double in size (Schoeberl and Hartmann 1991). Continued widening of the Antarctic ozone hole has potentially serious implications for mid- and high-latitude locations in the southern hemisphere. Australia, New Zealand, and southern Latin America may experience much greater ozone depletion during and immediately after the Antarctic ozone-depletion season as a result of further widening.

Arctic ozone loss is controlled less by the absolute amount of chlorine and more by the persistence of PSCs through the northern winter. Amounts of ozone lost over the Arctic are difficult to predict because of varying meteorological conditions. Winters in which stratospheric temperatures remain low beyond the middle of February and in which some denitrification occurs may lead to much greater ozone loss than has been observed to date (Salawitch et al. 1993). Furthermore, because the northern vortex lacks latitudinal and longitudinal symmetry, portions of PSC-processed air can occasionally travel through sunlit portions of high latitudes, giving rise to significant regional ozone losses in those places (Water et al. 1993; Proffitt et al. 1990; SORG 1993).

Aircraft

In the early 1970s, concern over the loss of stratospheric ozone centered around the proposal to fly subsonic and supersonic aircraft in the lower stratosphere and the potential atmospheric impacts of the proposed Space Shuttle.

In the case of commercial aircraft, it was feared that large emissions of nitric oxide (NO) and nitrogen dioxide (NO_2), together known as nitrogen oxides (NOX), would lead to significant catalytic losses of ozone in the lower stratosphere (Johnston 1971) via reaction sequence 1.7. The proposal for a fleet of high-flying aircraft was never pursued, and concern soon shifted to emissions of ozone-depleting compounds.

However, flying aircraft in the lower stratosphere is under consideration once again (Johnston et al. 1989). The impacts of this plan (which incorporates advances in the understanding of heterogeneous chemical reaction mechanisms achieved since the 1970s) were concisely stated in a World Meteorological Organization report:

Projected fleets of supersonic transports would lead to significant changes in trace-species concentrations, especially in the North Atlantic flight corridor. Two-dimensional model calculations of the impact of a projected fleet (500 aircraft, each emitting 15 grams of NO_X per kilogram of fuel burned at Mach 2.4 in a stratosphere with a chlorine loading of 3.7 ppb, implies additional (i.e., beyond those from halocarbon losses) annual-average ozone column decreases of 0.3–1.8 percent for the Northern Hemisphere. There are, however, important uncertainties in these model results, especially in the stratosphere below 25 km. (WMO 1994)

The impact arising from NO_X emissions in the lower stratosphere that this implies is significant, thought not as large as the original estimates.[12] The net effect may also change with varying amounts of chlorine in the stratosphere. We believe that further laboratory measurements and model refinements are necessary in order to conclusively determine the net effect of high-flying aircraft on stratospheric ozone.

Solid-Fuel Rockets

In 1973, a year before Molina and Rowland published the theory linking chlorinated halocarbon emissions to ozone depletion, two National Aeronautics and Space Administration scientists investigated the possibility that emissions from the proposed Space Shuttle would affect the ozone layer. Stolarski and Cicerone (1974) discovered that the exhaust from the shuttle's solid-fuel rockets would deposit hydrochloric acid (HCl) in the stratosphere, and that the HCl, after being broken down by UV radiation, would release reactive chlorine that might lead to the catalytic destruction of ozone. The discovery of chlorinated chemicals as a large source of chlorine soon overshadowed this relatively small source.

Concern over increasing ozone loss and the continued use of solid-fuel rockets has prompted a reassessment of this stratospheric chlorine source (WMO 1991, chapter 10). The largest rockets in this category, the American Space Shuttle and Titan IV rockets and the European Ariane 5, all use solid fuel composed of ammonium perchlorate, aluminum, and a polymer matrix.

Use of a chemical-dynamical model led to the conclusion that continual rocket launches could lead to a buildup of chlorine in the stratosphere, the magnitude of which would be governed by the frequency of launches and the rate of stratosphere-troposphere circulation. For a scenario of nine Space Shuttle and six Titan IV launches per year, the total chlorine added to the stratosphere was computed to be less than approximately 0.25 percent of the amount added by ground-based ODC emissions each year. Local depletion in the exhaust wake (within a few kilometers of the wake axis) of individual ascending rockets was found to be nearly total at altitudes of approximately 30–40 km. Owing to rapid horizontal mixing, this effect dissipated within hours. At no time were local column ozone losses computed to be greater than 10 percent.

Global impacts due to the added burden of chlorine in the stratosphere were much smaller. Stratospheric ozone loss was calculated to be less than 0.2 percent locally (at an altitude of approximately 40 km in winter), total column ozone loss to be less than 0.1 percent. None of these calculations took into account heterogeneous reactions resulting from existing aerosol or from aerosol in the exhaust.

Although this source appears small at the present time, the continuing unexpected ozone loss in the stratosphere emphasizes the need to reduce chlorine levels in every possible way. The amounts and the ozone-depletion implications of heterogeneous mechanisms must be determined, and observations must be intensified during launches.

Ozone-Depleting Compounds and Global Warming

"Global warming" and "greenhouse effect" are expressions commonly used to describe the increase in global average temperature due to the continued release of greenhouse gases to the atmosphere. The principal gas of concern, carbon dioxide (CO_2), a product of the burning of fossil fuels and of nonrenewable biomass, has consistently exceeded the pre-1850 levels. Other greenhouse gases are methane, nitrous oxide, ozone, water vapor, and ozone-depleting compounds. Although projections about temperature increases and potential nonlinear effects are still subjects of considerable uncertainty and debate, most studies indicate that the average surface temperature of the Earth may be from 1.5°C to 4.5°C higher around the year 2030 (WMO 1990).

The greenhouse effect is a relatively well-understood natural phenomenon, described as early as 1827 in a paper by the physicist-mathematician Jean-Baptiste Fourier. The Earth receives a constant amount of energy from the sun mostly in the form of visible or "incoming" radiation. Some of this energy is reflected directly back to space by the atmosphere and the surface, but most of it (approximately 70 percent) is absorbed. The absorbed energy raises the temperature of the surface and that of the atmosphere. This same amount of energy is re-emitted to space as thermal (infrared) or "outgoing" radiation. Thus, the amount of energy entering the Earth-atmosphere system is equal to the amount leaving. This energy balance is necessary to maintain the Earth's temperature at an approximately constant average level, which is currently about 15°C.

The essence of the natural greenhouse effect is the ability of particular trace gases in the atmosphere, primarily water vapor and carbon dioxide, to absorb some of the infrared radiation on its way out to space from the surface, thereby reducing the energy outflow and raising the temperature of the Earth-atmosphere system above what it would be without the presence of the atmosphere. It is this "blanketing" effect of the atmosphere that is referred to as the greenhouse effect. Without the presence of the atmosphere, the global average surface temperature would be approximately 43°C below the present 15°C. The trace gases in the atmosphere have an effect similar to that of the glass panes of a greenhouse, letting in visible (short-wave) radiation but impeding the exit of long-wave, infrared radiation (thermal radiation) and thereby increasing the temperature inside the "greenhouse."

The present concern centers not on the natural greenhouse effect of the atmosphere, which is essential to life, but on the effects resulting from increases in the concentrations of "greenhouse gases." Such increases could give rise to two deleterious effects: an increase in the Earth's average temperature (with severe consequences for many areas) and a rate of change in temperature that would be far more rapid than any rate attributable to natural variations. A higher rate of change could make it difficult or impossible for many biological species to survive.

It was recognized in the 1970s that ODCs behave in a manner similar to carbon dioxide, trapping heat that would otherwise be radiated into space (Ramanathan 1975). For some ODCs, each molecule is thousands of times more effective than a single carbon dioxide molecule in trapping infrared radiation (Lashof and Ahuja 1990). The ODCs emitted during the 1980s are calculated to account for approximately 25 percent of projected global warming due to all greenhouse-gas emissions during that decade (Hansen et al. 1989).

In a 1990 scientific assessment of potential climate change, it was estimated that CFC-11, CFC-12, and HCFC-22 would account for approximately 12 percent of the warming estimated to occur by the year 2025 (WMO 1990, p. 61).[13] This implies that these compounds would be responsible for approximately 0.3°C of the 2.5°C increase calculated to occur by 2025. These estimates did not incorporate the influence of methyl chloroform, carbon tetrachloride, and CFC-113, which together may have a substantial global-warming effect; nor did they consider feedback mechanisms involving the parallel effects of ODCs on ozone depletion, which may produce cooling.

Rather intensive investigation of the potential feedback mechanisms has now begun. Normally, much of the ultraviolet energy of incoming sunlight is absorbed by ozone in the upper atmosphere. When ozone is depleted, however, less UV radiation is absorbed in the upper stratosphere, and much more continues down into the lower stratosphere and the troposphere. The heating that would normally take place in the upper stratosphere is reduced, and that leads to cooling of the stratosphere. Since the troposphere receives some energy from the stratosphere, cooling of the troposphere may result. At the same time, the increased transmission and subsequent absorption of UV in the troposphere heats that layer. In addition, a portion of the UV absorbed in the troposphere can contribute to greater tropospheric ozone production. Since ozone is a greenhouse gas, this may lead to tropospheric warming.

Tropospheric temperature effects due to increases in UV have been difficult to detect because of the many human influences affecting climate and the large amount of visible sunlight that is absorbed by the troposphere. However, stratospheric cooling has in fact been observed in some regions. In a World Meteorological Organization assessment, scientists found that "in the upper stratosphere, the satellite data for the period [1979–1986] indicate a global temperature change of -1.5 ± 1°K [about −3°F]. It is compatible with the −3 ± 4 percent change in ozone concentration around 40 km as observed by SAGE [satellite measurement] instruments." (WMO 1989, p. xiii)

Cooling in the stratosphere may imply changes in the ozone-depletion process. For instance, a U.K. study pointed out how such effects might occur at other latitudes:

> It has long been recognized that the ozone destruction reactions in gas-phase chemistry proceed more slowly at lower temperatures. Thus the expected stratospheric cooling due to enhanced concentrations of carbon dioxide may reduce ozone loss, partially offsetting that from chlorine-catalyzed ozone depletion. This mechanism remains of potential importance in the middle to upper stratosphere.
>
> In the lower stratosphere, on the other hand, a decrease in temperature could lead to ozone depletion. PSCs would be expected to increase in frequency and duration, leading to a greater activation of reactive chlorine. Similar reactions on sulfate aerosols could also proceed more rapidly at the lower temperature. It is not known whether the observed trends have resulted in an increased occurrence of PSCs. *However, the fact that PSC occurrence is determined by a threshold temperature means that an ozone response may be sensitive to temperature change.* (SORG 1993, p. 18; emphasis added)

Future PSC occurrence may also be enhanced by an expected increase in stratospheric water vapor (Mahlman 1992). The stratosphere's internal water source is the oxidation of methane, which is currently increasing. Greater amounts of water vapor will allow PSCs to form at a higher temperature. The formation of PSCs in the Arctic vortex is currently limited by temperatures which are typically higher than in the Antarctic vortex. This could aggravate ozone depletion via the heterogeneous reactions described above.

The link among PSCs, temperature, and ozone may also initiate a feedback cycle. Less ozone in the stratosphere can lead to cooling, which induces the formation of more PSCs, which liberate more active chlorine, leading to greater ozone depletion, and so on.

An attempt to model these potential interactions between global warming and ozone depletion resulted in the occurrence of an Arctic ozone hole (Austin et al. 1992). Doubling carbon dioxide and including the radiative impacts of ozone loss in a three-dimensional model reduced ozone levels over a portion of the Arctic region to under 200 Dobson units, the defining line for the Antarctic ozone hole.

Changes in stratospheric temperature may also affect the ozone-depletion processes at the poles by altering the motions of air. According to research performed by the U.K. Stratospheric Ozone Review Group, changes in wind patterns occasioned by changes in temperature may extend the period of ozone depletion over the poles by preventing transport of ozone-rich air into the region (SORG 1991, p. 10).

There are many other unknowns. For example, changes in the global climate may also affect the dynamical processes that contribute to the global distribution of stratospheric ozone. According to the Stratospheric Ozone Review Group:

> Changes in the temperature structure of the troposphere and stratosphere are also likely to affect wind patterns in the lower stratosphere. This will directly affect column ozone amounts, since the lifetime of ozone is relatively long in the lower stratosphere and its distribution is significantly affected by the circulation of the atmosphere. Furthermore, changes in circulation patterns may indirectly affect ozone amounts throughout the stratosphere by altering the exchange of air between the troposphere and stratosphere, thus affecting the time taken for source gases to reach the stratosphere. The relative importance of these effects is not well understood. (ibid., p. 11)

In sum, there are a number of processes associated with ODC emissions that influence the temperature of the Earth's surface. The largest direct heating effects are those due to the increasing ODC concentrations. There may also be a cooling effect due to an overall decline in the amount of ozone. It is a matter of considerable debate at the present time whether the net effect of these competing phenomena is heating or cooling, although most studies have shown that cooling as a result of ozone loss very nearly offsets the warming induced by ODCs (Ramaswamy et al. 1992; WMO 1991, chapter 7; Lacis et al. 1990). In a manner of speaking, cooling due to ozone depletion may currently be masking some of the heating due to carbon dioxide and other greenhouse gases.

It would be unwise to delay efforts to reduce the concentration of greenhouse gases because of postulated cooling due to ozone depletion. First, ozone depletion is a serious health and environmental problem. Thus, the efforts to reverse ozone depletion must be increased, especially since the depletion is worse than anticipated. Second, it is important to pay attention to the timing of the various cooling and heating effects. Policies to protect the ozone layer now being rapidly implemented may

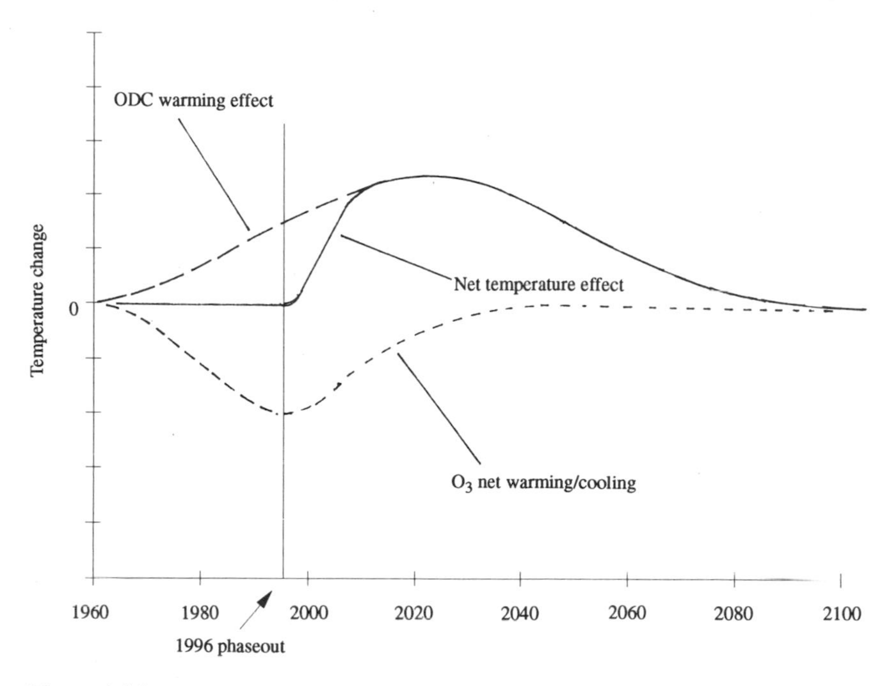

Figure 1.15
Effect of delayed phaseout of ODCs on global warming.

result in a restoration of the ozone layer by the middle of the 21st century. At that time, when the warming effects from pollutants such as carbon dioxide and methane may be at their peak, the cooling effect of ozone loss will disappear, and its disappearance may be as abrupt as the sudden appearance of the Antarctic ozone hole in the late 1970s. *The result may be rapid warming as the heating effect of ODCs is unmasked.*

Delaying the phasing out of ODCs would only increase the extent of this rapid warming. (For a qualitative rendering of this, see figure 1.15.) Many of the worst ecological effects of global warming are postulated to arise from the speed of the temperature rise, not just from its magnitude. The sudden unmasking of ODC heating may aggravate these problems considerably.

2

The Effects of Ozone Depletion

2.1 Biological Effects of UV-B Radiation

The primary motivation behind policy addressing the ozone-depletion issue is the fact that decreases in column ozone allow greater amounts of biologically damaging ultraviolet radiation to reach the surface of the Earth.

Not all radiation is equally damaging to all living systems. Most living things coexist with UV radiation. In fact, some exposure to UV radiation is needed for the formation of Vitamin D in humans. UV-A (radiation with wavelengths above 320 nanometers) does cause some biological damage, but it also stimulates repair mechanisms, so that such exposure is, at least to some extent, self-correcting in most people, if it remains within natural limits.

UV-B (290–320 nm) and UV-C (40–290 nm), in contrast, can cause severe biological injury because of their higher energy per photon, and because terrestrial life evolved with very little UV-B and UV-C radiation. This is particularly true of UV-C and of UV-B below 300 nm.[1] Although only a small amount of UV-B penetrates the atmosphere, it causes damage even at present levels. Almost all the UV-B that penetrates the stratosphere is in the less energetic 300–320-nm range. The results of UV-B radiation at the surface, even at pre-ozone-depletion levels, include skin aging and skin cancer in humans as well as damage to plants and animals.

Even moderate increases in UV-B can result in substantial harm to human beings and other forms of life. Four factors must be examined when considering the effects of ozone depletion on individual organisms and ecosystems:

the range of UV wavelengths to which organisms will be exposed as a result of ozone depletion,
the range of UV wavelengths that cause the greatest damage,
the sizes of the organisms considered, and
the damage done to individual organisms and ecological systems by increased UV radiation and their ability to repair the damage or to adapt to the increased radiation.

Optical density is a measure of how much incoming radiation of a particular wavelength a substance will absorb. As figure 2.1 shows, ozone absorbs UV efficiently, with a maximum absorption at about 255 nm. Molecular oxygen and ozone ensure that essentially no UV with wavelengths shorter than about 280 or 290 nm reaches the surface.[2] By the time solar radiation with wavelengths of 260 nm and 278 nm reaches the surface, its intensity has been reduced by factors of (respectively) 10^{40} and 10^{17}.

Since nucleic acids, the building blocks of deoxyribonucleic acid (DNA), absorb the greatest amount of radiation at a wavelength of 260 nm, and proteins absorb the greatest at a wavelength of 278 nm, it is believed that life on land could not have evolved without the protection of ozone (Smith 1977, pp. 146–147). Life first appeared underwater because the ocean screened out lethal ultraviolet radiation emitted by the sun. In fact, at that time there were no molecules in the atmosphere that could absorb UV (Levine 1991, p. xxv).

It is thought that early forms of underwater life existed at depths greater than 10 meters. This thickness was sufficient to screen out any UV with wavelengths below 300 nm (Smith 1977, p. 146). Life forms suited to land developed only after the levels of oxygen and ozone in the UV-B. As plant life on land increased, oxygen and column ozone levels also increased, allowing for further evolution of life on land shielded from UV-B radiation.

Living things have limited mechanisms to protect themselves from natural UV-B exposure. However, a wide variety of damage to particular species, as well as to ecosystems, can be expected with the increased UV-B exposure that will result from reduced stratospheric ozone. Such dam-

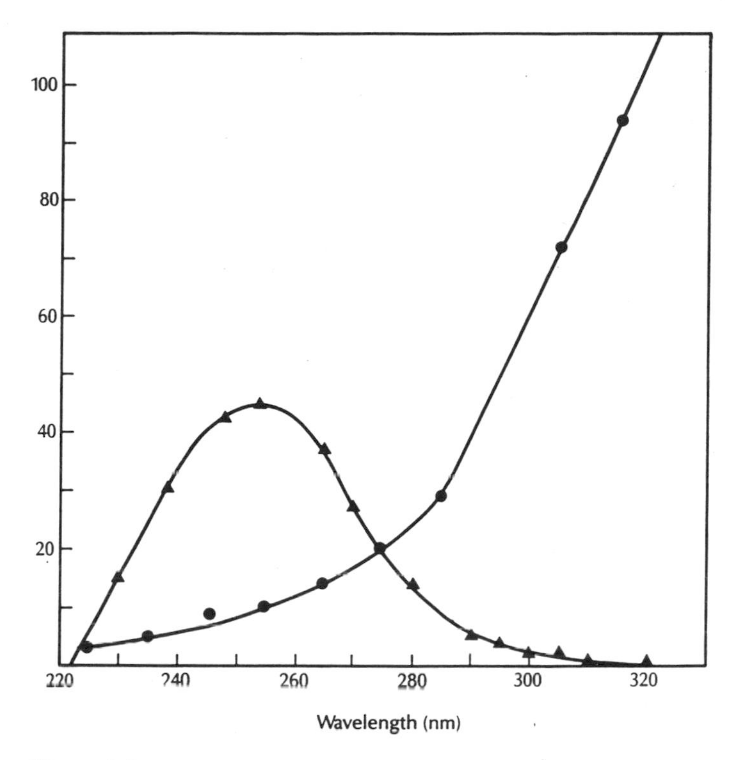

Figure 2.1
(●) UV irradiance in absence of ozone (photons $sec^{-1}cm^{-2}nm^{-1} \times 10^{12}$) and (▲) optical density of atmospheric ozone layer (3 mm) (Smith 1977, p. 147).

age can occur in single organisms as a result of damage to the defense mechanisms and/or as a result of those mechanisms' being overwhelmed. Such damage has already been documented in the case of marine phytoplankton during the Antarctic ozone-depletion season. In addition, there may be considerable harm to ecosystems as a whole in which injury to one class of creatures propagates through one or more ecosystems.

To be harmed by UV-B, an organism must absorb the radiation. Different organisms absorb UV-B to varying degrees. In some cases the radiation simply passes through organisms; such organisms are called **transparent**. For example, *Escherichia coli* bacteria absorb 25 percent of incident radiation at 250 nm but are almost transparent above 350 nm. In contrast, blue-green algae, which are as small as *E. coli*, absorb strongly at 350 nm (Smith 1977, pp. 21–22). Further, among

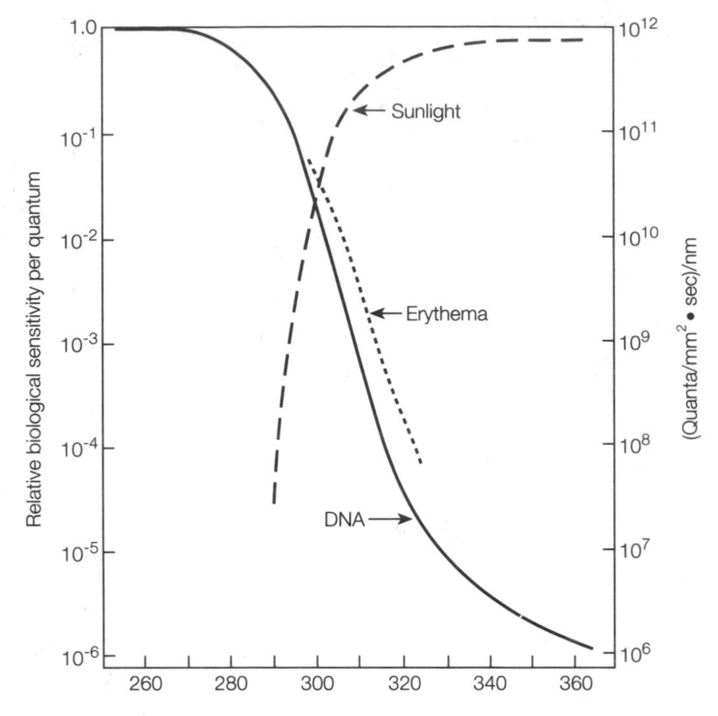

Figure 2.2
Average DNA action spectrum (EPA 1987, vol. IV, appendix A, p. 2-10).

those creatures that do absorb radiation, sensitivity to UV-B varies considerably.

One measure of an organism's UV-B sensitivity is its **action spectrum**, which shows how well a system absorbs and is affected by radiation of different wavelengths. Peaks in the action spectrum occur at wavelengths at which the amount of biological damage per unit of incident energy is the greatest (ibid., p. 25).

Figure 2.2 shows the action spectrum of DNA. The genetic material of cells is contained in long molecules of DNA which are coiled in the form of a double helix. DNA alteration is responsible for mutations and cancers. The DNA-altering effectiveness of UV-B increases by a factor of about 10,000 from 320 nm to 290 nm. For example, in human beings, the action spectra for both skin erythema (skin burn caused by radiation) and DNA are high in the range 290–300 nm (ibid., p. 151). Thus, an increase in UV-B in this range would produce more biological harm than a considerably larger increase at 320 nm.

Besides variations in UV sensitivity between organisms, size is important in determining the extent to which an organism will be affected. UV-B is absorbed by a few layers of cells. Thus, the larger the biological system, the more protected its interior; the smaller the organism, the more UV will penetrate its vital areas. As a result, unicellular organisms, such as those at the base of aquatic ecosystems, are among the most severely affected by UV.

For an organism as large as a human being, the direct effects of UV-B are primarily on and via exposed organs such as the skin and the eye. The internal organs (the reproductive organs, for example) of human beings are protected from direct UV-B, but they may be indirectly affected. For example, immune response may be weakened by UV-B exposure.

The mechanism that organisms have developed to protect themselves from the UV-B to which they are exposed naturally include avoidance of exposure, shielding (e.g., pigmentation), repair of DNA, and repair of tissue damage (from burns, for example). Many sensitive organisms, particularly aquatic ones, are relatively inactive in the middle of the day, when UV is at its maximum intensity. Almost all light-exposed organisms, including humans, have developed some form of external pigmentation to protect them from UV exposure. All organisms have the ability to repair damaged DNA. However, for many organisms the above mechanisms may not be sufficient to protect adequately against increased levels of UV-B, as is demonstrated by the reduction in the primary productivity of Antarctic phytoplankton associated with ozone depletion (see below). For some organisms, increases in UV may not result in increased tolerance through adaptation because radiation increases may be too rapid to allow the organisms to adapt. Rather, the result may be severe depletion of populations or complete extinction.

2.2 DNA Damage

Exposure to UV-B can cause damage to DNA. Since the genetic information of living beings is carried in DNA, they have many ways of repairing such damage. Yet exposure to substances that damage DNA increases the probability that some DNA will not be repaired and that cells will reproduce this altered DNA. Such exposure is classified as **mutagenic**, because it can mutate genes. Production of cells injured in this way can lead to cancer and other diseases.

Mutagenic effects of UV-B occur in a number of different ways. Eisenstark (1989) has described the effects of DNA damage by "near ultraviolet radiation," defined to include both UV-B and UV-A. According to Eisenstark's account (ibid., p. 104), absorption of near ultraviolet radiation may (1) directly alter DNA; (2) occur first in special cells called photoreceptors (since their function is to absorb photons of electromagnetic energy), which then transfer the energy to adjacent DNA strands, causing damage; (3) increase reactive varieties of oxygen in cells, causing mutagenic and toxic effects; (4) destroy critical enzymes; (5) destroy thiolated tRNA, a chemical important to growth of cells. There appears to be a sudden drop in "lethality, mutagenesis, and single-strand DNA breaks" above 330–340 nm (ibid., p. 105). Effects 2–5 listed above are indirect ways in which UV damages DNA. In these cases, radiation is first absorbed outside DNA and then imparted to the molecule, producing damage in diverse and complex ways (ibid., p. 104; Huselton and Hill 1990, p. 38).

Since all life has evolved essentially in the absence of the more energetic portion of the UV-B spectrum (below 300 nm), it is not known what damage to the genetic structure of living beings would result from sustained high exposure to such radiation.

Increased UV-B is known to cause damage to the DNA of phytoplankton, a single-cell plant at the base of the marine food chain that is a vital part of the carbon cycle, converting a considerable amount of the carbon dioxide in the atmosphere to organic carbon.

2.3 Effects on the Human Immune System

The deleterious effects of UV radiation on the immune system may be among the most important aspects of severe ozone depletion. Besides the direct effect of increased susceptibility to disease on individuals, severe ozone loss may cause widespread immune-system impairment and thereby increase the risk of epidemics. The medical, epidemiological, social, and economic aspects of such large scale problems have not yet been carefully addressed, apart from problems arising from increased skin cancer. In this section we discuss the effects of UV-B on the human immune system. We will discuss cancer, infectious diseases, and damage to disease resistance in plants and animals in separate sections.

The Human Immune System[3]

Immunity is the capacity of the human body to protect itself from foreign substances, such as disease-producing organisms. Immunity derives from a number of cell types, each with a specifically designated function, which allow the body to rapidly protect itself. Principal among these are **B lymphocytes**, derived from the bone marrow, and **T lymphocytes**, derived from the thymus gland. These cells and their products, notably **antibodies** and **lymphokines**, are largely responsible for the body's protective immunity, though other cells also play important roles. Other cells that provide immunity to the body include monocytes, macrophages, and neutrophils.

Antigens are the components of foreign substances that trigger the immune system. The introduction of antigens into a healthy person causes an activation and proliferation of specific T and B cells to fight the antigens. Stimulated B cells differentiate into antibody-secreting cells, which produce large molecules called **immunoglobulins** (also known as **antibodies**). These immunoglobulins are released into the lymph and travel to various parts of the body to destroy antigens. Immunity deriving from B cells is called **humoral immunity**.

Stimulated T cells work somewhat differently. Instead of releasing chemicals such as immunoglobulins, some T cells, called "killer" T cells, develop **cytotoxic** properties (that is, properties that are toxic to living cells), which enable them to destroy target cells bearing antigens. This aspect of immunity is called **cellular immunity**. There are also other types of T cells, namely "helper" T cells, which enhance the expression of immunity, and "suppressor" T cells, which reduce the immune response. Working together, these three types of T cells control cellular immune response. Helper and suppressor T cells also regulate the working of the immune system. T cells also regulate the production and the efficacy of both T and B cells. Thus, the functioning of the immune system is highly orchestrated, and disorders in one part can have profound effects on the whole.

Structure and Function of the Skin

The human skin is the organ most affected by UV. It consists of two distinctive layers, the outer one called the **epidermis** and the inner one called the **dermis**. Cells of the epidermis include resident melanocytes, immigrant Langerhans cells, lymphocytes, monocytes, macrophages, and neutrophils. Melanocytes synthesize brown, red, and yellow melanin pigments that give each person distinctive skin and hair coloration. Langerhans cells, derived from bone marrow, contain surface receptors for immunoglobulins and are able to capture external antigens that contact the skin; after doing so, they circulate to the lymph nodes to stimulate T lymphocytes.

Both layers of the skin contain proteins and nucleic acids that can absorb solar energy. Less than 10 percent of the UV-B incident on the surface of the skin penetrates to the deeper parts of the epidermis. UV-B appears to be most efficient in affecting immune response by interacting with Langerhans cells and activating suppressor T cells. The net result is systemic immune suppression, because suppressor T cells spread throughout the body and diminish the efficiency of natural killer T cells in destroying antigens.

Effects of UV on the Human Immune System[4]

The study of the effects of ultraviolet radiation on the immune system is a relatively new field, most of the work having been done since around 1980. The degree of knowledge varies a good deal from one area of the field to another. Much of our knowledge of the immune system's response to UV is derived from laboratory studies of animals. According to a recent review: "In experimental animals, exposure to UV-B radiation produces selective alterations of immune function which are mainly in the form of suppression of immune responses. This immune suppression response is important in the development of nonmelanoma skin cancer, may influence the development and course of infectious disease and possibly protects against autoimmune reactions." (Morison 1989, p. 515) The suppression of the immune system observed in laboratory animals appears to be selective. Cellular immune response is suppressed, but humoral immunity is mostly unaffected (ibid.).

The first realization that exposure to UV affected the human immune system came from observations relating to conditions, such as skin rashes and allergies, that occurred in some individuals after exposure to sunlight These observations led to the hypothesis that exposure to UV (mainly UV-A) triggered immune responses through the formation and release of antigens. However, experimental work on animals led to the conclusion that the selective suppression of cell-based immune function observed in the laboratory also occurs in humans (ibid., p. 516).

The entire spectrum of UV radiation reduces the number of Langerhans cells, which detect the invading antigens first. This adversely affects the immune system's functioning, since detection, capture, and transport of antigens are essential preliminary steps toward the ultimate destruction of the antigen. According to one researcher in the field, "Of great interest ... are recent studies by Streilein and colleagues demonstrating that applying contact allergy-producing chemical to skin given suberythemal doses [doses insufficient to cause skin rash] of UV radiation fails to induce the expected immune response in about 40 percent of normal subjects; in patients who have had one or more skin cancers, the proportion of nonresponsive individuals approaches 100 percent." (Kripke 1991, p. 5)

Some of the clearest evidence of actual UV-B-induced damage to the human immune system comes from the study of lymphocytes. Studies of T cells in exposed humans (*in vivo* studies) and studies of laboratory lymphocyte cultures (*in vitro* studies) have demonstrated that UV-B depresses the levels of lymphocytes, which rid the body of invading antigens, while there is an increase in suppressor T cells. Similar evidence derives from studies of allergic reactions of the skin to **haptens**—relatively light molecules, such as poison ivy toxin. According to Kripke: "mice treated with very low (suberythemal) doses of UV-B radiation fail to make a contact allergy reaction to chemicals applied through irradiated skin. Instead suppressor lymphocytes are produced. The mechanism involves an alteration in the activity of . . . the epidermal Langerhans cell, which normally picks up foreign substances in the skin, carries them to the lymph node, and stimulates the lymphocytes to initiate an immune reaction. Exposure of Langerhans cells to UV radiation renders them unable to stimulate the appropriate type of lymphocytes and instead, suppressor lymphocytes are activated." (ibid., p. 3)

In sum, there is considerable, even decisive evidence that UV-B radiation adversely and substantially affects the cellular portion of the immune system. Severe, persistent depletion of the ozone layer would mean corresponding persistent and large increases in exposure to UV-B. The effects of such increases on the immune system's response are at present a matter for conjecture rather than scientific prediction.

Whereas the skin-cancer-causing properties of UV radiation appear to be related to pigmentation, some evidence indicates that such immune suppression may be independent of pigmentation. According to Kripke: "A . . . study compared the same immunosuppressive effect [involving the Langerhans cell] of UV radiation in different ethnic groups having different amounts of skin pigmentation. There were no differences in the immunosuppressive effect of UV radiation in persons with white, brown, or black skin, indicating that pigmentation is not protective against this form of immunosuppression. Thus, the population at risk of immunological damage is much larger than that at risk of developing skin cancer." (ibid.)

However, a detailed study of changes in Langerhans cells at low levels of UV-B did find differences in immune response between individuals with dark and light pigmentation. Dark-skinned Australians of Aboriginal descent showed some damage to these cells in that there was "an organized death of scattered cells" of the type that "commonly occurs during the physiological atrophy of tissues" (Hollis and Scheibner 1988, p. 29). However, Celtic Australians under the same low doses of UV-B exhibited greater and qualitatively different damage. Light-skinned Celtic subjects appeared to suffer suppression of immune response; dark-skinned subjects did not lose immune-system function, and in some of them exposure even stimulated it (ibid., p. 30). However, since this experiment was conducted at low doses, such retention may not continue at higher doses associated with severe ozone depletion.

It appears that even among individuals with the same pigmentation there are genetic differences in the amount of damage to the immune system resulting from exposure to UV-B. For instance, there are considerable differences in skin cancer rates among light-skinned individuals of identical pigmentation. Further, the amount of damage to Langerhans cells, as determined by hapten sensitivity, also appears to vary from one individual to the next. It appears that 35–40 percent of individuals without a history of skin cancer are susceptible to immune suppression of this type (Streilein 1991, p. 886). In contrast, essentially all individuals who have had basal-cell and/or squamous-cell skin cancer are susceptible to UV-induced immune suppression, which indicates a link between the two (ibid.).

It should also be noted that damage to the immune system, as determined by hapten sensitivity, appears to be considerably greater at wavelengths between 290 and 300 nm than at wavelengths between 300 and 320 nm (Elmets et al. 1985). This is similar to the wavelength dependence of other kinds of biological damage.

Clear evidence of UV-B-induced immune-system damage was provided by joint research conducted by the University of Michigan, the U.S. Veterans Administration, and the U.S. Environmental Protection Agency (EPA). The work included studies of both erythemal and suberythemal doses. The results show reduced immune-system response, indicated by reduced sensitivity to the hapten dinitrochlorobenzene among 95 percent of individuals at levels of UV-B radiation to which humans are commonly exposed. Further, the degree of reduced sensitivity was highly correlated with UV in the range of doses high enough to produce erythema, the radiation-induced skin rash. The research indicated that the reduction in immune response was due to a reduction in Langerhans cells in the areas of skin exposed to UV. The reduction was small but statistically significant at suberythemal doses; there was a far higher (71 percent) decline in skin exposed to erythemal doses. In addition, the radiation appears to induce an increase in $CD1a^{-}DR^{+}$ macrophages, which in turn activate suppressor cells. These suppressor cells also reduce immune function (Cooper et. al. 1992).

2.4 Infectious Diseases

Increases in UV-B may also aggravate the spread of infectious diseases, in part due to damage to the immune system; but there has been little research on the link. Morison (1989, p. 520) writes: "Given the importance of infectious diseases in determining human health, both quantitatively and qualitatively, it might be expected that the effects of UV radiation on these conditions would occupy a central position in recent advances in photoimmunology. On the contrary, we are almost totally lacking any information in this field." There have, however, been some experimental studies on laboratory animals, and there have been epidemiological investigations for a few diseases. The evidence indicates that, as a result of damage from increased UV-B exposure, diseases caused by microorganisms that attack the skin (parasites, bacteria, viruses, etc.) could be aggravated. This appears to be caused by the adverse effects of UV on the immune system, notably on cellular immunity provided by T cells.

It is well established that exposure to sunlight sometimes triggers "cold sores" on the lips caused by the herpes simplex virus. Laboratory work has confirmed that exposure to UV-B in amounts that would cause a rash (called erythema) produces such sores in most instances. Exposure to UV-B has also been found to trigger active infections in the genital region due to herpes simplex type 2 virus, which might be dormant without such exposure. It is thought that local changes in immune-system function may be responsible for converting latent infections into active ones. Experimental evidence indicates a definite role for UV-B in triggering such infections. Moreover, studies on mice show that the severity of infection is dependent on the UV-B dose and that the rate at which the infection heals appears to diminish as the dose increases (Morison 1989, p. 520).

There is also evidence that infections caused by the protozoan parasite leishmania could also be aggravated by exposure to UV-B (ibid., p. 521). Leishmaniasis is a disease that occurs in tropical and subtropical countries. The vector for the spread of this infectious disease is the sand fly, whose bite delivers the protozoan to the skin. Visceral leishmaniasis can be lethal. Laboratory experiments have shown that exposure to UV-B increased the vulnerability of mice to leishmania infections by reducing cellular immunity. Further experiments indicate that 290-nm UV-B radiation, which is at the higher-energy end of the UV-B spectrum, is more effective than lower-energy 320-nm radiation in spreading leishmania in mice (ibid.).

Other diseases may also be aggravated. One of these is schistosomiasis, a debilitating condition caused by parasites that enter through the skin. Already epidemic in Egypt and Ghana, it is also prevalent in Puerto Rico. Leprosy, caused by a species of bacteria called *Mycobacterium leprae*, continues to be epidemic in much of the Third World. There are millions of cases in India alone. Increased UV-B radiation could exacerbate it and slow the healing of its lesions.

In sum, all populations would be at risk as a result of increased UV-B radiation, but among those particularly threatened would be people living in the tropical and subtropical regions where infectious diseases are prevalent and high UV exposure is the norm (EPA 1987, vol. III, pp. 9-1–9-3) and people whose immune systems are already depressed (for example, those undergoing organ transplants or suffering from immune-deficiency diseases, of which the best known and most serious is AIDS) (ibid.).

The number and variety of organisms whose attacks upon the human body would be made much more powerful by the reduction of immunity caused by exposure to UV-B radiation is not yet known. Laboratory data indicating that UV-B makes mice more susceptible to many bacterial and fungal infections are not reassuring. According to Morison (1989, p. 521):

> Taken together, all these groups of reports indicate that infectious disease due to bacteria, fungi, viruses and protozoa can all be influenced, at least in experimental systems, by exposure to UV-B radiation and that this is often due to a systemic effect on immune function. Perhaps more importantly, in *each reported instance the exposure to radiation has resulted in a suppression of natural defense mechanisms* and there are no reports of either a failure to detect an effect of radiation or the findings of a positive or beneficial influence from such exposure. [emphasis added]

Since the effects of sustained exposure to high levels of UV-B radiation are unknown, and since the adverse effects of such radiation on the immune system seem to be clear and generally dose dependent (particularly the effects of high-energy 290–300-nm UV-B), it is possible that there may be severe epidemics of various diseases, not only in the tropics, but all over the world. It is not possible at present to say whether such outbreaks are likely or how severe they might be. We can only point to clear warning signs that diverse epidemics are possible.

2.5 UV-B and Skin Cancer

There is scientific consensus that ultraviolet radiation has a role in causing two of the three types of skin cancer: squamous-cell cancer and basal-cell cancer.[5] The third and most frequently deadly type, malignant melanoma, is linked to solar radiation mainly by epidemiological evidence, though one particular form of melanoma, lentigo maligna melanoma, appears to be related to cumulative UV exposure (Jones 1987, p. 444).

Roughly 95 percent of solar radiation incident upon human skin will penetrate the outer layer; the remaining 5 percent will be reflected (EPA 1987, vol. III, pp. 7–13). At all wavelengths, most of this radiation is absorbed into the melanin pigment, which lies just below the outer layer of the skin. However, some radiation will penetrate to lower layers of the skin, the amount depending on the wavelength of the radiation, the thickness of the upper layers of skin, and how much melanin is found in upper layers. Radiation that penetrates to these lower levels will expose basal cells, squamous cells, and the melanocytes that produce melanin, possibly leading to mutation and cancer.

Melanin appears to play a dual role in causing skin cancer:

> The primary function of epidermal melanins is believed to be passive screening of harmful UV radiation. However, their chemical structure, physical nature, and anatomic distribution suggest a more complex role. . . . Since melanins can both quench and generate photoinduced active oxygen species, they may play both a preventive and a causative role in solar carcinogenesis, depending on their location and circumstances. (Huselton and Hill 1990, pp. 37–38)

Huselton and Hill analyzed this dual nature of melanins in experiments conducted with UV-C radiation and concluded that the reason melanins have different effects may be due to variation in the density and structure of melanosomes (the granules within melanocytes that synthesize melanin) in people of different pigmentation. In light-skinned people, melanosomes are "aggregated, round, and less densely packed with melanin"; in people with darker pigmentation the melanosomes are "single, oval, overlapping and densely melanotic [filled with melanin]" (ibid., p. 38). As a result, Huselton and Hill hypothesize, more photons reach the nuclei of melanocytes in light-skinned individuals, and more active oxygen species, which can cause mutagenic chemical reactions, can escape when melanin is loosely packed. They also conclude that the risk of melanomas to individuals with dark skin, while considerably lower than that for light-skinned individuals, also increases with ultraviolet radiation. As a result, ozone depletion may cause unanticipated increases in melanomas among dark skinned individuals who spend extended periods of time in the sun (ibid., p. 42).

However, the incidence of basal-cell and squamous-cell cancers in dark-skinned individuals is very low. This includes dark-skinned persons who live in areas of high solar radiation (Streilein 1991, p. 885). This does not necessarily mean that there is no increase among individuals with dark pigmentation, but it would seem to suggest that such increases are not particularly noticeable, if they exist. Thus, further work would appear to be necessary to verify the hypothesis that these populations are at an increased risk of these cancers.

Nonmelanoma Skin Cancers

Basal-cell cancer and squamous-cell cancer are infrequently lethal, but quite common. There is a clear correlation between these cancers and lifetime exposure to solar radiation. There is some indication that squamous-cell cancer may be much more common than is generally believed, because in the United States it is normally treated on an outpatient basis and is "customarily not reported to tumor registries" (Glass and Hoover 1989, p. 2097). A health-maintenance organization's registry of squamous-cell cancer indicates that the incidence of this cancer among the HMO's 300,000 members increased by factors of 2.6 in men and 3.1 in women in the period 1960–1986. Increases in incidence occurred among young and old and appear to be associated with increased exposure to the sun (Glass and Hoover 1989).

About 700,000 diagnosed cases of basal-cell and squamous-cell skin cancer were expected in the United States in 1994 (Microsoft Encarta, 1994 edition). Of these about 2000 cases are expected to be fatal, accounting for about one-fourth of skin-cancer fatalities (Skolnick 1991).

The incidence of nonmelanoma skin cancer among certain populations is much higher than the rate implied by statistics for the light-skinned population as a whole. This is clearly illustrated by statistics from Australia, where the early-1990s estimate for these cancers is 242,000 per year in a population of 17.3 million (Ewan et al. 1991, p. 555). This represents an increase of 1.4 percent per year, which is about six times the U.S. rate (unadjusted for age or gender). These differences may be due in part to some combination of differences in pigmentation, genetic susceptibility, and lifestyle.

One group of researchers concluded that a 1 percent decrease in ozone would produce a 1.56 percent increase in annual "carcinogenic radiation," leading to a 2.7 percent increase in nonmelanoma skin cancers (Kelfkens et al. 1990, p. 819).[6] There is, however, a considerable range cited for these relatively common cancers. Increases of anywhere from 1 to 10 percent in nonmelanoma cancers for each 1 percent ozone loss are reported, increases between 1 and 3 percent being reported most often (MacKenzie 1990, p. 152).

Malignant Melanoma

There appears to be a more complex link between UV-B exposure and malignant melanoma, a relatively rare but frequently lethal form of skin cancer. A recent estimate put the lifetime risk of an individual's developing malignant melanoma in the United States at 1 in 105. This risk may reach 1 in 75 by the year 2000 if present trends persist (Skolnick 1991, p. 3217).

Although malignant melanoma represents only 3 percent of all skin cancers, it is responsible for about three-fourths of all skin-cancer deaths (Skolnick 1991; EPA 1987, vol. IV, appendix A, pp. 3–33). For the United States in the year 1991, Skolnick (ibid.) predicted 32,000 cases of malignant melanoma and 6500 deaths (Skolnick 1991). The number of malignant melanoma cases around the world has been increasing for several decades and is expected to continue increasing.

Malignant melanoma does not appear to be directly related to total solar exposure; there is no one-to-one correlation between increases in exposure and melanoma. Yet there is widespread agreement that exposure to solar radiation, particularly UV-B, is related to malignant melanoma in some way. However, there is some debate as to the details of the relationship.

The most important piece of evidence linking skin cancer to solar radiation is that UV, particularly UV-B, has led to cancerous growth in animals. It has not, it should be noted, induced malignant melanoma. However, as we have noted, UV-B is known to damage cell DNA. According to an EPA study: "... it clear that UV is carcinogenic and that UV-B wavelengths are most effective.... It is also clear that the shorter wavelengths of UV-B, those that would be increased by decreasing levels of ozone in the atmosphere, are more carcinogenic than the longer wavelengths" (EPA 1987, vol. IV, p. 16-12).

In general, the risk of developing malignant melanoma, as opposed to the milder forms of skin cancer, has been shown to be directly related to the sensitivity of an individual's skin to sun exposure. Almost all individuals affected with malignant melanoma have very little melanin pigment in their skin. Light-skinned persons are from seven to ten times as likely to contract malignant melanoma as dark-skinned persons (LeJeune 1986, p. 94). Figure 2.3 illustrates the different patterns of UV absorption for light-skinned and dark-skinned individuals. Although genetic factors other than skin color may reduce the susceptibility of dark-skinned individuals to skin cancer, there is a significant gradation in coloring even among the light-skinned. In general, those with "Celtic" characteristics (red hair, pale and freckled skin, green eyes) are the most susceptible. Some studies show that those who sunburn and freckle easily are four times as likely as other light-skinned individuals to suffer from malignant melanoma (ibid., p. 95).

A particularly important piece of evidence linking malignant melanoma to solar exposure can be found in people who suffer from xeroderma pigmentosum (XP), a rare recessive genetic disorder that makes an individual highly sensitive to the sun. Persons with XP are 2000 times as likely as others to develop malignant melanoma before they are 20 years old, because they have difficulty repairing solar-induced lesions to DNA

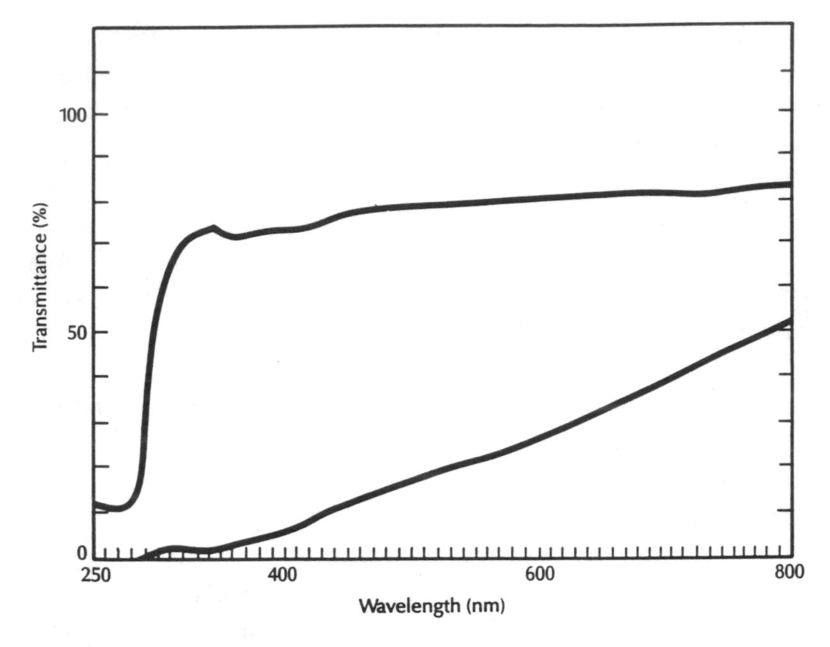

Figure 2.3
Measured epidermal transmittance in Caucasians (upper curve) and dark-skinned blacks (lower curve) (EPA 1987, vol. IV, appendix A, p. 3-19).

(EPA 1987, vol. III, p. 8-29). This indicates that increases in UV-B-induced cancer are linked to genetic factors and to immune-system damage.

The incidence of malignant melanoma is significantly higher among light-skinned persons who reside in areas with high solar intensity. For example, there is some evidence that incidence of malignant melanoma among light-skinned individuals is greater among those living nearer the equator, where UV radiation is higher.

Despite such latitudinal data on malignant melanomas, there is widespread agreement that the relationship between solar exposure and malignant melanoma is not as simple as that between solar exposure and basal-cell and squamous-cell cancers.

Malignant melanoma usually occurs in younger people, which indicates that cumulative exposure to the sun is not as important as in basal-cell and squamous-cell cancers. The most prevalent form, superficial spreading malignant melanoma (60–70 percent of cases), which is also the most dangerous, affects those with a median age of 40 on parts of the body less frequently exposed to sunlight (the torso in men and women, the backs of the legs in women) (LeJeune 1986, p. 94). Furthermore, malignant melanoma tends to in persons of higher socioeconomic status, who are unlikely to have spent long hours working in the sun. Some indoor workers, in fact, seem to be more susceptible to malignant melanoma than outdoor workers (Elwood and Hislop 1982, p. 168; Jones 1987, p. 444).

The above anomalies have led many scientists to hypothesize that malignant melanoma is tied to intermittent sun exposure, such as individuals of higher economic status may suffer during sporadic sunbathing or other recreational activities. A mild suntan such as is acquired working in the sun, some argue, protects the skin from UV exposure (Armstrong 1986, p. 153; Jones 1987, p. 444). However, the evidence for the intermittent exposure hypothesis is not conclusive. Other evidence suggests that malignant melanoma may be related to childhood exposure, particularly childhood sunburn. Yet other studies hypothesize a complex link between nevi (moles), solar exposure, and malignant melanoma (EPA 1987, vol. IV, appendix A, pp. 19-3 and 19-5). This debate is complicated by the fact that there are four different forms of malignant melanoma, each of which may arise in different ways.

The incidence of malignant melanoma has been increasing dramatically among light-skinned populations around the world for several decades (table 2.1). In several countries there has been a more than 100 percent total increase in the incidence of malignant melanoma over 10 or 15 years. In the United States here have been increases of 4–5 percent per year in both incidence and mortality for the last 10–20 years. These increases are found for all age groups and probably cannot be attributed solely to improved diagnoses.

The increase in malignant melanoma appears to be associated with increasing exposure to the sun due to leisure activities, such as sunbathing, and with the changes in dress styles that have increased exposure of the skin to solar radiation (EPA 1987, vol. IV, appendix A, p. 4-4).

Table 2.1
Increases in incidence and mortality from malignant melanoma (EPA 1987, p. 4-2).

			First period of observation		Second period of observation				
	State or country	Sex	Time	Rate per 10^6	Time	Rate per 10^6	Total increase (%)	Number of years	Annual percent increase
Incidence	N.Y.	M	1941–1943	1.2	1967	3.4	176	25	7.0
		F	1941–1943	1.8	1967	2.9	65	25	2.6
Incidence	Norway	M	1955	1.8	1970	6.3	264	15	17.6
		F	1955	2.6	1970	6.8	195	15	13.0
Mortality	Norway	M	1956–1960	1.6	1966–1970	2.7	69	10	6.9
		F	1956–1960	1.3	1966–1970	1.8	36	10	3.6
Mortality	Canada	M	1951–1955	0.7	1966–1970	1.4	93	15	6.2
		F	1951–1955	0.6	1966–1970	1.2	107	15	7.1
Mortality	U.K.	Both	1950	0.5	1967	1.0	100	16	6.3
Mortality	Australia	M	1931–1940	1.0	1961–1970	3.6	267	30	8.9
		F	1931–1940	0.8	1961–1970	2.5	267	30	7.6
Mortality	Denmark	M	1956–1960	1.6	1966–1969	2.4	49	10	4.9
		F	1956–1960	1.6	1966–1969	2.1	32	10	3.2

Mortality	Sweden	M	1956–1960	1.7	1966–1968	2.1	30	9	3.3
		F	1956–1960	1.1	1966–1968	1.5	40	9	4.4
Incidence	Conn.	M	1935–1939	1.1	1975–1979	8.2	645	40	16.1
		F	1935–1939	0.9	1975–1979	6.8	656	40	16.4
Incidence	U.S.	WM[a]	1974	6.7	1983	9.6	43	10	4.3
		WF	1974	6.0	1983	8.3	38	10	3.8
Mortality	U.S.	WM	1950	1.0	1977	2.6	160	27	5.9
		WF	1950	0.8	1977	1.6	200	27	7.4

a. W indicates white.

Recent increases in exposure to UV-B may be related to ozone depletion. It seems unlikely that changes in leisure patterns alone can account for the observed increases in malignant melanoma.

Scientists given scenarios of further ozone depletion have predicted increases in the incidence of malignant melanoma. One study by the National Cancer Institute, analyzing the UV-B level in seven U.S. cities, predicted that a 10 percent change in UV-B radiation would be associated with about a 7 percent increase in malignant melanoma incidence in males and a 7.5 percent increase in females (ibid., p. 19-12). A second study found that a 10 percent change in UV radiation would result in a 2.5–8.5 percent increase in mortality, depending on the assumptions (ibid.). More recently, McKenzie (1990, p. 152) and others have reported a 3.5–9 percent increase in melanoma cases, corresponding to a 10 percent increase in ultraviolet radiation.

Overall, the melanoma rate has been doubling approximately every decade, worldwide. Most of this doubling is probably due to causes other than ozone depletion, since it is likely that increases in ultraviolet radiation exposure, in bursts such as those which seem to be associated with melanoma, have been mainly due to other causes in the past (ibid., p. 154). However, this large rate of increase in melanoma would be exacerbated by increases in UV-B radiation due to ozone depletion. At the present rate, about 10 percent of all cancer deaths in the United States in the year 2010 are projected to be due to melanoma (Huselton and Hill 1990, p. 37).

2.6 Effects on the Human Eye

The human eye's structure enables us to see things that reflect or emit the visible portion of the solar spectrum (i.e., radiation with wavelengths in the range 400–700 nm).

Ultraviolet radiation reaches the eye not only when we look at the sun directly but also through scattering by the atmosphere and reflection from the Earth's surface in a manner similar to visible light. Most surfaces reflect less ultraviolet light than visible light. Snow and dry sand reflect the highest proportion of incident UV radiation. Snow reflects 50–85 percent of incident UV-B, dry sand 15–18 percent, concrete pavement

7–12 percent, ocean water 8 percent, boat decks 6–9 percent, soil 4–6 percent, and grass 1–5 percent (Charman 1990, p. 335).

The cornea (the outer tissue of the eye, containing live cells), the aqueous humor (the liquid within the eye), and the lens protect the retina from the deleterious effects of UV-B. The cornea screens out essentially all UV radiation with wavelengths shorter than 290 nm. Its effectiveness decreases at wavelengths longer than this, but it is still quite effective up to 300 nm. Less than 1 percent of UV-B is transmitted to the lens at wavelengths between 290 and 300 nm, and the lens itself screens out this radiation. The portion of UV-B not effectively removed by the cornea is remarkably coincident to the portion of UV-B filtered by stratospheric ozone, indicating a clear adaptation of the eye's structure to natural sunlight. The most effective range for producing damage to the lens is the 295–315-nm range of UV-B wavelengths (Yannuzzi et al. 1989, p. 38). Damage can occur to the conjunctiva (the mucous membrane covering the eye), the cornea, the lens, and the retina.

The Conjunctiva and the Cornea

The best-known effect of UV radiation on the eye is acute photokeratitis. This condition, also known as "snowblindness," is prevalent in environments where a great deal of light enters the eye after being reflected from snow. As we have noted, snow reflects 50–85 percent of incident UV-B, whereas vegetation reflects very little. Photokeratitis is like a sunburn of the cornea and the conjunctiva, with most damage occurring around wavelengths of 288 nm. Unlike the skin, however, the eye cannot adapt to repeated doses of UV-B (Emmet in EPA 1986, vol. I, p. 142).

The symptoms of photokeratitis are a gritty sensation in the eyes, tearing, and spasmodic closure of the eyelids. Symptoms begin after exposure, and healing occurs 48 hours later. Permanent injury is rare. However, in the wavelength region 288–290 nm, which is screened out almost totally by stratospheric ozone, death of corneal cells occurs. Exposure to these wavelengths causes painful and disabling, though usually temporary, injury to the eye. An increase in UV-B would put people living in many countries, and particularly outdoor workers, at higher risk of serious eye damage. Since snow and sand reflect a considerable amount of UV-B, vacationers (among others) would also be affected.

Benign tumors of the conjunctiva and the cornea are caused by ultraviolet radiation. UV can also trigger the development of certain corneal diseases, among them herpes simplex.

The Lens

Epidemiological studies indicate possible links between UV exposure and the occurrence of senile cataracts (an opacity of the lens in the elderly) (EPA 1987, vol. III, pp. 10-28–10-33). Although cataracts can be caused by both UV-A and UV-B, the latter is much more biologically effective—perhaps 250 times as effective at wavelengths below 290 nm as at wavelengths above 320 nm. A 1 percent increase in UV-B is expected to result in a 0.6–0.8 percent increase in cataracts (UNEP 1991g, p. 15). If not operated upon, cataracts result in blindness. In the United States, where there are about a million cataract operations each year, cataracts are still the number-three cause of legal blindness.

Increases in UV-B radiation will put an additional burden on older people's eyes. It will also be of significance in the Third World, where UV-B flux is higher and where surgical care is not routinely available to the poor. Increases in blindness will result.

The quantitative relationship between increased incidence of cataracts and increased radiation in the 290–300-nm range has not been well established. However, there is considerable evidence that a higher incidence of cataracts occurs in those parts of the world where there are greater amounts of natural UV radiation (for instance, the tropics) (Charman 1990, pp. 339–340). Although the causes of this higher incidence have not yet been clearly established, the known sensitivity of the lens to UV-B makes this radiation a likely cause. Thermal exposure may also be contributing.

The Retina

The retina is particularly sensitive to shorter wavelengths of UV-B radiation. Stable retinal disorders and retinal degeneration are two forms of blindness which may be caused by UV-B. According to a recent review, there is a serious dearth of research on this issue (Yannuzzi et al. 1989, p. 40).

The condition that puts an individual at the greatest risk is removal of the crystalline lens in a cataract operation, which exposes the retina to more UV-B unless it is otherwise protected. In the wealthier countries, it is now the normal practice to insert an artificial lens when the human lens is removed during a cataract operation. This provides some protection. However, in Third World countries most cataract operations are not accompanied by lens insertion. Partial protection of the retina occurs only while glasses are worn. Older people are also at greater risk.

The retina continues to develop after birth. Although experimental data on the exposure of the developing retina to UV are scant, there is concern that such exposure may cause disorders of visual development. Retinal damage leading to blindness in premature babies is known to be caused by an interaction of oxygen and light.

2.7 Effects on Plant Life

Knowledge concerning the effects of UV-B radiation on plant life is particularly deficient. Only four out of ten terrestrial plant ecosystems (temperate forest, agricultural, temperate grassland, and tundra and alpine ecosystems) have been studied. Among these four ecosystems, the most studied has been the agricultural crop ecosystem, and this only in the middle latitudes. The EPA has instituted a considerable research program to broaden the understanding of ozone-depletion impacts, including those on plant life and ecosystems. This research program is in its initial stages (Worrest 1991).

Though it is known that increased UV-B radiation damages DNA, the specific effects this damage has on plants are not well understood. According to the proceedings of an international conference held in February 1992 to discuss the state of knowledge on biological damage of UV-B radiation: "Experiments indicate that the level of free radicals in plant tissues increases under UV-B radiation. Little is known regarding the nature of these radicals, their role in UV-B-induced damage, or of the radical quenching systems in plants. It is suspected that free radicals play a role in radiation damage to membranes." (SCOPE 1992, p. 13)

In vitro studies have been performed on the interaction between UV-B and molecular systems in plants, but it turns out that action spectra for intact plants are quite different from isolated molecular systems. This indicates that laboratory research on molecular systems will not be enough to fill the vast gaps in our knowledge about the response of whole plants to UV-B radiation (ibid.).

Some findings of research on effects of UV-B on plants, as summarized at an international scientific workshop in October 1992, are as follows (SCOPE 1993, pp. 12–13):

In some cases there is a reduction in photosynthesis, and consequently in biomass production, as a result of increased UV-B exposure.

Some sensitive species exhibit decreased leaf expansion, which is linked to reduced biomass production.

Several UV-B receptors are responsible for interaction of plants with UV-B radiation and these receptors may interact.

Plants cultivated as monocultures may respond differently than plants in mixed-species stands.

Evidence of the effects on flowering is so far inconclusive.

Some plants may respond to increased UV-B by increasing their production of UV-B-absorbing pigments, thereby preventing reduction of photosynthesis. In a few cases, there may even be desirable effects. For instance, some plants used as food spices exhibit increases in production of ethereal oils.

Interactions of plants with UV-B are affected by nutrient and water availability. Water-stressed plants show a lower sensitivity to increased UV-B than well-watered plants. Phosphorous-deficient plants also seem to be less UV-B sensitive. Much work remains to be done in this area.

Some soil pollutants may interact with UV-B. In one experiment, the effects of cadmium on reducing the biomass of Norway spruce seedlings were exacerbated by increased UV-B.

Sensitivity of plants to increased UV-B radiation is highly species specific.

Crops

It is difficult to analyze the results for the crop species that have been tested. Most of the existing data are from greenhouses, where conditions differ from those in the field. Plants growing in growth chambers are more sensitive to UV-B than those grown outdoors. Also, the action spectra for most plants are unknown. Until 1991, only a dozen studies on about twenty crop species had been performed outdoors (Teramura 1991).

Despite these shortcomings, it is already clear that UV-B affects photosynthesis, nutrient content, and growth in many plants. According to Alan Teramura, who has performed extensive research on plant response to UV radiation: "Over the past 15 years, the scientific community has screened over 300 species and cultivars of crop plants and alarmingly, over one-half of these seemed to be sensitive to UV radiation. Among the most sensitive plant groups include peas, beans, squash, melons and cabbage." (ibid., pp. 1–2)

There are indications, furthermore, that weeds are generally more resistant to UV-B than crops. An increase in UV-B would thus likely alter the distribution and abundance of plants, disrupting ecosystems.

One important five-year study of the effects of UV-B on crops, performed at the University of Maryland, examined soybean yields. The United States is the world's leading producer of soybeans (its share was just over 50 percent in 1991), and soybeans are the third-largest crop in the United States. The United States provided almost two-thirds of the world's exports of soybeans in 1991 (DOC 1992, table 1107). A reduction in soybean yields would have serious consequences for the economies of the United States and the rest of the world.

In the Maryland study, a 25 percent depletion in ozone, simulated by artificial UV-B, reduced net soybean crop yields up to 25 percent; there were also reductions in the protein and oil contents of harvested soybeans (EPA 1987, vol. VIII, pp. 41–44). Currently, an estimated 56 percent of the global soybean crop is harvested; the remaining 44 percent is lost to weeds, insects, and diseases. A 25 percent reduction in stratospheric ozone could further reduce the crop yield to a mere 36 percent. It is also possible that increases in UV-B would increase crop losses due to weeds, diseases, and insects. Together, these two effects might cut net soybean yields in half (Teramura in EPA 1986, pp. 257–259). It should be noted that the soybean variety in the Maryland study, the Essex soybean, is particularly sensitive to UV-B. A combination of crossbreeding and switching to more durable strains could alleviate the problem. However, there are large uncertainties, particularly since UV-B-tolerant seeds seem to be of inferior quality with regard to yields and drought susceptibility.

One important result of research performed thus far has been to show that the response of plants to UV-B is very species specific, varying even among different cultivars of the same species (SCOPE 1993, p. 12).

Impacts on crop systems also need to be carefully considered. One important factor is the effect of increased UV-B radiation on interspecies competition in agriculture. Increased UV-B would preferentially reach taller plants in a mixed-species stand. This might affect intercropping, as well as competition between weed and crop species. Similar considerations apply to forest and grassland ecosystems (SCOPE 1992, p. 21).

Forests

The results from the extremely limited studies on tree species are similar to those from the crop studies. For example, of the fifteen species studied, more than half are UV sensitive. According to Teramura (1991, p. 3): "Preliminary evidence suggests that long lived plants such as trees can accumulate the damaging effects of UV radiation over many years.... Due to the long life span of trees and the lengthy time required for breeding, forests may be particularly vulnerable to increases in UV resulting from ozone depletion." One three-year study on the loblolly pine not only showed reduction of growth due to UV-B increases but also indicated that such damage was cumulative (SCOPE 1993, p. 12).

Such damage may be compounded by the postulated effects of global warming, since tree species migrate very slowly relative to the rapid temperature changes that may result from increases in greenhouse gases. Intense disruption and perhaps destruction of forest ecosystems may occur. The potential synergistic effects of global warming and ozone depletion on forest ecosystems are essentially unknown.

Noncrop Plants

About sixty noncrop plants have been screened for UV sensitivity. Though this is a minuscule number in relation to the great diversity of plant life in the world, the preliminary indications are that plants growing at high altitudes are more resistant to damage from UV than plants growing at sea level. For instance, plants from mountaintops in Hawaii are far more resistant to damage than plants that grow at sea level. This differential sensitivity means that the mix of species in natural ecosystems may change as a result of ozone depletion (Teramura 1991, p. 3). The consequences for complex ecosystems such as those in the Amazon forest are unknown. Indeed, the rich variety of life in the canopy of the Amazon and other tropical forests is only now beginning to be appreciated and studied. This would be the portion of the topical forest ecosystems most directly exposed to increased UV radiation.

Finally, there is some evidence that exposure to UV may reduce the ability of some plants to fight diseases (ibid.).

2.8 Aquatic Life

Ozone depletion poses a particular threat to marine life. Zooplankton and phytoplankton—animal and plant microorganisms found near the surfaces of water bodies—are highly sensitive to UV-B radiation. Much of the ocean's life is sheltered from UV, which penetrates only the upper 10 meters of water. However, zooplankton and phytoplankton spend either all or critical stages of their life cycles in sunlight. Since phytoplankton and zooplankton play crucial roles in complex ecological food webs, damage to these organisms may have important ramifications for all oceanic life.

Phytoplankton are the ocean's primary food producers, converting energy from the sun into organic matter through photosynthesis. They serve as the food source for zooplankton, and both phytoplankton and zooplankton are fed upon by larger animals. Some whales feed directly on plankton. In addition, many varieties of zooplankton are actually the larvae of larger animals, such as shrimp, crabs, and anchovies. This means that plankton is an important food source for humans, both directly (as some zooplankton will develop into full-size shrimp or crabs) and indirectly (plankton are a food source for larger fish). According to a 1987 report, fish from the seas provide about half of the world's protein, and the proportion is higher in many Third World countries (EPA 1987, vol. III, p. 12-3). However, important fisheries have experienced severe declines since 1987, and some are approaching collapse, because of overfishing (Pitt 1994).

UV-B radiation is known to damage both phytoplankton and zooplankton even at current levels. Particularly troubling is the fact that many plankton species may be at or near their maximum UV tolerance (Calkins and Blakefield in EPA 1986, p. 211). Thus, even limited increases in UV-B levels might have a disproportionate impact on plankton life and on marine ecosystems.

Phytoplankton

Damage to phytoplankton due to current increases in UV-B radiation is deduced from several sources. Increases in UV-B due to the springtime Antarctic ozone depletion have been found to cause genetic damage in phytoplankton. Phytoplankton can adapt to some extent by migrating deeper into the water, but DNA damage has been detected as deep as 37 meters (about 120 feet) (Weiler 1991).

UV-B seems to have the greatest impact on the primary productivity of plankton, probably because phytoplankton migrate deeper into the water to avoid UV-B exposure. This migration, however, simultaneously reduces the organism's exposure to visible sunlight, which is necessary for photosynthesis (Calkins and Blakefield in EPA 1986, p. 224).

There have been direct measurements of reduced phytoplankton productivity as a result of the formation of the Antarctic ozone hole. In 1992, an Antarctic expedition aboard a research vessel named *Icecolors* measured increased UV-B levels under the ozone hole, including underwater increases at levels where phytoplankton grow. The results showed a direct relationship between UV-B increases and ozone depletion as the ship traveled from lower-depletion areas to higher-depletion areas. It also found that phytoplankton productivity had declined by at least 6 percent, and that this reduction was linked to ozone depletion (Smith et al. 1992, p. 952).

The *Icecolors* expedition found photosynthesis inhibition down to depths of 25 meters. Moreover, the "results indicate that UV-B inhibition of photosynthesis increases linearly (within the accuracy of measurements) with increasing UV-B dose" (ibid. p. 956).

UV-B can damage phytoplankton by injuring the mechanisms that help the organisms move in response to visible sunlight. This can result in overexposure to UV-B or underexposure to sunlight. UV-B is also capable of directly damaging the photosynthetic capabilities of phytoplankton, although the organisms seem to be more resistant to this than to other kinds of injury (EPA 1987, vol. III, pp. 12-9–12-10).

One study estimates that a 25 percent reduction in ozone would lead to a 35 percent reduction in primary productivity of phytoplankton near the surface of the water and a 10 percent reduction throughout the euphotic zone (the warm, biologically rich upper layer of the ocean) (ibid.). A 10 percent reduction is within the range found by the *Icecolors* expedition.

Research on UV-B radiation and marine life suggests that limited ozone depletion is less likely to reduce the overall mass of living organic materials in the ocean's upper layers than to change its composition. This is because some types of phytoplankton are more sensitive to UV-B exposure than others. For instance, Smith et al. found that UV-B inhibited the growth rate of one of the two Antarctic species they tested experimentally much more than that of the other (Smith et al. 1992, p. 955). Even this kind of change, however, could have serious implications. The variety of plankton that survived enhanced UV radiation was found, in at least one study, to be less nutritious than the variety that perished (EPA 1987, vol. III, p. 12-10). Thus, predators dependent on the existing plankton mix might suffer. Furthermore, one researcher notes that changes in phytoplankton would lead not only to changes in populations or communities of animals but possibly also to changes in ecosystem processes (rates of primary production, oxidation of organic matter, etc.). This may result in changes to the "chemical broth in which organisms bathe," which could have devastating implications (Kelly in EPA 1986, p. 240).

Freshwater phytoplankton are also similarly affected by UV-B. Laboratory experiments on a particular freshwater phytoplankton (*Peridinium gatunense*) show that the organisms' natural ability to seek light when solar irradiance is low and to orient themselves away from it when solar irradiance is high is severely impaired by exposure to the entire spectrum of UV-B up to 320 nm (Häder et al. 1990).

It is difficult to estimate the net effects that increased UV radiation might have on phytoplankton productivity. According to one estimate, a 10 percent increase in UV-B is expected to lead to an effective loss of 2.5–5 percent of photosynthetically active radiation available for phytoplankton (EPA 1987, vol. III, p. 12-9).

Severe ozone depletion over Antarctic waters has been linked to a 6–12 percent decline in primary productivity of phytoplankton. Two things need to be borne in mind about this. First, it is based on short-term observations and does not give much indication of what the long-term effects would be. Second, the Antarctic ozone hole appears in the early spring, when UV-B is filtered by a much longer column of air in the troposphere. Since the ozone layer over the Antarctic recovers later in the spring and in the summer, the effects of persistent year-round depletion are, as yet, unknown.

Zooplankton

There are a variety of zooplankton, ranging from bacteria to the eggs and larvae of larger animals. We will briefly discuss some research on bacterial plankton and then go on to discuss more complex zooplankton.

Independent measurements of UV-B radiation underwater to the depth of 30 meters in tropical waters confirm the findings of the *Icecolors* expedition that UV-B radiation penetrates to considerable depths. In contrast to phytoplankton and other zooplankton, bacterioplankton appear not to have the adaptive capability to protect themselves from increased UV-B. Researchers at the Institute of Zoology of the University of Vienna found significant bacterial activity reductions due to UV-B exposure. This would result in reduced incorporation of dissolved organic matter by bacterioplankton (Herndl et al. 1993). How this would affect the oceanic ecosystem is presently not clear.

Zooplankton, like phytoplankton, need sunlight. Although they may not spend their entire lives near the surface of the water, they need solar exposure for important life stages, such as reproduction. Bacterioplankton appear unable to adapt to increased UV-B, as more complex zooplankton do. These zooplankton possess DNA-repair systems that can cope with a certain level of solar injury; at least one form of DNA-repair system (photoreactivation) has been shown to be specifically targeted at sun damage (Calkins and Blakefield in EPA 1986, p. 211). Furthermore, zooplankton are often pigmented for protection from the sun and are relatively inactive at midday, when solar radiation is greatest. Some have developed sensors to carefully regulate their position in relation to the sun (Häder et al. in EPA 1986, p. 197).

Zooplankton life cycles, specifically reproductive periods, have evolved to avoid UV exposure. Crab larvae have a relatively high tolerance for

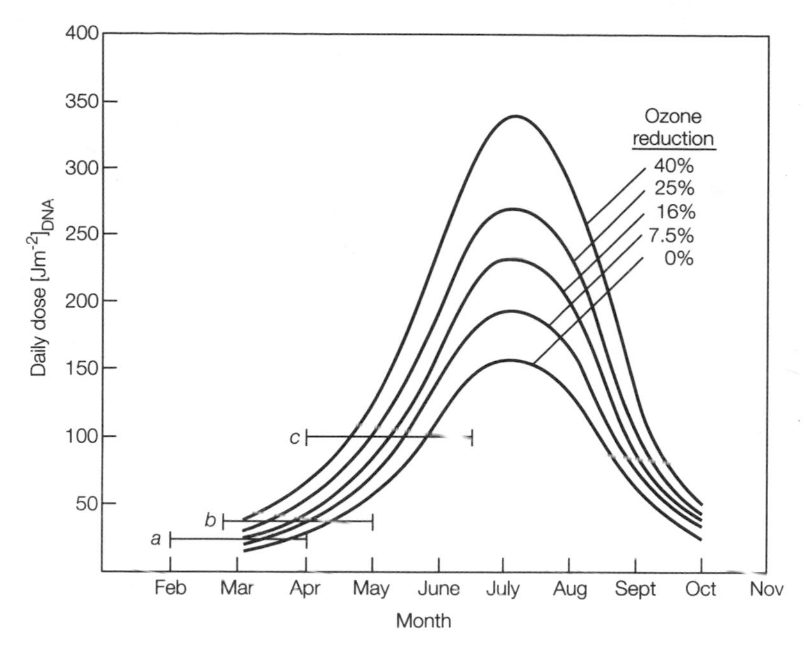

Figure 2.4
Estimated effective UV-B solar daily doses from sun and sky of (a) shrimp and euphausid larvae, (b) adult euphasids and crab zoea, and (c) crab megalopa at various atmospheric ozone concentrations, based on four-year mean of medians at Manchester, Washington, 1977–1980 (EPA 1987, vol. III, p. 12-14).

UV exposure and can sustain reproduction into June, when seasonal levels of UV begin to increase. On the other hand, shrimp, which have a relatively low tolerance for UV-B radiation, finish their reproductive season by April. Apparently, neither type of organism can sustain much more UV exposure at the surface than it currently receives. This has led researchers to question whether increased UV might not shorten these creatures' reproductive periods to the point where the integrity of the population is threatened. As can be seen in figure 2.4, it is estimated that an ozone reduction of only 7.5 percent would cut the breeding period of shrimp by approximately one-half (EPA 1987, vol. III, p. 12-12).

In general, zooplankton seem able to repair UV damage up to a certain dosage and period of exposure. However, beyond that threshold (which may be identical in similar species) researchers have found a sudden and irreversible drop in growth, productivity, and survival rate (ibid.). There is particular concern about sudden increases in UV exposure because, although organisms may have adjusted over evolutionary time scales to avoid UV exposure, most varieties of zooplankton do not have sensors for UV-B. In one study, shrimp were found to seek light even though it was accompanied by intolerable levels of UV-B. The shrimp stayed at the surface of the water until they ceased moving; they died within four days (ibid., pp. 12-21–12-22).

Measurements of precisely how much ozone depletion might injure aquatic organisms, and precisely which organisms it might injure, are complicated by many factors. Turbidity of water and the way particular organisms move within the water column need to be assessed to understand how much UV organisms actually receive. Furthermore, how the relationships between species would change is not fully understood. However, it is clear that there will be some damage from enhanced UV-B radiation. Table 2.2 shows biologically effective UV-B doses leading to significant population and reproductive effects in major zooplankton groups. We might expect a corresponding decline in higher-order species, which depend on the primary productivity of zooplankton and phytoplankton to sustain themselves.

Table 2.2
Estimated biologically effective UV-B doses leading to significant effects in major marine zooplankton groups (EPA 1987).

Zooplankton group	Dose rate, W/m^2 (DNA eff)[a]	Total dose, J/m^2 (DNA eff)	Time for effect
Shrimp/euphausiid larvae	6.0×10^{-2}	2.6×10^{3}	4 days
Euphasiid adults	9.9×10^{-2}	6.5×10^{3}	6 days
Copepod larvae	3.4×10^{-1}	1.4×10^{3}	1.0 hour
Copepod postlarvae	1.6×10^{-1}	2.7×10^{3}	4.5 hours
Copepod adults	3.4×10^{-1}	3.0×10^{3}	2.5 hours
Crab larvae	9.9×10^{-2}	6.5×10^{3}	6 days
Crab postlarvae	2.8×10^{-1}	6.0×10^{3}	20 days
Anchovy/Mackerel	6.0×10^{-2}	2.5×10^{3}	12 days
Oyster and mussel larvae	1.2×10^{-1}	2.2×10^{3}	5.0 hours

a. DNA action spectrum is referenced to 300 nm = 1.00.

It is possible that certain organisms and animals could adapt to higher levels of year-round ultraviolet radiation by changing their behavior patterns. As we have noted, some aquatic organisms could migrate to greater ocean depths. However, this will not be possible for many species, and certainly not for entire ecosystems. For instance, coral reefs cannot migrate. What might then happen to the rich marine life in and around these reefs?

Frogs

Precipitous declines in populations of certain frog and other amphibian species have been noted in many parts of the world. These declines have occurred in widely separated habitats. This suggests that common factors, such as increases in UV-B, could be responsible, at least in part.

Blaustein et al. (1994) studies the abilities of ten amphibian species to repair UV-B damage. In addition, they performed field experiments to test the relationship of hatching success to UV-B radiation in three frog species in the Cascade and Coast Mountains of Oregon. Two of these species, the Western toad (*Bufo boreas*) and the Cascades frog (*Rana cascadae*), have undergone such large population declines that they are candidates for the endangered-species list. The population of the third, the Pacific tree frog (*Hyla regilla*), is not known to be in decline. The objectives were to determine the extent of UV-B sensitivity and to understand some of the reasons for differential response among species.

Blaustein et al. examined an enzyme, photolyase, in the eggs of each frog species. Exposure to UV-B causes the production of cytotoxic and mutagenic chemicals, and photolyase helps repair the damage they cause. The experiments found a more-than-eightyfold difference in photolyase activity between the eggs of various species. This indicates that UV-B sensitivity is highly species specific. Further, the species that tend to have lower photolyase activity typically lay their eggs in places that have limited or no exposure to sunlight. High species-specific sensitivity to UV-B and adaptation to natural levels of UV-B (as indicated by photolyase activity relative to egg-laying locations) agree with findings for other plants and animals.

Blaustein et al. also tested the relation of photolyase activity in eggs to population declines. It was discovered that photolyase activity in eggs was far lower in the eggs of the two species in decline than in the eggs of the Pacific tree frog: the Western toad's level of photolyase was one-sixth that of the tree frog, and the Cascades frog's level was one-third that of the Pacific tree frog. The survival rates for the eggs of the Western toad and the Cascades frog were also far lower than that for the eggs of the Pacific tree frog. The survival rates for the eggs of the two vulnerable species were affected by the specific location in which the eggs were placed. Hatching outcomes of the two UV-B-vulnerable species improved when filters were used to block UV-B. Survival rates for eggs of the Pacific tree frog were high in all cases; this was attributable to the high repair activity of photolyase.

Repair mechanisms other than photolyase activity were not examined in this study. Further, the extent of the role played by UV-B in amphibian population declines is not yet clear, especially in relation to other factors, such as acidification of lakes and habitat destruction. However, the relation of UV-B exposure, species-specific sensitivity, and population declines of vulnerable species was dramatic enough that Michael Soulé, chairman of the Department of Environmental Studies at the University of California at Santa Cruz, went so far as to say "This paper provides something close to a smoking gun for the causation of the disappearances [of amphibian species]." (quoted in Yoon 1994)

A decline in frog and toad populations could have serious consequences for local and perhaps regional ecosystems in which frogs and toads are important regulators of insect populations.

2.9 Ecosystem Considerations

Ecosystems consist of complex interactions and dynamic balances within collections of living organisms and with the Earth's geochemical system. The entire ensemble of ecosystems together with the geochemical system has come to be called the **biogeosphere**. There are large gaps in the knowledge concerning the relationship of the few individual plants and animals that have been studied to increases in ultraviolet radiation. Even less is known about the effects of increased UV-B on specific ecosystems and on the biogeosphere. One indication of our limited knowledge is that "published estimates of possible effects of increased UV-B on various aquatic ecosystems have ranged from insignificant to catastrophic" (SCOPE 1993, p. 15).

As we have discussed, increased UV-B may decrease biomass production in terrestrial plant species. Reduction in phytoplankton productivity due to the Antarctic ozone hole has already been demonstrated. Since plant matter (including phytoplankton) is the principal vehicle for removing carbon dioxide from the atmosphere, increased UV-B could contribute to the ongoing increase in atmospheric carbon dioxide, and hence to global warming (SCOPE 1992, p. 13).

Increased UV-B could also enhance the rate of decomposition of plant litter and organic matter in the soil, further increasing concentrations of greenhouse gases. Enhanced decomposition rates would also result from increased temperatures due to global warming. Similar UV-B-induced increases in the decomposition of biological matter could occur in wetlands. This would result in greater methane emissions, providing further positive feedback to rising greenhouse-gas concentrations (SCOPE 1993, pp. 17–18).

Interactions between ozone depletion and greenhouse-gas increases are poorly understood. Overall, it is thought that the effects of increased UV-B on biogeochemical cycles could be considerable. As was noted at a recent international scientific workshop: "The biogeochemical cycles of various elements are linked like threads in a spider web. Thus, changes in the global carbon cycle are accompanied by changes in cycles of nitrogen, oxygen, sulfur, halogens, and biologically important transition metals." (ibid., p. 17) The workshop summarized some of these effects as follows (ibid., pp. 17–18):

> Damage to DNA and chlorophyll in plants and phytoplankton would result in many changes in their biological functioning and hence probably contribute to the increasing carbon dioxide in the atmosphere.
>
> Changes in marine and terrestrial geochemical processes would result in the formation of important trace gases in the atmosphere, including carbon dioxide, carbon monoxide, nitric oxide, carbonyl sulfide, and nonmethane hydrocarbons.
>
> Modification of chemical processes in the atmosphere could lead to increasing greenhouse gas concentrations, changes in aerosol concentrations affecting cloud formation, and changes in the oxidizing capacity of the atmosphere (which in turn affects a number of geochemical phenomena). Such chemical changes would interact with anthropogenic emissions, such as those resulting from biomass burning.

.ition, ecosystem-level consequences may result from adaptation .es to increased UV-B. For instance, in order to protect against skin cancer, animals might develop increased skin pigmentation. However, such a response may render some animals more visible and therefore more vulnerable to predators (SCOPE 1993, p. 23).

Finally, increased incidence of disease in animals, such as eye disorders, could cause serious disruption of agriculture, particularly in the large areas of the Third World that depend on cattle as the principal source of draft power.

2.10 Potential Impacts of Persistent Severe Ozone Depletion

Severe, persistent declines in stratospheric ozone could seriously damage life on Earth. For instance, disease epidemics could be triggered by a combination of decreased immunity, declining food production, and the direct disease-producing effects of UV-B radiation. These could be so serious as to greatly strain or even overwhelm the ability of economic and medical systems to respond, particularly in Third World countries. As we have discussed, severe depletion could also disrupt ecological systems considerably, with unknown effects.

With the present state of knowledge, quantitative analyses of the simultaneous interdependent changes in health, economic and ecological factors due to persistent severe ozone depletion are not possible. Serious disruption in any of these areas could by itself cause widespread societal harm. It is a challenge to even identify the potential combined effects, since little is known about the impacts of increased UV-B radiation on the ecosystem.

Some of the effects of severe ozone depletion have been studied in the context of hypothesized ozone loss as a result of a global nuclear war. While the possibility of such a war has diminished greatly with the end of the Cold War, the theoretical studies of its effects turn out to be a useful tool for making some qualitative observations about ozone depletion due to other causes.

Of course, an all-out nuclear war would have a great many horrendous and destructive effects, which would result in tens or hundreds of millions of human deaths within a few days. Our comparison does not refer to the explosive and high-level radiation effects of nuclear war or to the widespread radioactive contamination that would follow. It refers only to one of the most destructive long-term effects of nuclear war, which a 1975 study of the National Research Council of the U.S. National Academy of Sciences (NAS) identified as severe ozone depletion.

The NAS study was specifically directed at the long-term effects of nuclear war, stressing particularly those aspects that had not, until then, been studied in depth. Perhaps the most serious of these "frequently neglected effects" was severe depletion of the stratospheric ozone layer, which was hypothesized to occur as a result of the sudden injection of large quantities of nitrogen oxides into the stratosphere by the detonation of many large nuclear weapons. The NAS results have been reaffirmed by other studies (Turco and Golitsyn 1988).

The NAS study was used by the U.S. Arms Control and Disarmament Agency to prepare "An Assessment of Frequently Neglected Effects of Nuclear Attacks" in 1978. It summarized these effects as follows:

> ... the worst case of 70% ozone decrease ... would cause blistering sunburn after 10 minutes' exposure in temperate latitudes. The more probable lowering of 50% as a result of a nuclear war ... would cause blistering after 1 hour of exposure. This leads to the conclusion that outside daytime work in the northern hemisphere would require complete covering by protective clothing.... It would be very difficult to grow many (if any) food crops, and livestock would have to graze at dusk if there were any grass to eat. Even if ozone depletion only reached the best expected case of 25–30 percent reduction the effect on post-war recovery operations is difficult to imagine. (ACDA 1978, p. 7)

For such effects to occur, the study assumed that ozone depletion would be year-round and would persist for several years. Current chlorine-induced declines have not resulted in high levels of year-round depletion. However, there have been large seasonal declines over the Antarctic during austral spring, and phenomena similar to Antarctic ozone depletion seem to be reproducing themselves at a somewhat less intense level over the northern pole. Furthermore, ozone depletion of approximately 20 percent for periods up to one month has occurred over highly populated regions.

In Third World countries, most people depend upon farming for their existence, and in most of these countries cattle are the main source of power for tilling fields. This work must be done in the daytime. Though it may be possible to contemplate that cattle raised for beef production may be able to graze at dusk, there is no possibility that agricultural work could be carried out similarly. Moreover, during the peak agricultural season, most outdoor work by human beings must also be done in the daytime. The usual work day is long and extends throughout the daylight hours. Any substantial curtailment of the workday (occasioned by illnesses or disabilities caused by UV radiation) for cattle or humans could reduce crop production drastically.

The production of milk, already in extremely short supply in the Third World, could also be seriously curtailed. The ability of cattle to reproduce under the above-mentioned circumstances is essentially unknown at present.

Ecosystems would also be seriously harmed. In addition to the types of damage that we have already discussed, there might be large-scale blinding of many species of wildlife, since UV-B radiation induces cataracts. In such a case, entire land-based ecosystems might also be disrupted or destroyed, particularly if increased susceptibility to diseases and plant ecosystem damage are also taken into account.

The NAS report concluded that it was difficult or impossible to address these kinds of questions with the present knowledge. Although the work was performed in the 1970s, it remains true today that the effects of persistent ozone depletion of 25 percent or more are still essentially unquantifiable, mainly because of insufficient research. This relative paucity of research funds has persisted despite the reality that seasonal depletion larger than 25 percent has occurred every year over the Antarctic since the late 1970s (and has been generally known since about 1985), and despite the reported observations in the early 1990s of significant increases in ozone depletion over heavily populated areas in the northern hemisphere.

II

Ozone-Depleting Compounds: Sources, Uses, and Alternatives

3

Ozone-Depleting Compounds

Chemical compounds that destroy stratospheric ozone come from a variety of sources. Although some are of natural origin, maintaining the balance between ozone production and destruction, the decline in stratospheric ozone during the last two decades is due to human-derived chemicals and activities. Of the anthropogenic sources, a group of industrial chemicals used as solvents, foaming agents, fumigants, firefighting agents, coolants, and propellants are by far the most important. Other sources include nitrogen oxides, chlorine released from solid-fuel rockets, and chemicals released from the burning of biomass. In order to frame the options available to reduce the production and consumption of ozone-depleting compounds, it is useful to take a close look at their origins, evolution, and recent patterns of use.

3.1 Halocarbons

Halocarbons are a class of chemical compounds, both human-made and natural, containing carbon and one or more atoms belonging to the **halogen** group of elements. Halogens are the elements residing in the seventh column of the periodic table: fluorine, chlorine, bromine, iodine, and astatine.[1] They are naturally occurring elements found primarily in salts, such as sodium chloride (NaCl), and fluorspar (CaF_2). Chlorine and fluorine are the most common halogens found in halocarbons.

Halocarbons are derived from a larger class of chemicals, the **hydrocarbons**. Hydrocarbons, in their simplest form, are compounds that contain both carbon and hydrogen. A halocarbon is a hydrocarbon in which one (or more) hydrogen atoms has (have) been replaced by one (or more) halogen atom(s) in the hydrocarbon molecular structure. Halocarbons in which halogens replace all of the hydrogen in the hydrocarbon structure are commonly referred to as fully halogenated halocarbons. Halocarbons in which halogens partially replace the hydrogen in the hydrocarbon structure are commonly referred to as partially halogenated halocarbons.

Natural halocarbons are emitted into the atmosphere. However, the emission of a particular subset of halocarbons produced by humans, which we will refer to as **ozone-depleting halocarbons**, has been primarily responsible for the severe ozone loss experienced over the past two decades. A brief description of the synthesis and market penetration of industrial halocarbons can help in understanding the important properties and differences within this class of chemical compounds.

History

Halocarbon chemistry can be traced back to the 1890s, when a Belgian chemist named Frederic Swarts developed a method to bond carbon and fluorine in the laboratory (Manzer 1990). The significance of these compounds was not recognized until about 30 years later.

In 1928, a DuPont chemist named Thomas Midgely, Jr., in response to a request by the Frigidaire division of General Motors for a better refrigerant fluid, synthesized the first fully halogenated halocarbon. DuPont had majority control of General Motors stock at the time.

The young domestic refrigeration industry, which GM had recently entered through purchase of the Guardian Frigerator Company, had been hampered somewhat by cooling fluids that were potentially dangerous because of their toxicity.[2] The first halocarbon synthesized by Midgely, and many others that were soon developed by DuPont, came to be known commonly as **chlorofluorocarbons** (CFCs) (Midgley and Henne 1930). The name derives from the fact that these halocarbons contain only carbon, chlorine, and fluorine. As we will see, these early compounds are now a distinct class of the larger chemical family of chemicals pertinent to the ozone-depletion problem.

These structurally simple chemicals had some rather unusual properties, the most important of which was their nonreactivity. This led to two of their most favored characteristics from the perspective of the young chemical industry: low toxicity and low flammability. In fact, Midgely himself was reported to have given a demonstration in which he inhaled a mixture of chlorofluorocarbon and air and then exhaled over a flame, causing it to be extinguished. For the synthesis of the first chlorofluorocarbons, Midgely was later awarded the Perkins Medal, one of the highest honors in chemistry.[3]

Low toxicity to the immediate user, low flammability, and favorable thermodynamic properties led to the use of chlorofluorocarbons in a variety of industrial applications and consumer products. By the 1970s, they were used as propellants in aerosol cans, as solvents, as fire-extinguishing agents, as components of sterilizing fluids, as an expansion gas in the production of foams, as the heat-exchanging fluid in air conditioners, and, as originally intended, the working fluid in refrigeration systems.

Classification and Characteristics

Many halocarbons have been created and produced commercially. Figure 3.1 lists the halocarbons pertinent to the ozone-loss problem along with a relational hierarchy of the compounds as determined by their chemical structures and sources.[4]

Ozone-depleting halocarbons, as the figure shows, are referred to by various common names depending on differences in their chemical composition and industrial origins. The numbers associated with some of the names describe the relative proportions of elements in the molecular structure. Chlorofluorocarbons (also referred to by the trade name Freon) are the most established of the ozone-depleting halocarbons, the most versatile, and the first to have been regulated. **Halons** is the general name given to a class of halocarbons used primarily as fire- extinguishing agents. A common characteristic among the three listed in figure 3.1 is the presence of bromine in their molecular structure.

As it became clear in the 1980s that commercially successful halocarbons were primarily responsible for ozone depletion and that regulation was imminent, two groups of alternative halocarbons (also referred to as **transitional compounds**) were intensively developed to serve as replacements. These are referred to as **hydrochlorofluorocarbons** (HCFCs) and **hydrofluorocarbons** (HFCs).[5] As their names imply, the

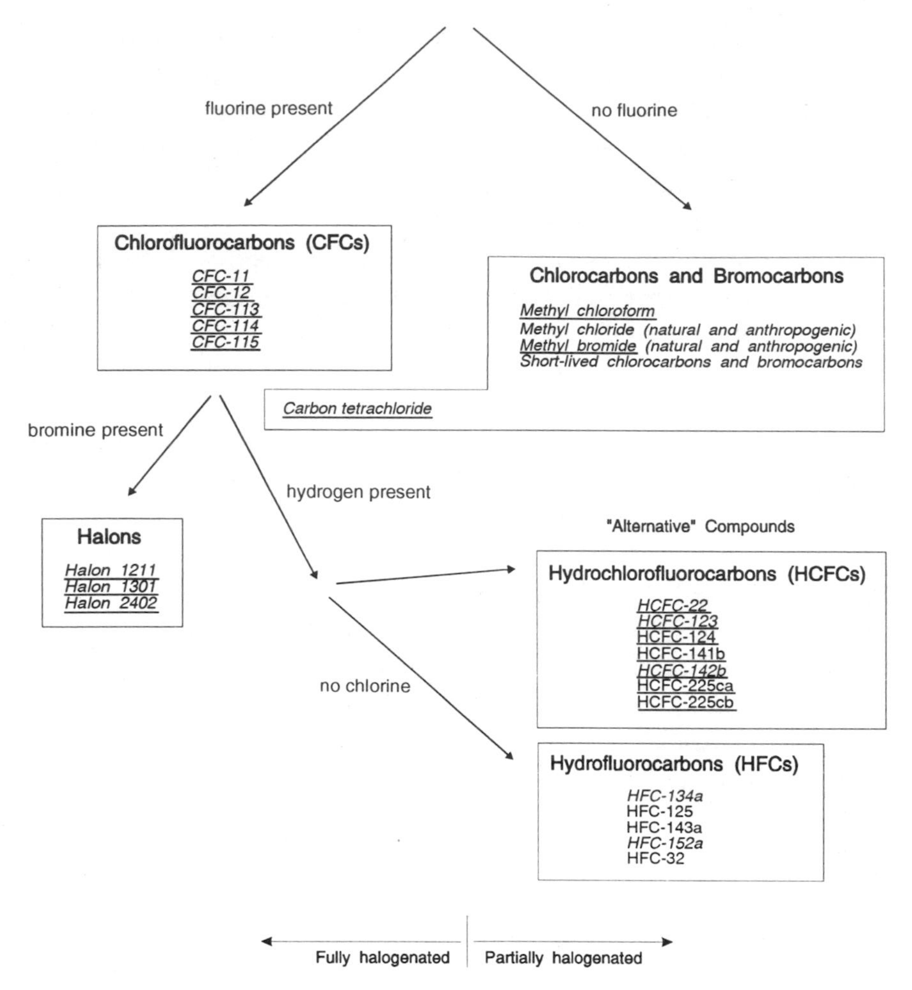

Figure 3.1
Relational hierarchy of halocarbons pertinent to the ozone-depletion problem. Those underscored are internationally regulated; those italicized are available in significant quantities. A letter at the end of a numerical sequence indicates the degree of spatial symmetry of the compound.

primary distinguishing characteristic of these partially halogenated compounds is the presence of hydrogen in the chemical structure, much as in their parent compounds, the hydrocarbons.

The presence of hydrogen in their molecular structure tends to increase the chemical reactivity of HCFCs and HFCs; they are therefore more likely to be chemically destroyed in the lower portion of the atmosphere (troposphere) than are the fully halogenated halocarbons. This means that, for comparable emissions, the partially halogenated halocarbons will not build up to as great an extent in the atmosphere as the fully halogenated halocarbons they have been designed to replace. Further, the presence of hydrogen in these compounds typically means that they contain fewer halogen atoms per carbon atom than CFCs. These combined aspects result in less ozone-depleting ability for a given amount emitted.

The HFCs contain no chlorine or bromine and are thought to pose no threat to stratospheric ozone. Recent modeling efforts have demonstrated that the impacts on ozone due to the CF_3 group, present on many HFC molecules, are likely to be negligibly small (Ko et al. 1994a; Ravishankara et al. 1994). As will be discussed later, these compounds may also contribute to global warming.[6]

Some alternative halocarbons are commercially available; these include HCFC-142b, HFC-134a, HCFC-123, and HFC-152a. Others are still being tested for suitability according to economic, environmental, safety, and toxicity criteria. One substitute compound, HCFC-22, is not of recent origin. This compound has been in production for approximately 40 years and was already commonly used as a working fluid in cooling systems when international agreements to phase out CFCs were being considered in the 1980s.

The remaining halocarbons pertinent to stratospheric ozone depletion are a diverse variety of nonfluorinated chemicals. Designated as chlorocarbons and bromocarbons in figure 3.1, they are chemically distinguishable from all the other ozone-depleting halocarbons in that they do not contain fluorine. Most of these compounds are used only in specialized applications, whereas many fluorinated compounds have been utilized in a far wider variety of applications.

There are a large number of compounds within the nonfluorinated group that we refer to as **short-lived** chlorocarbons and bromocarbons.

These compounds are removed from the atmosphere within a few days, weeks, or months after being emitted, unlike most of the other compounds in figure 3.1. They are commonly used as solvents, as pesticides, and as "feedstocks" (raw materials) in the production of plastics. Their ozone-depleting properties are currently being reassessed. Of particular concern are chloroform ($CHCl_3$); dichloromethane, also called methylene chloride (CH_2Cl_2); tetrachloroethene, also called perchloroethylene (CCl_2CCl_2); and 1,2-dichloroethane (CH_2ClCH_2Cl).[7] Because of considerable uncertainty associated with their lifetimes and their present distribution in the atmosphere, we do not include these compounds in our discussion of chemicals pertinent to ozone depletion. We will, however, discuss their potential role in stratospheric ozone depletion for completeness and to stimulate further research.

Mixtures or blends of two or more compounds are used in some industrial applications.[8] Their ozone-depleting properties are dependent on the constituents used and their relative proportions.

In addition to industrial activities, halocarbons are released from some non-industrial activities as well. Low-temperature biomass burning emits both methyl chloride (CH_3Cl) and methyl bromide (CH_3Br), though the exact amounts are difficult to estimate. Oceanic biological activity also results in emissions of both methyl chloride and methyl bromide. These compounds are also released as a result of industrial activity.

Human activities give rise to two additional ozone-depleting compounds not listed in figure 3.1: nitrogen oxides and chlorine from the exhaust of solid-fuel rockets. Mentioned here for completeness, these compounds will be not be considered in detail.

For simplicity and uniformity, all the compounds that contribute to stratospheric ozone loss will be referred to collectively as **ozone-depleting compounds** (ODCs) in the remainder of this book.

3.2 Applications

Ozone-depleting compounds are used in a variety of manufacturing processes and consumer products. Their lack of direct toxicity, their nonflammability, and their thermodynamic properties make them attractive for many applications. Their initial use as the heat-transfer fluids in refrigeration systems was the first in a long list of products either made with ODCs or containing them to varying degrees. The following are their principal applications.

Refrigeration and Air Conditioning

CFC-11, CFC-12, CFC-114, CFC-115, and HCFC-22 are used as heat-transfer fluids for mobile and stationary cooling systems of refrigerators, freezers, air conditioners, and heat pumps. Some CFC-113 is also used in special cases. More recently, HCFC-123, HFC-134a, HFC-32, and a variety of blends have been used as alternatives.

Foam Production

CFC-11 and CFC-12 are used as blowing agents to produce soft foam (such as that used in furniture, bedding, and car seats), packaging material, and rigid insulating foams (such as those used in refrigerators). In the last few years, HCFC-141b, HCFC-22, hydrocarbons, and water have been increasingly used as a substitutes in some of these applications.

Solvents

CFC-113 has been used, usually in combination with other chemicals, as a cleaning agent for electronic circuit boards and for metal parts and assemblies. Methyl chloroform (CH_3CCl_3) is used for metal degreasing, as an adhesive, as an ink solvent, and as a solvent in dry cleaning. Carbon tetrachloride (CCl_4), which is highly toxic, is still used in many parts of the world as a general solvent.

Firefighting

Halons are used in centralized firefighting systems (halon-1301) and in fire extinguishers (halon-1211). Halon 2402 is a relatively minor compound in this class. Carbon tetrachloride (CCl_4) has also been used as an extinguishing agent.

Aerosols

CFC-11 and CFC-12 have been used extensively as aerosol propellants, mostly in the United States, Europe and Japan. In the United States, Canada, Norway, and Sweden, most of these uses were banned in 1978 and replaced with other systems or chemical propellants. The use of fully halogenated halocarbons in aerosol products was reduced dramatically by the early 1990s; however, some were replaced with HCFC-22. Hydrocarbons have been the principal replacement chemical for ODC aerosol propellants. Medical inhalers, such as those used by asthmatics, are an exempted use of ODC propellants.

Sterilization
CFC-12 is used as the delivery vehicle for the sterilant ethylene oxide (EtO), which is used to sterilize hospital equipment, spices, and even books.

Soil, Structure, and Commodity Fumigation
Methyl bromide is commonly used as a soil fumigant and in quarantine and commodity fumigation. It is also used for rodent and termite control in homes and in some chemical processes. Recent research has suggested that methyl bromide is also emitted in vehicle exhaust. Some CFC-12 is also used as a fumigant.

Miscellaneous
There are a variety of small specialty uses of ODCs. For example, CFC-12 is used for quick freezing of perishable food and for leak detection; CFC-11 is used in the processing of tobacco.

Besides the use of ODCs in industrial applications, there are other anthropogenic activities that give rise to ODC emissions. The activities and the chemicals associated with them are as follows.

Biomass Burning
Methyl chloride and methyl bromide have been identified as by-products of low-temperature biomass burning. Methyl chloride is also used in industrial applications as a chemical feedstock, but emissions from this source are currently small. Methyl chloride and methyl bromide are also emitted from natural sources, such as the oceans. Anthropogenic biomass burning is responsible for approximately 20–30 percent (possibly more) of total methyl chloride emissions and 10–50 percent of total methyl bromide emissions.

Solid-Fuel Rockets

Solid-fuel rockets directly deposit chlorine in the form of hydrogen chloride into the stratosphere via the combustion of chlorine-containing fuel.

High-Altitude Aircraft

High-flying aircraft emit nitrogen oxides directly into the stratosphere, where they can catalytically destroy ozone. Nitrous oxide, which is stable in the troposphere, reacts to form nitrogen oxides in the stratosphere. Soil fertilization and the combustion of biomass and fossil fuels are the leading anthropogenic sources of nitrous oxide.

Table 3.1 lists the ODCs currently being produced and the recently developed alternative compounds, their concentrations in the atmosphere in 1989, their concentration trends, and their atmospheric lifetimes (also called residence times).

The **atmospheric lifetime** of a compound is related to the rate at which that compound is removed from the atmosphere. It reflects the combined effort of a variety of removal processes, such as chemical alteration, dissolution in rainwater and seawater, photodissociation, and surface adsorption. Quantitatively, the lifetime represents the amount of time required for a given quantity of a compound placed into the atmosphere to be reduced by approximately 63 percent.

The term "lifetime," though commonly used, is somewhat misleading in that it is *not* the amount of time required to completely remove an emitted amount of a compound from the atmosphere. The percent removal of a compound after a period of time equivalent to two, three, four, and five lifetimes is, respectively, 84, 95, 98, and 99. Thus, the time required for nearly complete removal of a compound from the atmosphere is equivalent to about 3–5 lifetimes. For example, one pound of CFC-12 emitted today will be reduced to just over one-third of a pound in approximately 120 years via removal processes, and will require about 500 years to be essentially purged from the atmosphere.

Lifetimes can be calculated from observations of concentration in the atmosphere or directly calculated from data on atmospheric removal. Both of these methods, however, contain significant uncertainties, even for well-studied ODCs like CFC-11. For simplicity, we have given only

Table 3.1
Atmospheric parameters of chemicals pertinent to ozone depletion (WMO 1991, pp. 1.4, 6.7, 8.7, 8.8; WMO 1989, p. 246; UNEP 1992, section II, p. 5).

Common name	Chemical formula	1989 concentration (pptv)	Concentration trend[a] (%/year)	Lifetime (years)
Fully halogenated				
CFC-11	$CFCl_3$	255–268	3.7–3.8	55
CFC-12	CF_2Cl_2	453	3.7–4.0	116
CFC-113	$CFCl_2CF_2Cl$	64	9.1	110
CFC-114	CF_2ClCF_2Cl	15–20	6	220
CFC-115	CF_2ClCF_3	5	?	550
Halon 1211	CF_2ClBr	1.8–3.5	20	11
Halon 1301	CF_3Br	1.6–2.5	15	77
Halon 2402	CF_2BrCF_2Br	negligible	?	20
Carbon tetrachloride	CCl_4	107	1.2	47
Partially halogenated				
Methyl chloroform	CH_3CCl_3	135	3.7	6.1
HCFC-22	CHF_2Cl	110	6.0–7.0	15.8
Methyl chloride	CH_3Cl	600	?	1.5
Methyl bromide	CH_3Br	9–13.0	>0[b]	2.0
Chloroform	$CHCl_3$	30[c]	?	~0.7
Methylene chloride	CH_2Cl_2	35[c]	?	~0.6
Perchloroethylene	CCl_2CCl_2	30[c]	?	~0.6
Dichloroethane	CH_2ClCH_2Cl	35[c]	?	~0.6

HCFC-123	$CHCl_2CF_3$	0	0	1.7
HCFC-124	$CHFClCF_3$	0	0	6.9
HCFC-141b	CH_3CFCl_2	0	0	10.8
HCFC-142b	CH_3CF_2Cl	0	0	22.4
HCFC-225ca	$CF_3CF_2CHCl_2$	0	0	2.8
HCFC-225cb	CF_2ClCF_2CHFCl	0	0	8.0
HFC-134a	CH_2FCF_3	0	0	15.7
HFC-125	CHF_2CF_3	0	0	40.7
HFC-143a	CH_3CF_3	0	0	64.6
HFC-152a	CH_3CHF_2	0	0	1.8
HFC-32	CH_2F_2	0	0	7.3

a. Owing to declining production of CFCs, methyl chloroform, carbon tetrachloride, and the halons, concentration trends are expected to decline from the 1989 values given. For CFC-11 and CFC-12, this has already been noted. See Elkins et al. 1993 and Cunnold et al. 1994.

b. Recent research suggests a global concentration trend of approximately 3.5% per year for the period January 1988 to December 1991 (Khalil et al.1993).

c. Owing to limited measurement and the short lifetime of these compounds, these values should not be considered global averages.

one number for the lifetime of each of the compounds listed in table 3.1. However, the estimated lifetimes from both measurements and model simulations depend on assumptions about a variety of fundamental variables, such as concentrations of other reactive species in the atmosphere, emission levels, and reaction rates. Further, the lifetime of a compound in the atmosphere is usually assumed to be approximately constant; however, changes in atmospheric chemistry or climate may change the lifetimes of some compounds.

As can be seen from the table, many of the ODCs now in use have long atmospheric lifetimes. These compounds tend to be the fully halogenated halocarbons. Their long lifetimes are primarily a reflection of the slow rate at which the photodissociation reactions occur. The partially halogenated halocarbons, many of which have been proposed as alternatives, generally have shorter lifetimes. Unlike the fully halogenated compounds, they tend to react in the lower atmosphere with a very reactive oxidizing agent, the hydroxyl radical (OH). Nevertheless, a significant portion of the many partially halogenated halocarbons emitted can reach altitudes in the stratosphere where they are photodissociated by UV radiation, the usual fate of the fully halogenated halocarbons. In this way, all the compounds listed in table 3.1 that contain chlorine or bromine can contribute to ozone loss in the stratosphere.

The efforts to produce alternatives to ODCs have focused on reducing atmospheric lifetime by constructing compounds that are more likely to be removed in the lower atmosphere. It is important to note that the resulting concentration of a compound in the atmosphere is a function of both the rate at which the compound is removed and the rate at which it is put into the atmosphere. In other words, a compound that is removed quickly and hence has a short atmospheric lifetime, can attain high concentrations only if it is being emitted in substantial amounts. Methyl chloroform is a useful example. It has a lifetime of only 6.1 years, but it is as abundant in the atmosphere as compounds with much greater lifetimes. This is due to large, uncontrolled emissions associated with its use as a solvent.

The high concentrations of many of the compounds are due to the combination of large emissions and long lifetimes. This is true for three commonly used CFCs: CFC-11, CFC-12, and CFC-113. The high concentration of methyl chloride reflects the fact that the concentration of this compound is dominated by a large, relatively constant source. In contrast, CFC-114, CFC-115, and the halons, though they have long atmospheric lifetimes, are emitted in relatively small quantities and remain at relatively low concentrations.

The concentration of the principal ODCs in the atmosphere were all increasing as of 1989. However, recent emissions data and atmospheric observations for 1990 and 1991 suggests that concentrations are beginning to level off and may soon decline, particularly for CFC-11 and CFC-12. This is due primarily to international agreements and national and local legislative actions in many countries. At what time concentrations begin to decline and how quickly that decline proceeds depend critically on the timing and the content of further regulations aimed at limiting the production and the consumption of ODCs and the thoroughness with which they are implemented.

3.3 Patterns and Evolution of ODC Use

The pattern of ODC use has changed rapidly in recent years, primarily because of the existence of an internationally agreed upon timetable aimed at ending the production and consumption of many ODCs in the major producing and consuming countries by the year 1996.

We will first examine the uses of ODCs in 1985, before the onset of regulation. We will then examine the evolution of uses since 1985—especially since late 1987, when the international regulations limiting the production of ODCs were signed by many of the producing and consuming countries. This historical examination of the pattern of ODC uses forms the basis for a discussion of the extent to which alternative technologies and chemicals are needed and serves to demonstrate the connections between regulation, innovation, and environmental protection.

Figure 3.2 is a chart of global ODC use, according to application, as it prevailed in 1985. Figure 3.3 shows the principal uses of these compounds in the United States, and figure 3.4 shows the uses in all other countries.[9] Until recently some striking differences were apparent in the international use patterns of ODCs. For example, in 1985 the use of ODCs in aerosol products was still quite common in many countries.

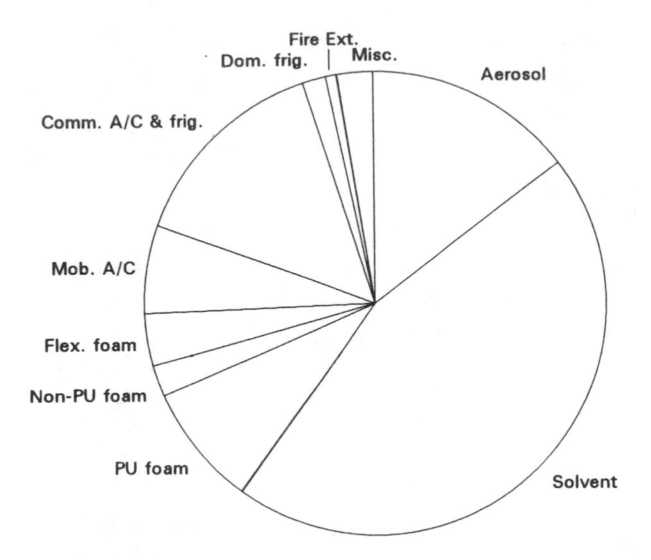

Figure 3.2
Estimated global ODC use, excluding methyl chloride and methyl bromide, 1985 (total: 1883 kilotonnes).

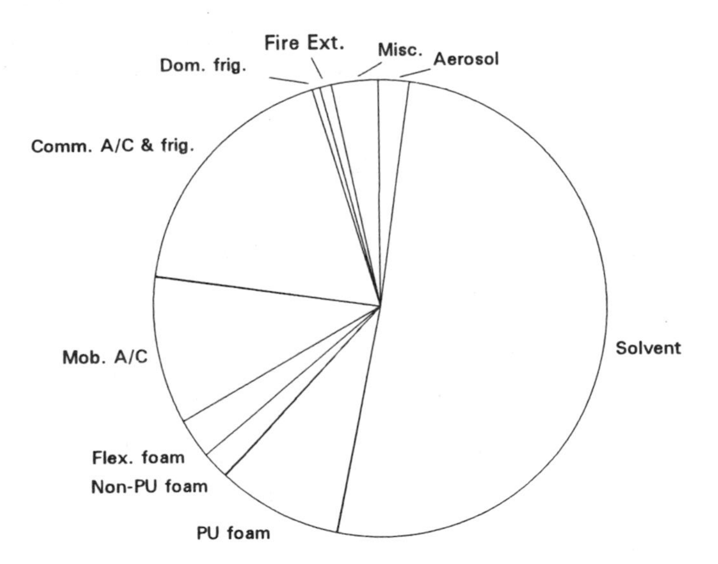

Figure 3.3
Estimated ODC use in the United States, excluding methyl chloride and methyl bromide, 1985 (total: 700 kilotonnes).

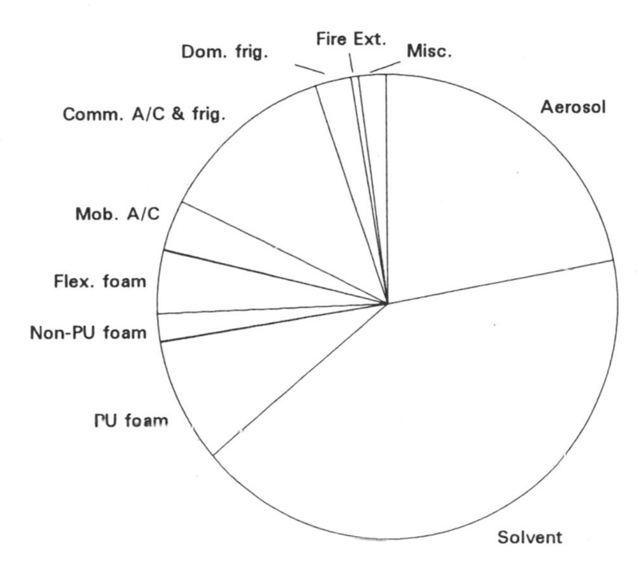

Figure 3.4
Estimated ODC use outside United States, excluding methyl chloride and methyl bromide, 1985 (total: 1184 kilotonnes).

The United States, Sweden, Norway, and Canada banned the use of CFC propellants for most applications in the late 1970s, while other countries continued to employ them. Another difference stems from the fact that commercial and industrial uses (particularly mobile air conditioning and refrigeration) accounted for a significant share of ODC use in the United States and Japan. This characteristic of international ODC consumption has likely not changed since 1985. For example, it is estimated that in 1991 approximately half of the world's air conditioned vehicles were in the United States (UNEP 1991d, pp. 166 and 175), 35 percent in the rest of the industrialized world (mostly in Japan), and 15 percent in the Third World.

Legislation banning the use of ODCs as propellants (e.g. in aerosol cans) has been quite common in recent years (UNEP 1991a, pp. 19–21). Recent data (figure 3.5) indicate that the aerosol component of world ODC use was reduced significantly between 1985 and 1990. Contrary to what is commonly reported, the use of ODCs as solvents constitutes the greatest share of global use. This is due to the inclusion of methyl chloroform and carbon tetrachloride in the global use estimate.

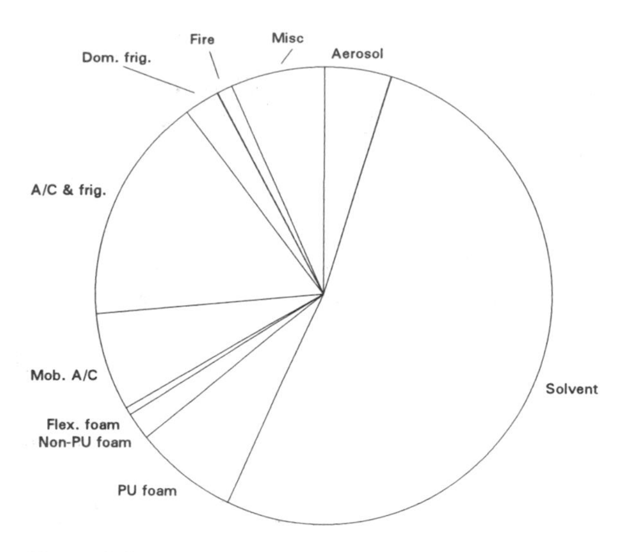

Figure 3.5
Estimated global ODC use, excluding methyl chloride and methyl bromide, 1990 (total: 1898 kilotonnes).

Table 3.2 presents the most recent application-specific estimates of reductions in ODC use for 1991 relative to 1986. In absolute quantities, the use of these compounds in aerosols was reduced from approximately 300,000 tonnes in 1986 to an estimated 115,000 tonnes in 1990.[10] The shift in overall ODC use has been away from aerosols and foams to cooling-system applications. HCFC-22 production increased during this period and accounts for much of the increase in commercial A/C and refrigeration represented in figure 3.5.

Figure 3.6 shows the global consumption of CFCs and halons by region in 1986. As can be seen, consumption takes place mainly in the industrialized regions of the world. Figure 3.7 shows per capita consumption for various industrial and Third World countries as of 1985. More current data on regional production and consumption are not available. However, the indications are that the Third World's share of consumption may be rising relative to the industrial countries' share. For example, it is estimated that Thailand and India increased their use of ODCs by approximately 300 percent between 1985 and 1991 (EIN 1991; Billimoria & Co. 1990).

Table 3.2
Worldwide estimates of ODC use in 1991 compared to 1986 (UNEP 1991f, p. ES-3; AFEAS 1993c; Pauline Midgely, personal communication, December 8, 1993).

ODC application	Percent of 1986 total uses	1991 vs. 1986
CFCs		
Propellants	28%	−58%
Cleaning	21%	−41%
All foam blowing	26%	−35%
Polyurethane		−30%
Phenolic		−65%
Extruded polystyrene sheet		−90%
Extruded polystyrene board		−32%
Polyolefin		−35%
Refrigerants	23%	−7%
Miscellaneous	2%	
Total		−40%
Other ODCs		
Methyl chloroform		19%
HCFC-22		30%
Carbon tetrachloride		NA[a]

a. Not available.

ODC consumption continued to grow between 1985 and 1988, despite the discovery of the Antarctic ozone hole. It reached a peak of 2,290,000 tonnes in 1988. But in late 1987, international regulations confronted users with the reality that the availability of these compounds would be sharply curtailed and that they would be far more expensive. At that time, most user industries began serious efforts to phase out ODCs. These efforts were intensified in 1989, when it became clear that the worsening news about ozone depletion would lead to a total production and consumption phaseout in approximately a decade. As a result, the production of ODCs has changed rapidly since 1988 (figure 3.8). The changes were especially dramatic during 1990 and 1991, when the efforts of industries to install new equipment, retrofit old processes, and use new chemicals began to have a large cumulative effect on ODC

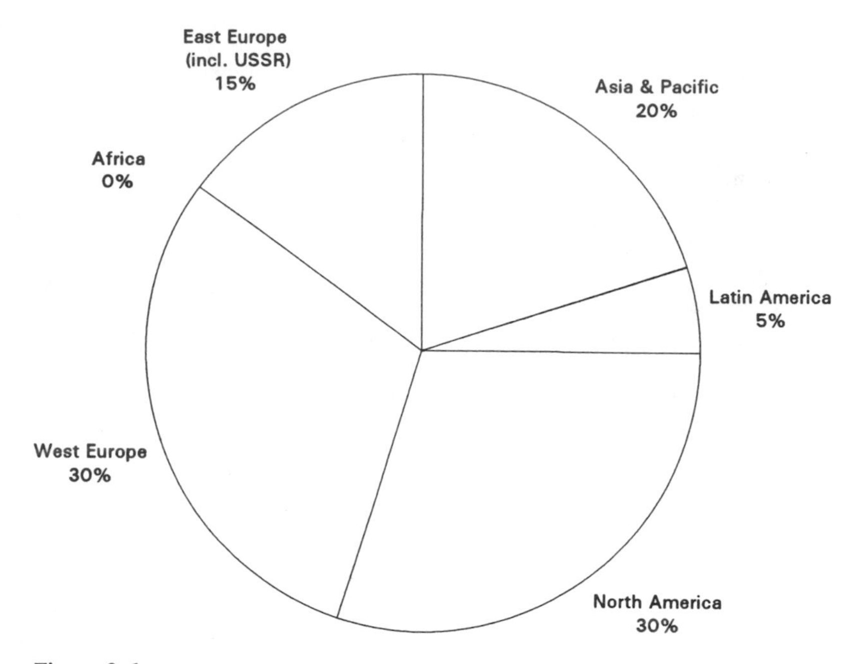

Figure 3.6
Regional consumption of CFCs and halons, 1986 (UNEP 1989g, p. 18, figure 2-3).

production. As depicted in table 3.2, emissions estimates indicate a worldwide decline of about 40 percent between 1986 and 1991.

In the United States (the largest user), a tax on many ODCs also helped reduce demand and encourage recycling, notably by the automobile air conditioning service industry and by manufacturers using ODCs as solvents. For example, the combined production of CFC-11 and CFC-12 in the United States declined by 42 percent between 1989 and 1991 (USITC 1991).

Owing to the limited historical data and the fact that both natural and anthropogenic emissions exist, the preceding data concerning the patterns and trends in ODC production and use do not include methyl chloride or methyl bromide. However, we will discuss the existing data and the procedures for estimating the recent emissions of methyl chloride and methyl bromide in detail in chapter 10.

Recent data indicate that approximately 63,000 tonnes of methyl bromide were produced for use as a fumigant in 1990 (UNEP 1992).

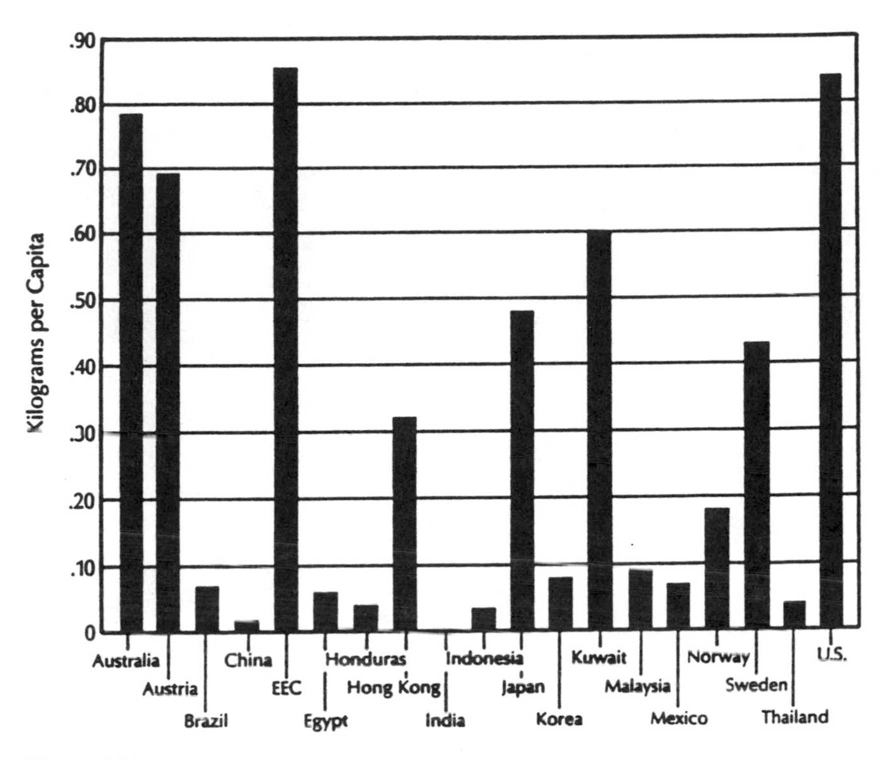

Figure 3.7
CFC-11 and CFC-12 production (kilograms per capita) in various countries (derived from exhibit A-5 on pp. A-8–A-9 of vol. II of EPA 1987).

From 1984 to 1990, this production grew at an average rate of approximately 7 percent per year (ibid., section II, p. 6). A regional breakdown of fumigant consumption is presented in table 3.3. The majority of sales are to North America and Europe.

The timing of ODC releases to the atmosphere is closely tied to the ways in which these chemicals are used. In many applications, ODCs are promptly emitted (defined as emissions in less than one year) during either the manufacture or the use of particular ODC applications. For example, ODC-based solvents or propellants typically result in releases within a year of production. However, in applications such as refrigeration and insulating foams the ODCs are still mostly or entirely contained within the equipment even after the useful life of the product (which can be on the order of decades). As a result, large quantities of

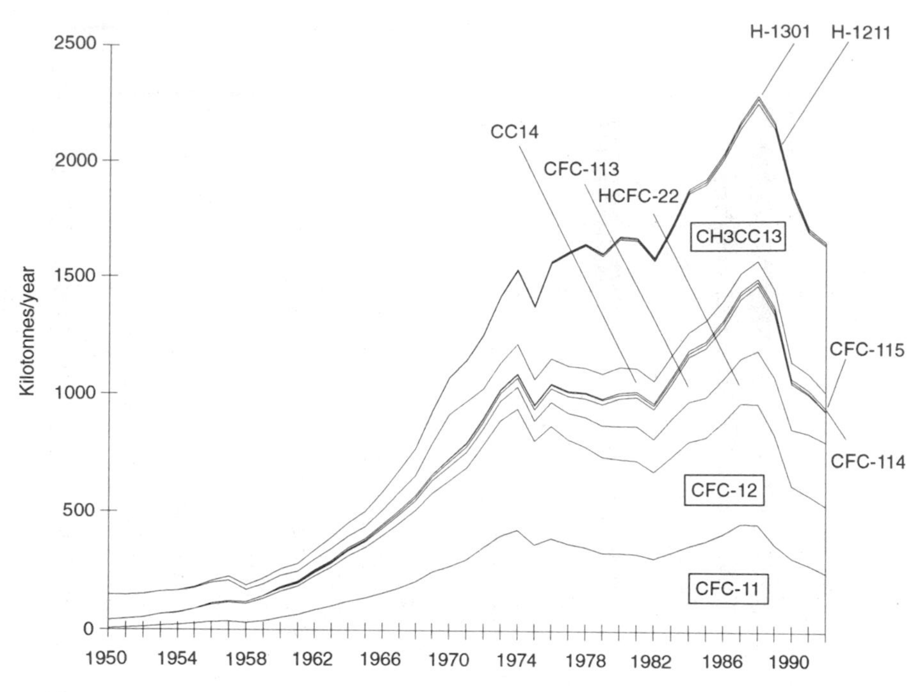

Figure 3.8
Estimated global ODC production, omitting CH_3Cl and CH_3Br. (See appendix for sources.)

ODCs (notably CFC-11, CFC-12, and HCFC-22) are stored in existing equipment. These ODC accumulations are called **banks**. The amount of CFC-12 contained within banks in 1990 was almost three times the amount of new CFC-12 produced in that year (AFEAS 1993a).

The following are significant sources of banked ODCs:

insulating polyurethane foams.
air conditioning systems in vehicles
residential and commercial refrigeration and air conditioning systems
stocks at the sites of processing and manufacturing industries that use ODCs
halons in existing fire-protection systems.

The rates at which various banks leak ODCs vary significantly. Releases of ODCs from insulating foams found in appliances, such as

Table 3.3
Kilotonnes of methyl bromide sold in 1990 (UNEP 1992, section II, p. 6, table 2). Presumably, 1990 sales exceeded production because of the sale of existing stockpiles.

Region	Soil fumigant	Space and structure fumigant	Chemical processes	Quarantine and commodity fumigant	Total	Share
N. America	22.7	1.4	2.8	1.2	28.1	42.2%
S. America	1.1	0.1	0.0	0.4	1.6	2.4%
Europe	16.6	0.6	0.9	1.0	19.1	28.7%
N. Africa	0.4	0.0	0.0	0.1	0.4	0.6%
Africa	1.4	0.1	0.0	0.3	1.8	2.8%
Asia	8.4	0.9	0.0	5.3	14.6	21.9%
Australia	0.7	0.1	0.0	0.2	0.9	1.4%
Total	51.3	3.2	3.7	8.4	66.6	100%

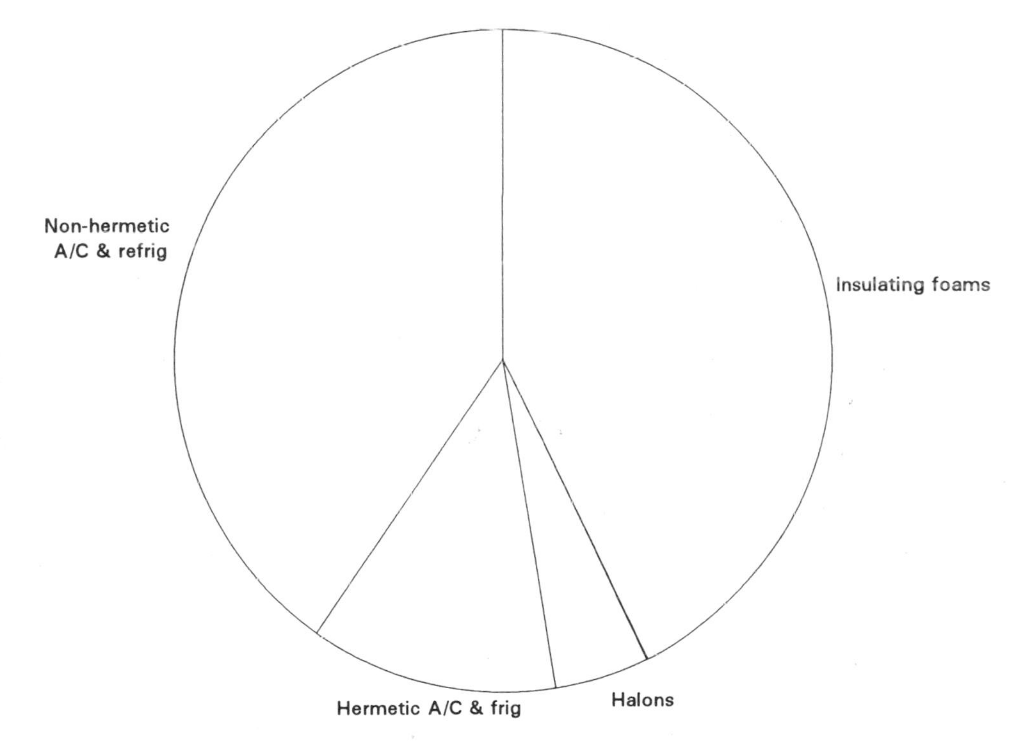

Figure 3.9
Global ODC banks, 1992 (derived from data in AFEAS 1993a, 1993c) (total: 3080 kilotonnes).

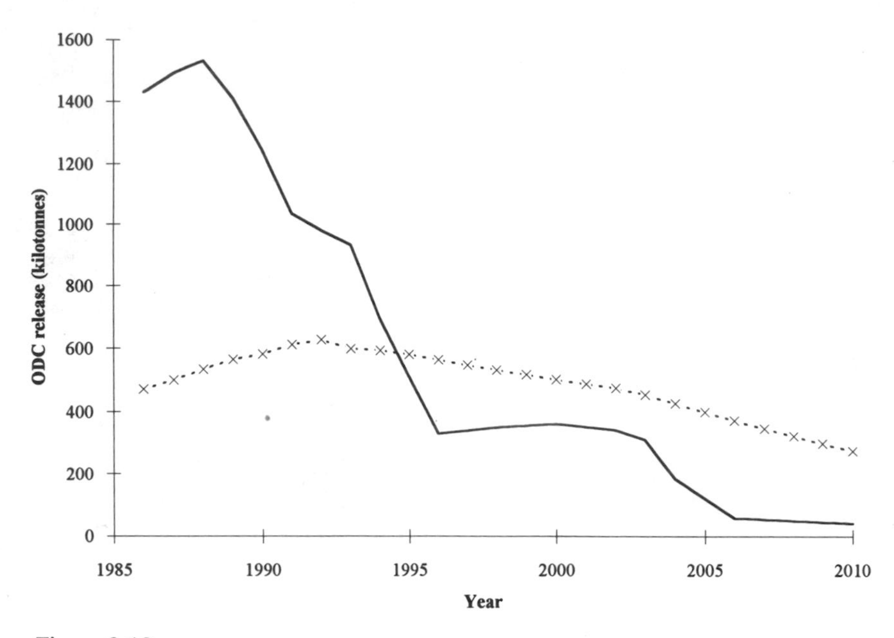

Figure 3.10
Projected global ODC emissions, based on IEER calculations (see chapter 12 below). Solid line: prompt release. Broken line: bank release.

refrigerators and freezers, can take decades. Similarly, ODCs used as working fluids circulating in hermetically sealed cooling systems (such as air conditioners, refrigerators, and freezers) will not leak unless the systems are damaged, or until disintegration occurs after disposal. In contrast, car air conditioners are not hermetically sealed and leak their coolant charge within a few years. Some large central air conditioning and refrigeration systems are also not hermetically sealed and can leak their charge, though more slowly than car air conditioners. The breakdown of banked ODCs in 1992 is given in figure 3.9. In view of the current size of ODC banks and the international controls on production and consumption, banked releases of these compounds are likely to exceed prompt releases in 1995 (figure 3.10).

4

Goals for Protecting the Ozone Layer

4.1 Environmental Philosophy

Establishing clear goals is a crucial step in formulating any effective policy: this is as true for a project to protect the ozone layer as it is for any other kind of program. Once the goals are agreed upon, policies to meet them can be put into place, and their effectiveness can be evaluated against commonly recognized standards.

Protecting the ozone layer from damage that would adversely affect human health or the environment has been the stated or implied goal of most programs to regulate ozone-depleting compounds. But the failure to establish more sharply defined aims has lead to serious gaps in policy, especially at the global level. There are also wide differences in the stringency of regulations in different countries, even among industrialized countries. This is because no explicit overall goal or philosophy for protecting the ozone layer has been incorporated into the global process as a guide to regulation. For example, until 1988, debate at the international level regarding emissions of ODCs focused almost exclusively on the regulation of CFCs and halons. Methyl chloroform and carbon tetrachloride were added in 1990. Regulation of HCFCs and the industrial uses of methyl bromide is still insufficient, and there is no policy at all regarding methyl chloride or methyl bromide emissions from biomass burning. Coordination of policies to protect the ozone layer with other environmental protection issues, such as regional pollution or global warming, is limited. Since 1990 there has been some progress in recognizing that policy must be linked to total stratospheric chlorine levels, notably the amount that triggered the Antarctic ozone depletion.

In order to specify the goals of our program to protect the ozone layer, a common framework incorporating all ODCs must be established.

4.2 Chlorine Concentration and Bromine Concentration as Policy Tools

Two approaches have generally been used to estimate the impacts of policies aimed at limiting the release of ODCs into the atmosphere. One approach involves the construction of complex computer models that simulate the effect each new potentially destructive compound has on the ozone layer. (See, e.g., Wofsy et al. 1975; Wuebbles 1983; WMO 1985, chapter 13; Hammitt et al. 1987; WMO 1989, chapter 3; Bruhl and Crutzen 1990; WMO 1991, chapter 8.) A simpler approach, adopted in our analysis and described later in this chapter, involves measuring the contribution of each ODC to total atmospheric concentrations of chlorine and bromine. (See, e.g., WMO 1989, chapter 3; Prather and Watson 1990; WMO 1991, chapter 8.)

ODCs themselves do not directly deplete stratospheric ozone. Rather, the chlorine and bromine released from these compounds engage in catalytic ozone loss.[1] The basic logic of this approach was noted in a 1988 U.S. Environmental Protection Agency study: "The merit of examining the potential for future ozone depletion by examining future levels of chlorine and bromine stems from the fact that chlorine and bromine abundances are thought to be the primary determinant of the risk of ozone depletion." (EPA 1988, p. 4)

The eventual human contribution to atmospheric chlorine and bromine concentrations can be approximately calculated without detailed knowledge of other atmospheric constituents and related uncertainties (a major problem in complex computer simulations of stratospheric ozone loss). This approach allows us to set specific goals for allowable atmospheric chlorine and bromine concentrations and, thereby, establish comprehensive policy goals for the entire class of ODCs.

Furthermore, the relative effectiveness of policies to reduce chlorine and bromine concentrations can be assessed in a straightforward manner. Atmospheric measurements, laboratory experiments, and theory can confirm and refine the scientific validity of this approach.

The computational aspects of this method are relatively straightforward (see chapter 12 and the appendix), in contrast with models that explicitly calculate and project stratospheric ozone loss.

This method also has potential drawbacks, since it does not explicitly link specific levels of atmospheric chlorine and bromine to ozone depletion. For example, chlorine and bromine concentrations do not explicitly reflect the potential nonlinearity in ozone loss. This approach is linear: it assumes that the effect of every additional molecule of chlorine will be the same as that of the previous molecule, irrespective of total concentrations. But natural processes tend to be nonlinear, with certain concentrations triggering sudden changes (on the model of the straw that broke the camel's back). Such nonlinearity produced the Antarctic ozone hole at chlorine concentrations of about 2 parts per billion by volume. Furthermore, each chlorine or bromine atom released into the atmosphere from the decomposition of an ODC is not equally likely to engage in stratospheric ozone loss. The altitude and the latitude at which the release occurs will, in most instances, determine the extent to which the particular chlorine and bromine atoms participate in ozone destruction. This is particularly true for HCFCs and other partially halogenated compounds (e.g., methyl chloroform, methyl chloride, methyl bromide) that release some of their chlorine into the lower part of the atmosphere, reducing their contribution to ozone depletion.

In addition, this method does not encompass the ozone-depleting effects of other chemicals. As the EPA study noted, "to the extent that other chemicals, for example NO_X [nitrogen oxides] from high speed airplanes, or climate-induced shifts in atmospheric dynamics threaten stratospheric ozone, chlorine and bromine are inadequate measures of the potential risk of ozone depletion" (ibid., pp. 4–5).

Finally, the approach does not directly take into account the greater ozone-depleting efficiency of bromine and the interactions between chlorine and bromine.

This last weakness can be overcome somewhat. The greater ozone-depleting efficiency of bromine can be accounted for by calculating its chlorine equivalent. Nonlinear chlorine-bromine interactions are more difficult to model, but even these can, in principle, be accounted for by

the parameter that is used to convert bromine to chlorine and
it dependent on total chlorine and bromine concentrations.

With appropriate cautions regarding nonlinear ozone-loss mechanisms, the total equivalent chlorine concentration (which includes the effect of bromine) can serve as an important policy tool for protecting the ozone layer. Nevertheless, in making chlorine and bromine concentrations the primary determinants of regulatory policy, we cannot lose sight of the fact that it is ozone depletion that is of concern, not chlorine and bromine as such.

4.3 Selecting Goals

As we have discussed, Antarctic ozone depletion was triggered in the late 1970s, when chlorine and bromine concentrations in the atmosphere reached approximately 2 ppbv. If we include bromine, weighted by its chlorine equivalent, the total concentration that triggered the Antarctic ozone depletion was approximately 2.4 ppbv.[2] Weighted equivalent chlorine concentrations as of 1993 were about 4.4 ppbv.

Long-Term Goals

One way to establish a long-term goal for protecting stratospheric ozone is to examine the amounts of natural chlorine and bromine in the atmosphere. We propose that acceptable concentrations of total equivalent chlorine from human activities be maintained far below the smaller of two values: the average natural level of equivalent chlorine in the stratosphere and the fluctuations in this level caused by natural phenomena (notably, volcanic eruptions). These can be of the same order of magnitude, since single volcanic eruptions can substantially increase stratospheric hydrogen chloride. The El Chichon eruption is estimated to have increased stratospheric hydrogen chloride by roughly 9 percent (Mankin and Coffey 1984). By this criterion, the total concentration of equivalent chlorine from human activities should be about 0.2 ppbv or less.

With a standard of 0.2 ppbv, we would have to eliminate all but the most essential uses of ODCs. Furthermore, regulatory control would have to be broadened to include all ozone-depleting chlorine and bromine that arise from human activities. It would probably be difficult to detect the effects of human activities on stratospheric ozone. Indeed, this is another way of stating the overall objective of ozone-layer protection: Human activities should not significantly alter natural stratospheric ozone levels and natural fluctuations in these levels.[3]

It might be argued that since significant ozone depletion might result from natural events, such as volcanic eruptions, relatively high concentrations of chlorine and bromine due to human activities should be permissible. From a number of viewpoints, this argument is not very sound.

We have already discussed the uncertainties and the possible nonlinearity that might give rise to Antarctic-ozone-hole-type losses over middle latitudes. To permit concentrations of chlorine and bromine comparable to natural fluctuations might mean permitting a doubling of natural levels of equivalent chlorine in the atmosphere. When natural events such as volcanic eruptions occurred, they might trigger unanticipated phenomena, which might not have arisen without the added burden of chlorine and bromine from human activities. Given the uncertainties and gaps in our knowledge of the functioning of the ozone layer, the changing burdens of other pollutants (e.g. methane and nitrogen oxides), and the potential nonlinear effects induced by aerosols from natural and human activities, we should set a long-term goal that eliminates so far as practicable the risk of significant ozone depletion.

We will not be able to achieve the goal of reducing equivalent chlorine concentration from human activities to a level of 0.2 ppbv for hundreds of years, even if all emissions of chlorinated and brominated compounds stop today. This is because levels of equivalent chlorine from human activities are already approximately 20 times this level and because some of the chemicals containing chlorine have lifetimes of 100 years or more. Recognizing this, we put forward 0.2 ppbv as an ultimate goal, derived from environmental considerations, and not as a goal that can be achieved by any policy prescriptions in the near future. At this stage it is not important whether this ultimate goal is 0.05 ppbv or 0.5 ppbv, since we are at least 200 years from anything in that range.

Transitional Goals

Short-term and medium-term goals must be set on the basis of far more urgent criteria. In the short term, we risk the expansion to heavily populated mid-latitude locations of the nonlinear ozone loss now beginning to occur over the northern pole.

Even at existing equivalent chlorine levels, the outbreak of nonlinear loss at locations other than over Antarctica is a constant possibility. Fluctuations in polar stratospheric temperatures can alter the formation of polar stratospheric clouds, thus accelerating ozone depletion over the northern pole and contributing to ozone loss over mid- and high-latitude locations. Such fluctuations occur frequently and are difficult to predict.

The 1991 eruption of Mount Pinatubo, which injected large quantities of sulfate aerosols into the stratosphere, has greatly increased the urgency of limiting chlorine concentrations. The surfaces of sulfate aerosols facilitate the onset of heterogeneous chemical reactions much as ice particles do over the southern pole. This injection of sulfate into the stratosphere is thought to be partially responsible for the unprecedented increase in worldwide ozone-loss rates during the early 1990s. With chlorine and bromine levels far above what is natural, future eruptions of greater magnitude could lead to severe, if not devastating, ozone losses.

Therefore, a medium-term goal should be the reduction of weighted equivalent chlorine concentrations below 2.4 ppbv. This is the approximate level that triggered the Antarctic ozone depletion. It has been recognized by the international scientific community as a reasonable medium-term goal.

These goals will require considerable effort and cooperation worldwide. However, they are achievable.

4.4 Comparing Stratospheric Ozone-Protection Policies

The methodology chosen to compare specific policies for protecting the ozone layer has a profound effect upon the efficiency and thoroughness of the results. Though the history of regulation pertaining to stratospheric ozone loss is short, various approaches have been suggested.

Regulating Individual Chemicals

To date, the approach of international ozone-layer-protection policy has been through regulation of individual chemicals or groups of chemicals. The predominant criterion used in evaluating particular chemicals has been a quantity referred to as the **steady-state ozone-depleting potential** (S-S ODP) (Wuebbles 1981; Fisher et al. 1990). It has been considered satisfactory to replace, at least for some period of time, chemicals with "high" S-S ODPs by chemicals with "low" S-S ODPs. The changing nature of the ozone-depletion problem and the reevaluations of the short-term effects of these compounds have made this approach inadequate.

S-S ODPs are calculated by assuming constant emissions of a particular compound for periods equal to several lifetimes of the chemical itself or a reference chemical, whichever is longer. CFC-11 has generally been used as the reference chemical. Because this ODC has a lifetime of 55 years, calculation times on the order of 200 years or more are required. After such a period, the annual amount of the compound of interest and CFC-11 are approximately equal to the amounts of each annually added, so the total atmospheric amount of each remains constant thereafter (provided that emissions remain constant). When such conditions occur, a compound is said to have achieved a steady state. The steady-state concentration is then used to calculate the S-S ODP of the compound relative to CFC-11.

For instance, consider annual one-tonne emissions of two ODCs currently being produced, CFC-11 and HCFC-22. After about 220–275 years (4–5 CFC-11 lifetimes), CFC-11 will build up to steady-state levels in the atmosphere of approximately 55 tonnes. Each molecule of CFC-11 contains three atoms of chlorine. An annual one-tonne emission of HCFC-22, which has a lifetime of about 16 years, will reach steady-state levels of about 16 tonnes after approximately 64–80 years. Each molecule of HCFC-22 contains one atom of chlorine. To compare the steady-state effects of HCFC-22 with those of CFC-11 implies a waiting period of about 250 years, even though HCFC-22 has achieved a steady state in a far shorter time period. After this waiting period, the number of chlorine atoms in the atmosphere due to CFC-11 emissions is approximately 20 times (accounting for differences in molecular weight) the number of chlorine atoms due to HCFC-22 (figure 4.1). The calculated steady-state ozone loss due to CFC-11 is also approximately 20 times that due to HCFC-22. It is on this basis that the S-S ODP of HCFC-22 is calculated to be approximately 5 percent of CFC-11 (Fisher et al. 1990).[4]

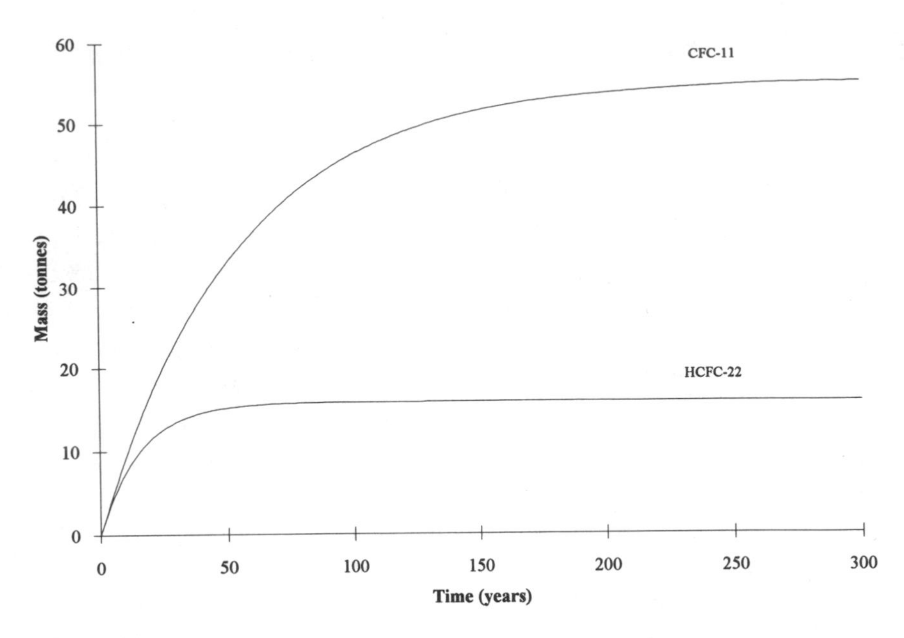

Figure 4.1
Mass of compounds in the atmosphere, with annual emissions of 1 tonne assumed.

From the perspective of protecting the stratospheric ozone layer, there are several things wrong with the use of the S-S ODP. For instance, in the above example the relative impacts of HCFC-22 emissions are considerably different prior to steady-state conditions. Such effects are not characterized in the steady-state determination of the ODP. The World Meteorological Organization's 1989 report on ozone depletion described the problem as follows: "Just as valuable as knowing the applicability of ODP, we should also recognize the limitations of ODP definition in order not to extend interpretation beyond its valid scope. For instance, since ODP is defined at a steady state, it is not representative of transient effects, especially during the early years of emissions." (WMO 1989, vol. 2, p. 308)

Since the incorporation of the S-S ODP into international regulations, the projections of ozone-depleting chlorine and bromine have changed dramatically. According to the current understanding, the worst depletion is expected to occur shortly after the year 2000, when atmospheric chlorine and bromine concentrations reach their maximum amounts

(WMO 1991, chapter 8). Between now and then, unforeseen events such as large volcanic eruptions and persistent cold stratospheric temperatures threaten to initiate serious ozone depletion over middle latitudes. Efforts to protect stratospheric ozone must now focus on limiting this upcoming maximum depletion and restoring the ozone layer to its pre-ozone-hole levels as rapidly as possible. Regulating on the basis of the steady-state impact of chemicals is not an appropriate way to meet this crisis.

Furthermore, the S-S ODP values currently used in regulations protecting stratospheric ozone do not take into account heterogeneous ozone-loss mechanisms. Heterogeneous ozone-loss mechanisms are considered responsible for the severe ozone loss observed over the Antarctic and are likely responsible for the increasingly severe loss over the northern pole and the middle latitudes.

One way to overcome the temporal bias of the S-S ODP is to look at the relative impact on stratospheric ozone over time. For instance, in the first decade or two after the start of HCFC-22 emissions, their contribution to atmospheric chlorine (and, hence, to ozone depletion) was far larger relative to CFC-11 than in later years. The time dependence of the atmospheric chlorine release can be used to calculate a time-dependent or "transient" ODP (T-ODP) (WMO 1989, volume 2, pp. 365–374). The T-ODP can then be evaluated at any point between the time at which release of the chemical began and the time at which it achieves a steady state.

Attempts to include heterogeneous ozone-loss mechanisms in estimates of T-ODPs have used a combination of observations and model calculations for many of the important ozone-depleting compounds (Solomon and Albritton 1992). Thus, although the S-S ODP of HCFC-22 is about 0.05, the estimates for 5-year, 10-year, and 20-year T-ODPs incorporating heterogeneous ozone-loss mechanisms are 0.19, 0.17, and 0.14, respectively (ibid., table 2). Figure 4.2 shows T-ODP as a function of time for a number of currently used ODCs and a number of proposed alternative ODCs.

No single number can capture the complexity of a chemical's contribution to ozone-layer depletion; however, 10-year and 20-year T-ODPs may indicate what impact ODCs may have during the most crucial years for stratospheric ozone loss. Table 4.1 presents a T-ODP calculated

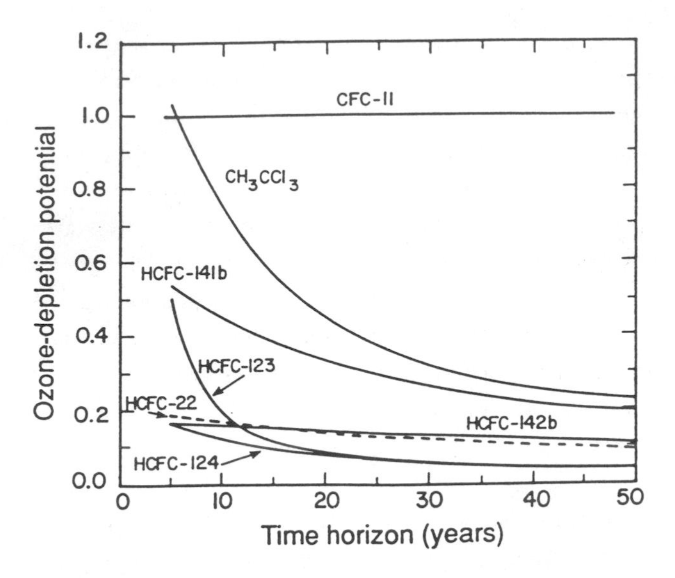

Figure 4.2
Semi-empirical time-dependent ozone-depletion potentials (Solomon and Albritton 1992).

at various time horizons for both the ODCs in current use and those proposed as alternatives. (Although the numbers in table 4.1 represent a combination of polar-ozone-loss observations and calculations, they are considered nearly equivalent to global values (WMO 1991, p. 6.14).)

The short-term T-ODP values for many of the HCFCs are significantly higher than their steady-state values.[5] These compounds are being proposed as alternatives to the ODCs currently scheduled to be phased out by 1996 in the industrialized countries (CFCs, halons, methyl chloroform, and carbon tetrachloride). Because the projected maximum atmospheric chlorine content is expected to occur soon after the turn of the century, the ozone-depleting ability of these compounds during that period should be considered an important element in their regulation.

Table 4.1 has significant implications for the regulatory structure of one of the largest ODC-producing countries, the United States. The stratospheric-ozone-protection regulations embodied in the amended U.S. Clean Air Act (CAA) group ODCs by ODP value. Compounds with ODPs of 0.2 or greater are classified as "class I substances" and are subject to the most rapid phaseout schedule (U.S. CAA 1990). A number

Table 4.1
Semi-empirical time-dependent ozone-depletion potentials (T-ODPs)(IEER calculations).

Compounds	Cl+Br atoms	Lifetime (years)[a]	Time horizon (years)						
			5	10	20	30	50	100	500
Fully halogenated									
CFC-11	3	55	1.00	1.00	1.00	1.00	1.00	1.00	1.00
CFC-12	2	116	0.46	0.47	0.50	0.52	0.56	0.66	0.96
CFC-113	3	110	0.55	0.56	0.59	0.61	0.66	0.78	1.1
CFC-114	2	220	0.17	0.18	0.19	0.21	0.23	0.30	0.62
CFC-115	1	550	0.039	0.040	0.044	0.047	0.055	0.075	0.23
CCl_4	4	47	1.3	1.3	1.2	1.2	1.2	1.1	1.1
Halon 1301	40	77	10.3	10.4	10.7	10.9	11.4	12.4	14.4
Halon 1211	41	11	11	9.6	7.2	5.7	4.1	2.9	2.4
Partially halogenated									
CH_3CCl_3	3	6.1	1.0	0.75	0.45	0.32	0.22	0.15	0.12
CH_3Cl	1	1.5	0.59	0.21	0.089	0.059	0.039	0.027	0.022
CH_3Br	40	2	17	7.0	3.0	2.0	1.3	0.92	0.76
HCFC-22	1	15.8	0.19	0.17	0.14	0.12	0.09	0.066	0.055
HCFC-123	2	1.7	0.51	0.19	0.082	0.055	0.036	0.025	0.021
HCFC-124	1	6.9	0.16	0.12	0.079	0.057	0.039	0.027	0.022
HCFC-141b	2	10.8	0.54	0.45	0.33	0.26	0.19	0.13	0.11
HCFC-142b	1	22.4	0.16	0.15	0.14	0.12	0.10	0.080	0.067
HCFC-225ca	2	2.8	0.42	0.21	0.10	0.067	0.044	0.030	0.025
HCFC-225cb	2	8.0	0.21	0.17	0.11	0.084	0.058	0.040	0.033

a. Lifetimes are from WMO 1991, pp. 6.7, 8.7 (H-1211 and H-1301), and 8.8 (CH_3Cl), and from UNEP 1992, section II, p. 7 (CH_3Br).

of compounds not now included in this group, such as HCFC-22, HCFC-141b, HCFC-123, and both isomers of HCFC-225, may have to be reclassified because of their large short-term T-ODPs. For instance, HCFC-141b has a T-ODP value greater than 0.2 for all periods less than 40 years, the critical period for ozone-layer protection. Many of the HCFCs have ODPs close to 0.2 for shorter periods. The relative lifetime uncertainty of 10–30 percent embodied in these values would put an upper-bound estimate for the 10-year T-ODP above 0.2 for every HCFC except HCFC-124 (Solomon and Albritton 1992).

The time-dependent ODPs overcome many of the limitations of their steady-state counterparts. However, they still do not take into account nonlinear effects due to atmospheric chlorine and bromine from all compounds combined. This is crucial, because the effect of a particular chemical depends not only on the emissions of that chemical but also on the total concentrations of ODCs present when that chemical is emitted. Just as the Antarctic ozone hole was an unforeseen phenomenon, we cannot yet say what surprises may await us if increases in chlorine concentration continue. T-ODPs cannot take this into account .

Although we do not emphasize comparisons of ODCs, there may be a need to consider priorities in recycling and the interim use of compounds. This will require a comparative index. We recommend the use of the time-dependent ODP, with heterogeneous mechanisms incorporated. Such a measure has some limitations, but it would be far superior to homogeneous models for the purposes of public policy, since it is oriented toward limiting the peak of atmospheric chlorine and bromine and toward rapidly reducing chlorine and bromine concentrations.

Technological Approach

In order to eliminate the production and the consumption of ODCs, alternative technologies and chemicals must be rapidly identified. To a certain extent, this was done during the formation of existing regulations. However, many promising non-ozone-depleting alternatives have yet to be examined fully or encouraged sufficiently. In searching for alternatives, a number of issues must be considered.

First, technologies, processes, and chemicals used as replacements must be evaluated for their effects on atmospheric chlorine and bromine concentrations. Second, their impacts on environmental and human health (such as groundwater pollution or cancer rates must be closely examined. Third, the radiative properties of the substitute chemicals must be examined to determine the degree to which they will contribute to global warming. Fourth, since a number of the processes that now use ODCs affect a substantial portion of energy consumption, the net effect of replacing currently used ODCs on energy use—and hence on the accumulation of greenhouse gases—must be considered. Finally, we should examine what new problems alternative technologies, processes, and chemicals may occasion. For example, some new perfluorinated compounds are being tested as substitutes in many of the applications currently using ODCs (UNEP 1991f). Perfluorinated compounds tend to have lifetimes in excess of 1000 years, so they have the potential to accumulate in the atmosphere even when emitted at modest rates. Any possibility that environmental harm might result from their presence should be thoroughly investigated.

In many instances, there is potential not only to do away with the use of ODCs but also to greatly improve the environmental and economic performance of the processes now dependent upon them. For example, many present-day technologies were designed when energy consumption was not as great a concern as it is today. There is, therefore, the opportunity to replace technologies that rely on ODCs with far more efficient technologies. This is particularly true in the sphere of stationary air conditioning and heating in commercial and large-scale residential applications. In the long term, some of these technologies can be adapted to use solar energy.

Similarly, there are technologies that obviate the cleaning of electronic circuit boards, a major use of ODCs. These technologies give rise to products of the same quality as those now cleaned with ozone-depleting chlorinated solvents. Finally, vacuum insulation panels, based on the same principles as Thermos flasks, can be developed for refrigerator and freezer insulation, in place of the polyurethane foam manufactured with CFCs.

Some of these technologies are available today; others must be developed. The rate at which these alternatives can be implemented or developed is a matter of some debate. In those instances in which the most desirable alternative technologies remain to be developed, it may be necessary to adopt interim measures which will have some short-term deleterious effects, but less so than emissions of ODCs.

As we will discuss in the chapters that follow, nontoxic, non-ozone-depleting alternative technologies, processes, and chemicals with comparable characteristics in other areas, such as energy use, are now available and could be fully utilized within a relatively short period of time. Substitution with HCFCs is not required. These compounds should therefore be phased out, along with the fully halogenated ODCs, on a timetable appropriate for individual applications.

Banked ODCs

Many of the ODC-using applications release ODCs very slowly to the atmosphere. As a result, large quantities of ODCs, notably CFC-11, CFC-12, and HCFC-22, are "banked" within existing equipment. In order to keep the peak of ODCs in the atmosphere as low as possible, it will be necessary to recover these ODCs for reuse and eventual destruction.

Some research on existing refrigerant-recovery systems in the United States suggests that in many cases much of the CFC charge goes unrecovered or is released as a result of poor recovery-equipment performance and poor procedures (Kolar 1993). A thorough reevaluation of the available recovery equipment is necessary in order to ensure that banked compounds are properly recovered before destruction or reuse.

Mere recovery and recycling will not be enough in most cases, since repeated reuse in applications such as automobile air conditioning results in the eventual release of the entire quantity of the chemicals to the atmosphere. Recovery, recycling, and reuse of ODCs for existing equipment should be allowed for strictly limited periods, and only when such use is necessary and no substitutes are available. As we will see, there is considerable scope for modifying existing systems to accommodate different chemicals and processes. The goal of minimizing the peaks of atmospheric chlorine and bromine will require that much of the existing stock of ODCs be removed from existing equipment and destroyed.

The principal method of destroying ODCs is incineration. Most of the ODCs currently in use are not flammable, since this is one of the properties used to screen these compounds for commercial use. Thus, high-temperature incineration such as is typically used in destroying hazardous wastes, or some equivalent process, is necessary to destroy ODCs.

Incineration of CFCs used as refrigerants has been the subject of two experimental investigations in Denmark. In one investigation, only the CFC-12 was incinerated; in the other, the entire refrigerator was incinerated, which would destroy both the CFC-12 (averaging 140 grams per refrigerator in Denmark) and the CFC-11 in the insulating foam (averaging 500 grams per refrigerator in Denmark) (Pedersen 1991). One reason that the Danes have pursued this research is that CFCs destroyed in this way can be taken as credits toward new CFC production under the Montreal Protocol, since the destroyed CFCs cannot contribute to ozone depletion. We do not recommend such credits at this stage, but it is interesting to note the manner in which regulations have brought forth technological development.

ODCs destroyed in incinerators appear not to give rise to toxic byproducts under test conditions. The principal reaction in the particular case of CFC incineration is (Pedersen 1991, p. 517)

$$CF_XCl_Y + 2H_2O \rightarrow CO_2 + Y \cdot HCl + X \cdot HF, \quad (4.1)$$

where CF_XCl_Y is the CFC molecule being destroyed, containing X atoms of fluorine and Y atoms of chlorine; H_2O is water (in the form of steam in this case); HCl is the hydrochloric acid which results from the presence of chlorine in the CFC being destroyed; and HF is the hydrofluoric acid that results from the presence of fluorine in the CFC being destroyed.

The acids in the exhaust gases can be removed by scrubbing with calcium hydroxide, leaving essentially only carbon dioxide to be released to the atmosphere.

In addition to impurities in the ODC being incinerated, trace contaminants from combustion will arise (Pedersen 1991). Further complications would arise from the incineration of an entire refrigerator rather than just the halocarbon refrigerant. In such instances, the simple reaction depicted above would not adequately model the actual incineration chemistry.

Measurements of the exhaust gases revealed that about 99 percent of the CFC-12 and essentially all of the CFC-11 were destroyed in the

Danish study. The data for various pollutants emitted by the facility were compared against "threshold values" specified under Denmark's air-pollution regulations. All the pollutants reported were below the threshold values. However mercury was up to 86 percent of the threshold value (Pedersen 1991, pp. 518–519). Additional filtering with activated-carbon filters would reduce the heavy-metal content of both the exhaust air and the ash (Cleve 1989).

No amount of pollution control can completely remove all pollutants from incinerator exhaust. We are unfortunately confronted with the choice between allowing banked ODCs to escape into the atmosphere and recovering them so we can destroy them. Our recommendation that these compounds be incinerated with the best available pollution-control technology stems from the essentially complete destruction that can be achieved and the benefits of reducing the risk of severe ozone depletion. Further, this recommendation allows most of the risk of ODC use in existing equipment to be borne by our generation rather than future generations.

This recommendation makes sense only in the context of a phaseout of ODCs. It is a one-time sacrifice that our generation, which has benefited from the use of ODCs, must make to protect future generations. If production of these compounds were to continue, incineration and the risks it entails would not make sense—at least, not in the same straightforward manner.

Research by Japanese chemists indicates that a method other than incineration may be developed to destroy CFCs (this method has not been tested on other ODC classes such as chlorocarbons or halons) (*New Scientist* 1992b). The process involves passing a mixture of CFC and ethanol over a catalyst of activated-charcoal pellets coated with iron chloride ($FeCl_3$). The main by-products are carbon dioxide (CO_2), carbon monoxide (CO), hydrogen fluoride (HF), and a hydrogenated CFC whose exact nature depends on the compound initially introduced into the process. It is believed that the process can be improved to eliminate this last unwanted by-product. While questions remain concerning the effectiveness and safety of this method, the preliminary results are promising.

5

Air Conditioning and Refrigeration

Several ozone-depleting compounds possess properties that make them good refrigerants; that is, they can efficiently transfer heat from one location to another. They are used in household and commercial refrigerators and freezers, in stationary and mobile air conditioners, and in refrigerated trucks, water coolers, and similar applications. (Mobile air conditioners are those found in cars, trucks, and other vehicles.) ODCs are also used as working fluids in heat pumps (which are essentially air conditioners operated in reverse) and in chillers used to keep certain industrial equipment at low temperatures. The generic term for the working fluids in all these applications is **refrigerant**. The ODCs most commonly used as refrigerants are CFC-11, CFC-12, and HCFC-22. CFC-113, 114 and 115 are also used in some applications, although in relatively small amounts. These compounds may be used alone or combined and used as mixtures.[1] Since about 1990 a number of other halogenated chemicals have come into use. These include some ozone-depleting compounds, such as HCFC-123, and some non-ozone-depleting compounds, the most notable of which is HFC-134a.

Nonhalogenated refrigerants are also used in some air conditioning and refrigeration systems. For instance, ammonia is used in significant quantities in cold storage and in large absorption air conditioning systems. We will discuss these refrigerants, along with emerging technologies, as options that might facilitate the phasing out of ODCs in air conditioning and refrigeration. We will also consider the ODCs that are banked in these systems, since the recovery and destruction of ODC banks can make a crucial contribution to the protection of the ozone layer.

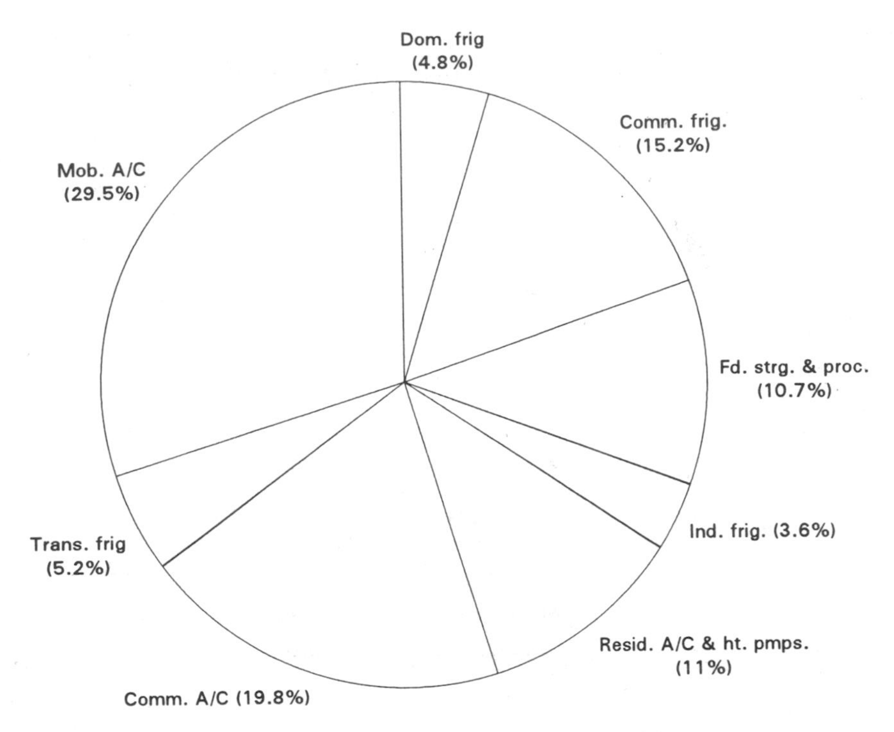

Figure 5.1
Global ODC use for refrigeration and air conditioning, 1990 (derived from data in AFEAS 1993b and UNEP *1991*d)

5.1 Use of ODCs as Refrigerants

Figure 5.1 presents the worldwide 1990 use of ODCs in refrigeration and air conditioning applications. A total of 482,220 tonnes were used in that year.[2] In 1985 approximately 420,700 tonnes were used.[3] While total consumption increased until 1990, there were modest declines in ODC use in 1991 (see table 3.2), indicating a downward trend. In 1990, the consumption of ODCs for home refrigeration, commercial and industrial refrigeration, and mobile air conditioning was 2.1, 38., and 26.5 percent, respectively, of the total for refrigeration and air conditioning applications. The balance was consumed primarily in commercial and residential air conditioning (24.9 percent and 7.5 percent of the total, respectively).

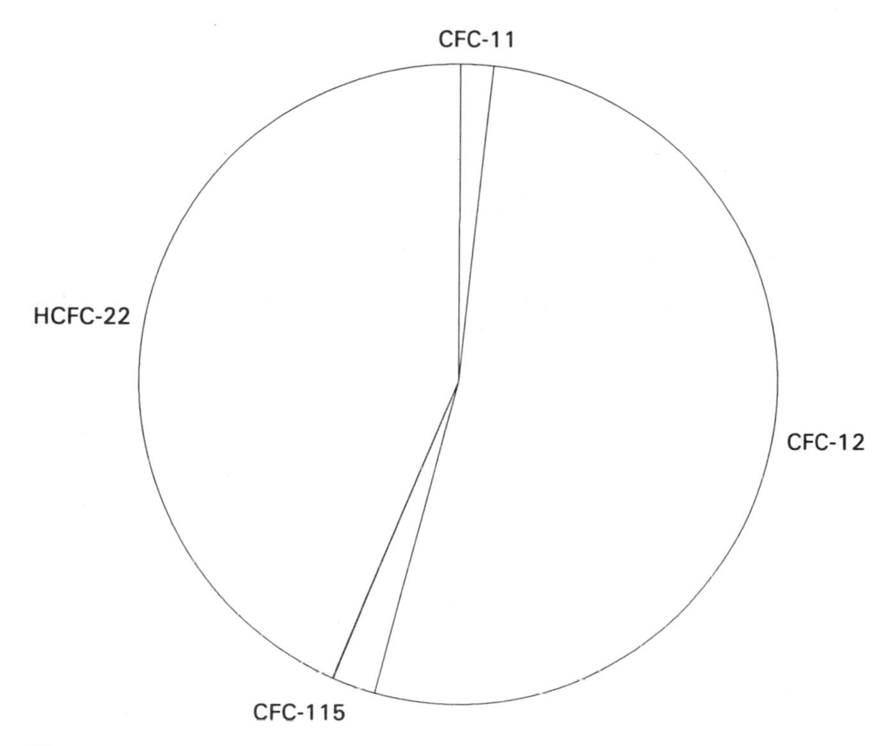

Figure 5.2
Use of various ODCs in refrigeration and air conditioning, 1990, relative to the total of 497,760 tonnes (UNEP 1991d).

The most significant change since 1985 has been the increase in the proportion of ODCs used in all refrigerant and air conditioning applications, relative to other uses, that has resulted from the rapid reduction in the use of these compounds as propellants and solvents and in the blowing of foam.

Most central home air conditioning systems, room air conditioners, and small-scale commercial systems, and many commercial refrigeration and air conditioning systems, use HCFC-22 as a refrigerant. Indeed, Richard Ertinger of the Carrier Corporation, which makes central air conditioners, calls HCFC-22 "the world's most widely used refrigerant" (Ertinger 1991, p. 340). It is used in both medium-temperature and low-temperature applications. Home, commercial, and industrial refrigeration and freezing use both CFC-12 and HCFC-22. Mobile air conditioners also use CFC-12. CFC-11 is also a common refrigerant for large non-unitary commercial and industrial air conditioning and chilling systems. Specialized systems use CFC-113, CFC-114, and CFC-115, often in combination with other chemicals. CFC-114 has been the refrigerant of choice of the U.S. Navy for shipboard chilling and air conditioning applications (Smith et al. 1993).

In terms of releases, the differences among home, commercial, and mobile refrigeration and air conditioning are considerable. Refrigerants in home refrigerators and freezers are contained within hermetically sealed systems; generally, they are released only when the systems are discarded without refrigerant recovery. Even these releases can be largely prevented by recovering the refrigerant at the end of life of the system. In contrast, car air conditioning systems leak continually, because engine-mounted air conditioning systems require flexible hoses. As a result, such systems require coolant replacement every 3–4 years. For example, CFC-12 emissions from automobile air conditioners were approximately 200,000 tonnes in 1990, versus about 5000 tonnes for domestic and commercial refrigeration.[4] Releases from mobile air conditioning systems can be greatly reduced, as discussed below.

Equipment for air conditioning and refrigeration contains a very large bank of unreleased ODCs (primarily CFC-12 and HCFC-22), totaling approximately 1.6 million tonnes in 1990.[5] These banked ODCs, which are now approximately equal to the global production of all ODCs, will contribute significantly to releases in the future unless appropriate actions are taken to collect and destroy them.[6] In the United States and some European countries, regulations that came into force in the early 1990s require refrigerant recovery. They have probably contributed to the reduction in emissions since 1990.

Mobile Air Conditioning

Next to solvent use, mobile air conditioning, principally in passenger vehicles, represents the largest specific end use of ODCs worldwide. In 1987, 57 percent of the world's air conditioned vehicles were in the United States.[7] Mobile air conditioning is also common in Japan and among the wealthy in many Third World countries. Approximately 85 percent of the air conditioned vehicles are located in the industrialized countries (UNEP 1991d, p. 166). CFC-12 was the main refrigerant used in mobile air conditioners until 1992.

The capacity of a car's air conditioner is determined primarily by the requirement that the air inside a car, as well as the solid parts such as the steering wheel, receive a blast of cool air moments after the vehicle is started. This design criterion applies to a car started after having been parked in direct sunshine for a long period. In other words, almost all car air conditioners are designed for the conditions one might expect in the hot, arid southwestern United States. As a result, car air conditioners are quite large: each has about 24,000 Btu/hour of cooling capacity. (In the air conditioning industry this is said to be equal to 2 "tons" of cooling capacity.[8])

Even with such a large air conditioner, the solid parts of the car cannot be cooled very fast ,since they store a considerable amount of heat. When the outside temperature is 110°F, the inside of a car with untinted glass parked in the sun without shade or ventilation can heat up to approximately 225°F (above the boiling point of water) (Chiou 1986).

With a large air conditioner, a relatively large amount of coolant must be used: about 2 kilograms of CFC-12 per car. This also means that large amounts of this compound are needed for leak detection and recharging. Therefore, reducing air conditioner size is a factor in reducing ODC use. Smaller air conditioners offer the added, if relatively modest, benefit of better fuel mileage.

Four-fifths of the ODC use in mobile air conditioning occurs during servicing and recharging of existing units (UNEP 1991d, p. 167). Refrigerant loss in a typical car air conditioner occurs within 5 years, necessitating service and recharging. The loss occurs because car air conditioners are not hermetically sealed units. Instead, they are engine-mounted, the most convenient and initially inexpensive design. Engine mounting of the compressor subjects the system to considerable vibration, necessitating the use of flexible hoses and connectors. Typical causes of refrigerant loss are the permeability of the hoses, the loosening of connections from vibrations, and the deterioration of the hoses.

When an air conditioned car is taken to a garage for service, typically only about 40 percent of the original charge of up to approximately 2 kg of CFC-12 is left in the vehicle. As a result, the annual servicing requirements per vehicle average about 0.4 kg (Baker 1991). This is more than an order of magnitude greater than the requirements for domestic refrigerators, which typically contain only about 0.25 kg (or less) of CFC-12 and are hermetically sealed, with an overall leak rate (including that from disposal of refrigerators) of about 5–10 percent of the charge per year.[9] Thus, a domestic refrigerator, without refrigerant recycling, would emit about 0.02–0.04 kg per year, including the loss after disposal. Recycling greatly reduces these emissions since it eliminates the main source of loss from refrigerators.

A typical service routine for a leaking car air conditioning system when there is no refrigerant recovery is as follows:

Any refrigerant remaining in the system is vented into the atmosphere prior to checking for leaks.
The system is flushed with CFC-11 after the remaining CFC-12 in the system is exhausted. (With tighter restrictions on CFC-11 disposal, U.S. service stations are now favoring filter and component replacement over flushing.)
CFC-12 is introduced into the system, and a detector is used to locate the leak.
The system is repaired.
The system is evacuated and, for 10 minutes, reloaded with CFC-12 vapor. It is then reevacuated.
The system is recharged.

At each step in which refrigerants are introduced, they are eventually vented into the environment. In some cases, where small cans are used to recharge the systems, the residual CFC-12 in the can is discarded also. Small coolant containers are now prohibited in the United States; refrigerant recovery has been mandated. There are, however, some concerns about the effectiveness of the program.

Commercial and Industrial Refrigeration and Air Conditioning

HCFC-22 is widely used, especially in air conditioning. It has been estimated that there were 332,000 tonnes of HCFC-22 banked in installed unitary air conditioning equipment worldwide in 1990 (UNEP 1991d, p. 27). (Unitary air conditioners cool air directly, while non-unitary systems cool other fluids, typically water, first.) The motors that drive these sys-

Table 5.1
HCFC-22 use in refrigeration and air conditioning, 1990 (derived from data presented in UNEP 1991d).

	Unitary	Non-unitary	Commercial and industrial
Charging and service	104,300	3,200	small
New	32,700	2,700	74,000
Total	137,000	5,900	74,600

tems have a total installed electrical capacity of 1.385 billion kilowatts. Unitary systems require approximately 0.24 kg of HCFC-22 per kilowatt of installed capacity. Given average cooling efficiencies, this amounts to a charge of about 0.86 kg per tonne of cooling capacity (Ertinger 1991, p. 341). Additional quantities of HCFC-22 banked in other refrigeration and air conditioning equipment amount to approximately 250,000 tonnes.[10]

In 1990 the total estimated consumption of HCFC-22 was about 218,000 tonnes for all refrigeration and air conditioning applications.[11] As table 5.1 shows, 137,000 tonnes were used to charge and service new and existing unitary air conditioning equipment, while 5900 tonnes were used to charge and service new and existing non-unitary equipment (UNEP 1991d, p. 143). The remainder of the HCFC-22 consumed for refrigeration and air conditioning applications, approximately 74,600 tonnes, was used in applications such as industrial cold storage and transport refrigeration. It is further estimated that unitary products (the charging of new products, plus the service of existing equipment) will consume about 220,000 tonnes by the year 2000 if there is no refrigerant recovery. This requirement can be reduced to about 75,000 tons if there is refrigerant recovery (Ertinger 1991, pp. 344–347).

Many commercial and industrial refrigeration and air conditioning applications still use CFC-12 as a refrigerant. For example, it is used in many food service and food stores and cafeterias and in the processing and storage of dairy processing and storage of dairy products. In 1990, CFC-12 consumption in non-automobile refrigeration applications totaled approximately 110,700 tonnes.[12] Many of these applications are in the midst of a conversion to HFCs, notably HFC-134a.

Domestic Refrigeration
While a relatively small use within the broad category of refrigeration and air conditioning, ODC use in domestic refrigeration gets a significant amount of public attention. In addition to being used as refrigerants, ODCs are commonly used to produce the insulating foam found in the walls and doors of refrigerators. The average charge of coolant within domestic refrigerators varies: in North America it is typically about 180 grams (~0.4 pound) per unit, whereas in Europe it is about 140 grams (~0.3 pound) per unit (UNEP 1991d, p. 79). North America and Western Europe account for just under 50 percent of the CFC-12 used in domestic refrigerators and freezers (ibid.). Asia and Eastern Europe (including the former Soviet Union) each account for approximately 23 percent of the CFC-12 used in this application.

5.2 Alternatives to CFCs in Refrigeration and Air Conditioning

Broadly speaking, the following technologies and chemicals are available for use today to replace CFCs in air conditioning and refrigeration:

HFCs (see chapter 3), which are close in their thermodynamic properties to CFC-11 and CFC-12 but which contain no chlorine or bromine. Chief among these is HFC-134a. Other products are HFC-152a, HFC-32, and HFC-125.

Partially halogenated compounds containing chlorine, known as HCFCs. These compounds have shorter lifetimes than CFC-11 and CFC-12 and are largely destroyed in the troposphere. They are therefore less ozone-depleting than CFC-11, but proportionately more so in the near term (see chapters 4 and 12). Chief among the HCFCs used as refrigerants are HCFC-22, HCFC-123, and HCFC-142b, all of which are commercially available. HCFC-124 is under investigation.

Mixtures of various halogenated chemicals.

Nonhalogenated refrigerants, which are used in some compressor systems similar or somewhat akin to halocarbon-using systems. Among the refrigerants that can be used are hydrocarbons (notably propane), fluoroiodocarbons, and carbon dioxide.

Absorption refrigeration and air conditioning systems, which use working fluids such as ammonia and lithium bromide. In commercial and large residential applications (such as in large apartment buildings), these systems can be built to operate on waste heat from electricity generation.

In addition to these chemicals and systems, there are categories of chemicals and systems which are not yet on the market but which could be important in a few years, either for a broad range of applications or for niche markets that are served today by chlorine-containing chemicals and mixtures. Among the chemical candidates are fluorinated ethers, fluorinated propanes, and other organic compounds containing fluorine but not chlorine or bromine (IAC 1991, pp. 385–415). Helium can also be used as a refrigerant in the Stirling-cycle system in some low-temperature applications, but it is not yet commercially suitable for mass-market refrigeration and air conditioning applications. Some new developments in compressor design may enable Stirling-cycle machines to be developed for refrigeration and freezing (Berchowitz, n.d.).

A possibility for some applications is the use of heat to generate high-intensity sound waves in a gas, which are then used for cooling. This refrigeration system, called thermoacoustic refrigeration, is based on the principle that sound waves can be used to create differences in temperature. A standing wave is set up in a tube which contains a set of plates called a stack. The alternating compression and expansion of the sound waves extracts heat from one end of the set of plates (the cold, or refrigerated end) and transfers it to the other, where it is removed. The device to be cooled is connected to the cold end of the stack via a heat exchanger. This principle has been used in specialized very-low-temperature applications where the amount of heat pumping required is modest. Thermoacoustic refrigeration is attractive for a number of reasons: it has no moving parts (the vibrating gas is the only moving element); the refrigerant, an inert gas such as helium, does not contribute to global warming or ozone depletion; and it can use a variety of energy sources, such as natural gas or hydrogen. However, it has poor efficiency for refrigeration and air conditioning applications in which temperature differences are small and heat-pumping requirements are high, such as comfort air conditioning. It cannot be used in these applications without major design changes (Garret and Hofler 1992).

Broadly speaking, HFC-134a is the most promising candidate in the short term for use in existing vapor-compression refrigeration systems using CFC-12 and in stationary and mobile air conditioning systems. This compound is nontoxic and nonflammable and possesses thermodynamic properties close to those of CFC-12. The initial concerns surrounding its use in domestic refrigerators and freezers focused on the lack of lubricants and on the potential energy-use penalty relative to CFC-12. Intensive development resulted in a resolution of concerns in both of these areas by about 1991 (Riffe and Dekleva 1991). In fact, there is a connection between the two concerns, since appropriate selection of lubricants can improve energy efficiency. With appropriate design, HFC-134a "can provide at least equal and possibly even better energy performance than [CFC-12]" (Riffe and Dekleva 1991, p. 465, referring to HFC-134a used in domestic refrigerators and freezers). Experience since the early 1990s has shown that the concerns over the potential increase in energy consumption associated with the use of HFC-134a were unwarranted once the necessary investments were made in changing equipment to fit the characteristics of HFC-134a.

In the early 1990s, HFC-152a was regarded as a substitute for CFC-12 that would have a significantly higher energy efficiency than HFC-134a. The disadvantage of HFC-152a that kept it from coming into widespread use was its flammability. The initial calculations indicating higher efficiency for HFC-152a refrigeration systems appear to be incorrect, at least for use in domestic refrigerator-freezers. Test results reported in 1993 indicate that if the components and materials (notably compressors and lubricants) of each refrigeration system are properly selected, and if the system is designed for the specific refrigerant, there is "no statistically significant difference between the efficiencies of HFC-134a and HFC-152a" (Vineyard and Swatkowski 1993, p. 86).

In commercial refrigeration and air conditioning applications, HFC-134a is also an option for many systems now using CFC-12 or R-500.[13] HFC-134a can be used in place of CFC-12 or R-500 in water chillers (large air conditioners that cool water and then air, instead of cooling air directly), both as a substitute in existing systems (with some minor retrofitting) and in new designs (Snyder 1992). HFC-134a is being used in new retail refrigeration.

For some applications, retrofitting of existing systems with HFC-134a appears to carry no energy penalty, even though, in general, operating conditions and equipment cannot be fully optimized when retrofitting occurs. For instance, a retrofit with HFC-134a was carried out in a large-scale district heating heat pump system that used R-500 as a refrigerant. There was essentially no change in the efficiency of the system; however, there was about a 2 percent decrease in its output (with a corresponding decline in energy input) (Dekleva et al. 1993).

Replacing CFC-11, HCFC-22, CFC-114 (in naval systems), and the blend R-502 (which consists of 51.2 percent CFC-115 and 48.8 percent HCFC-22) has been more complex. However, CFC-11, CFC-114, CFC-115 (including its use in the blend R-502) together constituted only about 5 percent of the total ODC refrigerant use in 1990. HCFC-22 is often cited by sections of the air conditioning and refrigeration industry as the most important refrigerant. We will therefore discuss it some detail. We will also discuss alternatives to HCFC-22 for use in existing systems. In general, the alternatives to R-502 are similar to those for HCFC-22.

CFC-11 is used in large commercial non-unitary air conditioning systems as well as similar industrial chilling systems. HCFC-123, which contributes considerably less to the buildup of ozone-depleting chlorine than CFC-11 or even HCFC-22, is one replacement for CFC-11 in these applications.[14] An azeotropic blend of HFC-32 and HFC-125 could replace CFC-11 in both applications. Allied-Signal has done extensive field testing and has submitted data to the EPA's Significant New Alternatives Policy Program for review.[15]

Other HFCs have been evaluated as replacements for CFC-11 and CFC-114 in large cooling systems. In particular, the properties of HFC-245ca and HFC-236ea correspond approximately to those of CFC-11 and CFC-114, respectively. From tests and calculations in non-optimized systems it appears that HFC-236ea is well suited to replace CFC-114 (Smith et al. 1993; Kazachki and Hendricks 1993). However, there are still questions of efficiency and performance regarding the use of HFC-245ca as a CFC-11 replacement. Evaluations need to be done in optimized systems to further evaluate efficiency and capacity. Theoretical considerations indicate that HFC-245ca may be the most efficient HFC replacement for CFC-11 (Kazachki and Gage 1993).

In late 1993, a group of halocarbons containing iodine were identified as possible substitutes for ODCs. Fluoroiodocarbons (FICs), which contain carbon, iodine, and fluorine, have been suggested as alternatives in a variety of ODC-using applications (often in combination with HFCs, ethers, and hydrocarbons).[16] In a preliminary test, a mixture of 87 percent CF_3I and 13 percent HFC-152a was placed into a small CFC-12 refrigerator without an oil change. As of late 1993, the unit had run over 1500 hours without apparent difficulty and appeared to be equal to or better than a CFC-12 unit in capacity and efficiency (Lankford and Nimitz 1993). According to the FIC developers, the refrigerant blends developed so far are "nonflammable, noncorrosive, have low toxicity, low environmental impact, and display high energy efficiency and high enthalpy of vaporization" (Lankford and Nimitz, p. 141).[17] While a considerable amount of testing must be performed on these compounds, their potential as drop-in substitutes has garnered attention. The developers hope to have commercial blends available by 1996 and expect costs to approach that of HFC-134a (GECR 1994; Kimball 1993).

Alternatives to HCFC-22

As we have noted, HCFC-22 is used in unitary air conditioning applications and in medium- and low-temperature commercial and industrial applications. The lead in phasing out HCFC-22 is being taken by Europe, since tougher regulations to phase out HCFC-22 before the turn of the century are anticipated there (von Eynatten et al. 1993, p. 119).

In 1993, two German companies announced results of tests of non-azeotropic mixtures containing varying proportions of HFC-23 and HFC-134a. Solkane XF is the name under which one such mixture is marketed. Both HFCs are commercially available and have passed necessary toxicity testing. Performance tests showed that the mixture had "excellent refrigeration performance," with higher efficiency than HCFC-22 and similar capacity (ibid., p. 123).

Non-azeotropic HFC mixtures are still in the initial stages of development, so little operating experience has been acquired. Further, the non-azeotropic mixture means that condensation of the refrigerant takes place over a range of temperatures, as does evaporation (in contrast to a single refrigerant or an azeotropic mixture). This phenomenon is known as "glide" range of temperatures. Use of such a refrigerant would require redesign of existing cooling systems, though the basic components and principles remain the same as those of other vapor-compression systems.

In sum, this refrigerant could be used in new systems before the turn of the century, if further tests prove the initial reported test results. It cannot be used as a drop-in substitute in existing HCFC-22 systems.

HFC-32 (CH_2F_2) and a variety of blends incorporating HFCs are also viable alternatives to ODCs in new refrigeration and air conditioning applications. Blends such as HFC-32/HFC-125, HFC-32/HFC-134a, and HFC-32/HFC-125/HFC-134a have been tested as possible replacements for HCFC-22 in new air conditioning applications (Godwin and Menzer 1993). The same set of compounds is also under consideration for use in place of R-502.

A blend of 30 percent HFC-32, 10 percent HFC-125, and 60 percent HFC 134a appears to be comparable in performance to HCFC-22 in capacity and efficiency tests. This performance was obtained even though the tests were conducted with compressors that were, in general, not optimized. If earlier experience with HFC-134a is any guide, the performance of this and other mixtures of HFCs containing HFC-32 could be superior to that of HCFC-22 (ibid., pp. 109–113).

A ternary blend developed in Canada as a replacement for HCFC-22 and R-502 consists of HFC-32, HFC-125, and HFC-23. This blend is nonflammable and nearly azeotropic and has been certified as nontoxic (Sami 1993).[18] Not a "drop-in" substitute, it would require a different lubricant and different seals. Preliminary data from the inventor indicate sizable efficiency gains.[19]

In existing screw compressor chiller systems using HCFC-22, ammonia can be used as a drop-in substitute for HCFC-22. Its performance is approximately the same (Godwin and Menzer 1993, pp. 114–115). In 1995, the Eastman Chemical Company planned to replace one-third of its existing chillers with new chillers using HCFC-22. The remainder are to be replaced in 1999. One of the reasons given for choosing systems using HCFC-22 instead of going to other alternatives was that such a system "can be converted to a variety of refrigerants" (EPA 1993b). One can only conclude that had more stringent regulations to phase out HCFC-22 been in place, Eastman would have used a non-ozone-depleting refrigerant in the first place.

Propane, a highly flammable substance, also happens to be a proven refrigerant that can be used within the vapor compression cycle. Propane is widely available, not patented, and relatively inexpensive. Isobutane, another flammable hydrocarbon, is also a refrigerant. A propane-iso-butane mixture could be used as a refrigerant in household refrigerators. Between 50 and 100 grams of propane coolant (equivalent to the amount in a table-top cigarette lighter) is all that would be required.

In 1992 the German branch of the Greenpeace organization initiated a campaign to encourage the ongoing development of a hydrocarbon-based refrigerator. Foron, a company in the former East Germany, began manufacturing the "Greenfreeze," and other manufacturers (Bosch, Lebherr, Siemens) followed suit. The refrigerant is 24 grams of a propane-butane mixture. According to Greenpeace, the refrigerator has been certified as safe by the Überwachungsverein, the German safety and standards institution. This refrigerator also uses non-ozone-depleting insulation: its foam is blown with cylcopentane, which is a hydrocarbon. It has therefore won approval as an environmentally sound product from the German Environmental Protection Agency.[20] According to A. Meyer, the chief of Foron's testing laboratory, the small amount of refrigerant means that a flammable mixture "cannot occur outside the refrigerator" (Meyer 1993).

There has been some controversy over the efficiency of hydrocarbon refrigerators, notably those with freezer sections. A paper presented at the 1993 Greenpeace conference claimed that propane-butane refrigerators with freezers were as efficient as those using CFC-12 or HFC-134a. However, the paper asserted that for such efficiencies to be achieved, the entire machine must be designed to use the hydrocarbon refrigerant, so that the operating system can be optimized as a whole. HFC-134a was also initially thought to suffer from energy penalties, but with attention to system design these fears proved unfounded. Conclusions about the efficiencies of new systems that do not consider optimization of the whole system are likely to be misleading and sometimes even wrong.

The need for optimization was demonstrated in tests conducted by one of the manufacturers, Foron. Dropping pure propane into a CFC-12 refrigerator resulted in a 40 percent energy penalty. Changing the refrigerant to a mixture of propane and butane reduced the energy penalty to 22 percent. Initial design changes to the evaporator led to energy parity with the CFC-12 refrigerant; further evaporator design changes led to 10 percent less energy consumption than the CFC-12 machine (Meyer 1993).

Propane can also be used a substitute for HCFC-22 in air conditioning systems. With replacement of the compressor, tests indicate that propane is a good replacement for HCFC-22 in the following respects (Treadwell 1991):

zero ozone-depletion potential
slightly higher energy efficiency
considerably lower refrigerant cost
potential for longer compressor life (due to lower discharge temperatures and pressures)
compatibility with components of existing systems, including lubricants
far lower global warming potential.

The amount of propane in a typical home central air conditioning system would be rather small—about 3.5 pounds, or about 1.6 kilograms. Since propane must be present in a concentration of at least 2.3 percent in air to ignite, even a total and sudden leak into the vent system, dispersed in a typical residence by the system fan, would maintain a concentration well below this limit. This is not to say that there is no danger, but rather that the danger from such use is likely to be far lower than that arising from the use of many flammable and polluting materials in the home, including natural gas (and often propane) space heating and water heating systems, storage of oil and other flammable chemicals in the home, etc. Asphyxiation is not a concern when such small quantities of refrigerant are involved, since there is not enough refrigerant to displace most of the air.

In a 1991 United Nations Environment Programme review, the barriers to greater utilization of propane as a refrigerant were stated as follows: "While propane is an established refrigerant with operating characteristics not too dissimilar from CFC-12, it is not seriously considered for domestic application because of its poor perceived efficiency and its flammability." (UNEP 1991d, p. 81)

The existence of dangers from present practices is not a reason to compound them. However, the dangers arising from the threat to the global environment, the damage already occurring due to ozone depletion, and the potential damage from the buildup of greenhouse gases need to be assessed against the dangers posed by substituting new compounds. In other words, refrigerants such as propane and ammonia should be carefully considered in light of all the relevant concerns. As a practical matter, concern over flammability has limited the development of refrigerant systems using propane or other hydrocarbons.

Propane is used in industries in which flammability controls are routine, such as the oil, gas, and petrochemical industries. The hydrocarbon n-butane, which is similar to propane, can also be used as a refrigerant, but the system is still under development (Rice et al. 1993).

Approximately 3–4 percent of industrial refrigeration worldwide uses hydrocarbons as coolants. Ammonia represents approximately 34 percent ,with the remainder belonging to ODCs, primarily HCFC-22 (UNEP 1991d, p. 114).

In systems where large quantities of refrigerant are required, flammability is a more serious practical constraint. In such cases, the best system may not be based on the use of vapor-compression cycles which use halogenated refrigerants. Rather, the solution lies in cogeneration technology, which generates electricity on site and uses waste heat for both heating and air conditioning.

New Large Commercial Air Conditioning Systems

The phasing out of ODCs presents an opportunity to address a number of environmental, energy, and economic issues associated with energy use in large commercial structures and large residential apartment buildings. The approach we discuss here could also be extended to lower-density housing, but the economic and energy-efficiency aspects would have to be examined on a case-by-case basis.

Building design has generally not integrated economic, energy, and environmental considerations while minimizing environmental impacts within given economic constraints. While there are many reasons for this, two of the principal ones are that building designers and builders often do not pay the energy bills and that vendors of electricity and natural gas have traditionally seen themselves as being in competition with each other. Electric utilities have favored vapor-compression-based, electricity-driven air conditioning systems. In certain climates they have also promoted heat pumps.

Yet systems that use natural gas for generating electricity on site (at the point of use) and use the waste heat for heating and air conditioning can be far more efficient and far less polluting. The most common systems use electricity for cooling and natural gas or oil for heating. Efficiency can be further optimized by connecting such on-site generation systems to the electricity grids of utilities, so that the local system can buy power from and sell power to the centralized system. Such systems are in commercial use today, but only in large buildings. For air conditioning they use, rather than a vapor-compression cycle in which a refrigerant is compressed, expanded, and condensed, an absorption cycle in which a chemical like ammonia or lithium bromide absorbs energy from a fuel or from waste heat and is driven out of a fluid (usually water).

On-site electricity generation with waste heat used for heating and cooling has the following advantages:

zero ozone depletion and essentially zero global warming potential so far as the refrigerant fluids are concerned

higher energy efficiency compared to systems that use a primary fuel for heating and vapor compression for cooling, reducing the contribution to global warming

growth of electrical generating capacity in step with demand, so that the need for large power plants, with the associated environmental and financial risks, is reduced

reduction of peak electricity load due to new construction of vapor-compression systems

possible compatibility with a transition to solar hydrogen and solar absorption air conditioning.

Since generation would be dispersed in cities, the consequences for local air pollution must be carefully examined. This may not generate any more pollution than most existing systems, since local air pollution is also generated by heating systems that use oil or natural gas.

One very important facet of cogeneration is the contribution it can make to proper functioning of the energy markets, which work poorly at present, especially in the area of electricity production and consumption. This is important to reducing the concentration of ODCs and greenhouse gases in the atmosphere.

A principal reason for the poor functioning of the electricity market is that the time horizon for decisions about new building construction (and other consuming equipment) is much shorter than that for central-station power plants. Moreover, in times when economic growth is highly uneven and uncertain, the fluctuations of market signals to electric utilities are so great as to make the building of large power plants financially risky. This has resulted in large losses for many utilities relying on large plants, such as nuclear power stations, which require long lead times.

An approach to electric power planning that creates the supply of electricity in direct response to demand via a cogeneration system would make for an economically functional market in electricity. As we have discussed, on-site generation, with the system connected to the electricity grid, accomplishes a broad range of economic and environmental goals.

Another important facet of on-site generation is that many existing buildings can be converted from conventional systems. This would enable a rapid increase in energy efficiency, since the amount of energy used in existing buildings in any year is far larger than that used in new buildings and equipment. Another advantage of converting existing buildings is that this would reduce peak electricity demand, averting the need for new power plants.

Absorption air conditioning using waste heat from electricity generation is not yet economical in small units of the type that would be needed for existing single-family residences or very small commercial developments. However, in new developments it may be possible to go to this approach if the long-term economic and environmental advantages are taken into account. The smallest size at which such an approach may be economical will vary from one region to another, depending on fuel and electricity prices and on regulatory requirements regarding efficiency.

Where the use of absorption air conditioning based on waste heat from electricity generation is not possible, other approaches must be taken. One answer may be absorption air conditioning using natural gas (instead of waste heat) as the fuel. For example, there are approximately 200,000 absorption air conditioners in the United States, most of which use an ammonia/water absorption system (UNEP 1991d, p. 134). Ammonia/water absorption refrigeration is also used in mobile homes in the United States and in domestic refrigeration in many European countries (ibid., p. 83). Because of higher electric rates, the use of absorption air conditioning is growing rapidly in Japan. However, with some system designs, absorption cooling may carry energy penalties relative to efficient vapor-compression systems when these systems use a primary fuel (e.g. natural gas) instead of waste heat. In such cases, vapor-compression systems using HFC-134a or ammonia may be one answer that is available today. Ammonia constituted approximately 81 percent of the refrigerant used in cold storage and food processing in the United States in 1990, and in Germany ammonia accounted for approximately 70 percent of this use in 1987 (ibid.). Worldwide, ammonia use in industrial refrigeration accounted for about 55 percent of refrigerant use in 1990. It is expected that large water-chiller air conditioning systems can utilize ammonia as a working fluid to a greater extent. The use of flammable or toxic refrigerants may also be inhibited by local building codes and other regulations.

A water/zeolite absorption system may be adaptable to a wide variety of applications, ranging from domestic refrigeration to mobile air conditioning to large-scale stationary air conditioning running on waste heat. Zeolites are natural crystalline aluminosilicate minerals with millions of tiny pores that absorb water when they are cool and expel it when heated. Synthetic zeolites are commercially available.

Zeolites have a strong affinity for water. When they absorb water from another medium, such as air or a body of water, they cool it down. When the zeolite becomes saturated with water, the cooling action stops. The zeolite is then regenerated by heating, which expels the moisture from the pores, making the zeolite ready for another cooling cycle. Continuous cooling can be obtained by using more than one zeolite cartridge, so that one is being regenerated while the other is carrying out the cooling (Schwartz et al. 1993).

Zeolite/water systems have advantages over ammonia/water or lithium bromide/water systems in that zeolite systems can be used in freezing applications, whereas the other two can only be used for cooling to temperatures above the freezing point of water. Further, no volatile toxic materials need be involved in zeolytic systems. Finally, zeolites can be regenerated at temperatures that are equal to or below exhaust gas temperatures from diesel, gasoline, and natural gas engines. Therefore, they are well suited to a wide variety of applications: ice-making machines, car air conditioners, domestic refrigeration, heating and hot water systems, large-scale industrial and commercial chilling and air conditioning applications. In addition, zeolite/water systems can be used in conjunction with solar water-heating systems (Energy Concepts Co. 1993).

Zeolite/water systems are not yet widely available. For many applications, such as mobile air conditioners using waste heat, a considerable amount of research and development is still required (Schwartz et al. 1993). Other applications, such as ice making and industrial chilling, are closer to commercial use (Energy Concepts Co. 1993).

Evaporative air conditioning systems provide another route to cooling in which water is the working fluid. These systems are based on the principle that the evaporation of water from a surface causes the surface to cool. Air, after passing through the cooled porous surface, is used for air conditioning. Such a system is called direct evaporative cooling. These systems cause the moisture content of the air to increase and therefore can only be used in very dry areas.

Indirect evaporative cooling can extend the range of evaporative cooling systems. Here the primary air stream, which is wet and cool, is used to cool a secondary air stream with which it does not come into contact. Such a system is less efficient than a direct evaporative system because of the extra heat-exchange step. A combination direct/indirect system can work in a wider range of climates and provide relatively high efficiency. One manufacturer claims that its system reduces energy use by up to 80 percent (untitled pamphlet, Norsaire Systems, Denver).

Evaporative cooling systems can be used in humid areas if the incoming stream of air is dried by means of a desiccant. Desiccants are commonly used in industrial drying and in car air conditioners. The evaporative cooler is used on dry days, and the desiccant system is added to the cooling loop during humid conditions. This added step reduces efficiency relative to the evaporative system, but the overall seasonal efficiency would still be good if the number of dry days were sufficiently high (Heimann 1993).

A highly innovative evaporative solar cooling system, designed and built by the Davis Energy Group, was installed on the roof of the Office of State Printing in Sacramento in 1991.[21] This system uses the concept of a "cool storage roof." Light insulation panels installed on the building's roof float on water, which is the working medium. At night, the water is sprayed over the panels. The water cools as energy is radiated into the night sky and water is evaporated from the surface. This cool water then drains back below the insulating panels, where it is stored for use in the daytime. According to the Davis Energy Group, the system has better energy efficiency and service life than conventional vapor-compression machines, and the stored water provides additional fire protection.

New Mobile Air Conditioning Systems: Near-Term Approach

In 1988 the use of CFC-12 in automobile air conditioning was considered to be a problem of considerable economic and environmental importance, both because it is the largest source of ODC emissions within the refrigerant sector and because there was no non-ozone-depleting substitute commercially available at that time.

The leading non-ozone-depleting chemical alternative, HFC-134a, was still undergoing toxicity testing. Moreover, there were no commercial plants manufacturing HFC-134a, and it had not been extensively tested on the road in automobiles. Another significant problem which remained to be addressed at the time was the development of lubricants compatible with HFC-134a. All these problems have since been addressed. HFC-134a was tested and is now used by the automotive industry as the air conditioning refrigerant in new vehicles. It has also been tested successfully in older systems.

It has been estimated that about 80 percent of new vehicles used HFC-134a by the end of 1993. The conversion to HFC-134a for new vehicles will be complete by the end of 1994, a year ahead of the phaseout date for CFCs (Atkinson 1993, p. 269). These observations apply to the industrialized countries.

HFC-134a as used in car air conditioning systems has one principal long-term disadvantage: it is a greenhouse gas. Existing automobile air-conditioner designs have flexible hoses which permit significant refrigerant leakage. There are a number of ways to overcome this problem.

One relatively straightforward principle involves reducing the heat load on the air conditioner. As we have discussed, car air conditioning systems have traditionally been sized so as to provide the largest possible blast of cool air to lower the temperature of hot parts of a car's interior after it has been sitting in the sun for a considerable period. If cars were to be designed so as to prevent the interior from getting hot when standing in the sun, system size could be reduced considerably. Tinted windows and a ventilation system can reduce the heat load on a car considerably. Solar-powered ventilation systems have been designed and tested and are now in commercial production. These systems ventilate a car, preventing the interior from reaching very high temperatures. This reduces the demand for cooling when the car is started.

According to simulations done by J. P. Chiou (1986), the interior of a car parked in a sunny, dry place with an exterior temperature of 110°F can reach 225°F—a temperature rise of 115°F. Under such circumstances a typical air conditioning unit would take over 10 hours to cool the solid parts of a car back down to 110°F. With a solar-powered active ventilation system, the interior temperature reaches 120°F—a temperature rise of only 10°F (ibid.). The solar-powered system has an additional advantage in that the power available to run the fan increases with the increased solar radiation and thus can match the need for increased ventilation as the temperature rises.

A solar-powered ventilation system costing $25 can be retrofitted to most existing cars; it vents hot air through a small opening in the rear window. Systems with solar cells on the roof and larger fans providing better cooling capacity are available as options on some cars (Wald 1992). These systems should be adopted generally because they enable considerable fuel savings due to smaller air conditioning systems, require smaller charges of refrigerant, eliminate the need for air conditioning in temperate areas, and reduce the damage that high temperatures do to vinyl interiors.

New Mobile Air Conditioning Systems: Long-Term Solutions

Many of the environmental problems associated with HFC-134a-based systems stem from the use of flexible hoses. The elimination of flexible hoses and their replacement by a hermetically sealed system is one straightforward solution with some additional benefits. For instance, hermetically sealed systems could be driven electrically rather than mechanically, possibly allowing for greater motor efficiency.

Hermetically sealed systems would also permit near-complete recovery of refrigerants when systems are serviced or discarded, greatly reducing the emission of greenhouse gases for a given level of mobile air conditioner use. Mobile air conditioners could also be built to use the waste heat discharged from engines, thus increasing efficiency by as much as 70 percent. This is the same efficiency-increasing principle we discussed for large commercial air conditioning systems. These technologies could further be developed for larger vehicles such as buses (Mei et al. 1991, p. 591). The discussion of the zeolite/water absorption system above illustrates one approach that could eventually lead to a mobile air conditioning system that would make use of waste heat.

Hermetically sealed systems and systems that use waste heat are heavier than the traditional engine-mounted systems. The energy penalty of increased weight must be considered when developing such systems. Another concern with systems using waste heat is whether there is sufficient waste heat to supply the cooling needs. This may constrain system size considerably, since absorption air conditioning systems require large heat exchangers. Reducing cooling requirements through the use of solar-powered fans could make the use of waste heat, as well as hermetically sealed systems for mobile air conditioning, more feasible.

5.3 Recovery and Recycling of ODCs from Existing Systems

Recovery, reconditioning, and reuse of ODCs is now a well-established technology for refrigeration and air conditioning systems of all sizes. At one time the technology was used only for large systems. By 1989, the concerns regarding the use of CFC-12 recovered from existing equipment had been resolved, and such recovery is becoming standard for mobile air conditioning systems.

There remains one important difficulty with the recovery and recycling of ODCs from air conditioning systems: on average, 60 percent of the refrigerant in a mobile system leaks out before the vehicle is brought in for service. Even with complete recovery of what is left, the continued use of ODCs allows for substantial increases in atmospheric chlorine and bromine.

Further, recycling rather than destroying CFC-12 after recovery means that most of the recovered product will be lost within a few years because of the unavoidable losses from mobile air conditioning systems. The following approaches may overcome this problem:

Continue to collect CFC-12 during servicing until the year 2000. (Most car owners do not service the air conditioner after a car is more than about 5 years old; see Corr et al. 1991, p. 554.)
Retrofit existing vehicle air conditioners to use HFC-134a. This will require changing some components.
Allow decline in refrigerant charge in existing vehicles without replacement, forgoing the use of the systems in a few years.

The last option will probably not be acceptable to most people who have invested money in vehicles with CFC-12 air conditioning systems. There are various cost estimates for retrofitting existing CFC-12 systems to use HFC-134a (including changing hoses, lubricants, etc., and the necessary labor time). Industry discussants often include the costs of repairing existing systems in the costs of converting them to HFC-134a, on the ground that car air conditioners are often in some disrepair when they are brought in for service. However, such repair costs would accrue independent of conversion, so they are not part of the actual conversion cost.

Estimates of the conversion cost remained roughly constant in the 1991–1993 period—just over $200 (Baker 1991; Atkinson 1993). However, in March 1994, General Motors announced that Buick and Oldsmobile dealers would begin converting systems to HFC-134a for about $100 (Ozone Depletion Network Online Today, March 25, 1994).

A policy prohibiting the use of ODCs in existing automobile air conditioners should be implemented at the earliest possible date after 1995. Such a policy would require that recovered CFC-12 be sold to repositories of ODCs for destruction or essential use, so that as small a portion as possible of this bank will be released to the atmosphere.

The service requirements for non-automobile, non-hermetically-sealed systems and for hermetically sealed household refrigerators are far smaller than those for mobile air conditioning systems and can be met with ODCs recovered from stationary systems.

Existing HCFC-22 systems would pose the greatest difficulties were this compound to be phased out well before the end of the century. Improvement of service practices can minimize the amount of HCFC-22 needed to service existing systems. Using the statistics on HCFC-22 discussed above, we estimate that the service requirements of equipment in use by 1995 will be under 50,000 tons per year worldwide. It may, therefore, be necessary to allow very limited production of HCFC-22 for existing equipment. This amount can be minimized if it is announced in advance that it will be costly to use new HCFC-22 for existing equipment. A large tax on HCFC-22 in industrialized countries would provide the incentive to recover and store it for future use and would motivate the conversion of large-scale air conditioning systems now using HCFC-22 to alternative refrigerants. Substitutes for existing systems include ammonia for screw-compressor chillers, propane for small systems where flammability is not a major concern because only a small amount of refrigerant is used, absorption air conditioning with on-site generation of electricity, and conversion to one of the above-mentioned mixtures of HFCs. These are not "drop-in" alternatives; they would require changes in materials and/or components.

Recovery of ODCs from many refrigeration and air conditioning devices can be performed when the devices are retrofitted or retired. For domestic appliances, recovery can be performed at the end user's location or at waste facilities. For example, in the state of Wisconsin an electric utility company pays consumers approximately $50 each for retired refrigerators, freezers, and air conditioners in an attempt to upgrade the average efficiency of its customers' electricity-consuming appliances of their consumers (Malaspina et al. 1992, p. 79). This program has retired approximately 235,000 appliances since 1987. After ozone-depleting compounds are removed, the appliances are recycled. In some parts of Canada, ODC recovery and recycling from domestic, industrial and commercial refrigeration became mandatory in early 1992 (FOE 1991b). In the United Kingdom, it is estimated that approximately 13.4 percent of the potentially recoverable coolant from refrigerators is being recovered with the existing voluntary programs (ibid., p. 2). Far greater amounts could be recovered were legislation to be enacted.

It is imperative that recovery equipment function well and that it recover ODCs according to the specifications. Data indicate that some recovery equipment does not perform according to specifications. David Kolar, an engineer with the CFC Steering Committee of the New Jersey Environmental Federation, claims on the basis of limited tests and anecdotal reports that much of the existing stock of refrigerant-recovery equipment "is short-lived ... performs poorly at high temperatures or ... degrades when used at high temperatures" (Kolar 1993). However, tests also indicate that equipment that performs well is also available. These claims should be carefully evaluated by the EPA. If they are verified, stricter standards for equipment performance under expected conditions of service should be enacted.

5.4 Recommended Phaseout of CFCs and HCFCs

A review of the list of available chemicals and technologies makes it clear that CFCs and (with few relatively minor exceptions) HCFCs could have been completely phased out by the end of 1995. Today such a phaseout schedule would be impractical because of delays in enacting strict HCFC regulations. A review of the developments that led to the complete elimination of CFC-12 in mobile air conditioning leaves little doubt that the prospects of stringent and realistic regulation and large economic losses spurred innovation. The widespread industry sentiment in 1988, just after the Montreal Protocol was signed, was that a 50 percent cut in CFC production by 1998 might pose intolerable economic burdens. This gave way to the reality of a near-total phaseout of CFC-12 use in new automotive air conditioners by the end of 1994 in the industrialized countries.

Chemicals, technologies, and processes that could replace the overwhelming bulk of HCFCs exist today, either in fully developed form or in forms that are close to commercialization. That the latter applications are not yet commercial, in our judgment, is due largely to a lack of stringent, realistic phaseout dates. Such dates should have been set in 1992, when government representatives convened in Copenhagen to amend the Montreal Protocol.

Beyond the end of 1995 we recommend the Swedish approach, which provides for time-limited permits for the use of HCFCs in restricted applications where commercial alternatives are not available. The permits should be restricted so that the users are obliged to replace HCFC refrigerants with non-ODC alternatives at the earliest practical date.

6
Rigid Foam

Plastic foams were first manufactured in the 1930s using urea formaldehyde. Since that time they have replaced a variety of traditional products, including fiberglass, cork, and perlite. They are commonly used as insulating material for walls, roofs, pipes, and appliances such as refrigerators and freezers. Because of their protective, insulating and cushioning properties, they are also used in packaging. The two general categories of foam are **rigid** and **flexible.** Flexible foams are typically used as cushioning in car seats and in bedding.

Foam plastics are produced by injecting "blowing agents" into liquid plastic or plastic-forming material. The most common blowing agents are ozone-depleting compounds, but carbon dioxide and hydrocarbons are also used. ODCs are primarily used to create the tiny gas pockets in foams. In some instances, they serve both to absorb heat generated in the foam-production process and to lower the thermal conductivity of the resulting foam product.

Both rigid and flexible foams have porous structures. Rigid foams are characterized by closed cells in which gases can be trapped for varying periods of time. This cell structure is what gives rigid foams their characteristic structural integrity and insulating ability. Flexible foams' open cells, similar in structure to sponges, allow for greater cushioning and flexibility.

One important difference between most rigid foams and flexible foams is that in the latter almost all ODC emissions take place during the manufacturing process. The open cells retain none of the blowing agent used to produce the foam. Rigid foams, on the other hand, tend to trap the blowing agent in their closed cells. However, many rigid foams (such as polystyrene sheet, which is used for packaging) lose the blowing agent in their cells by diffusion to the atmosphere within several days or weeks after being manufactured. In contrast, insulating foams, such as those used in refrigerators, lose their blowing agent very slowly.

6.1 Current Use

Figure 6.1 shows the breakdown of ODCs used in foam production by foam application in 1990. The production of insulation for walls and roofs constitutes the majority of consumption. Figure 6.2 shows this same breakdown by compound type. The principal ODC used is CFC-11, of which approximately 140,200 tonnes were used in rigid foam production in 1990.[1] CFC-12, CFC-113, CFC-114, and HCFC-22 are also used bringing the total global amount of ODCs used in rigid foam production to approximately 174,450 tonnes.[2] In 1985, approximately 202,000 tonnes were consumed in this application.[3] In the late 1980s

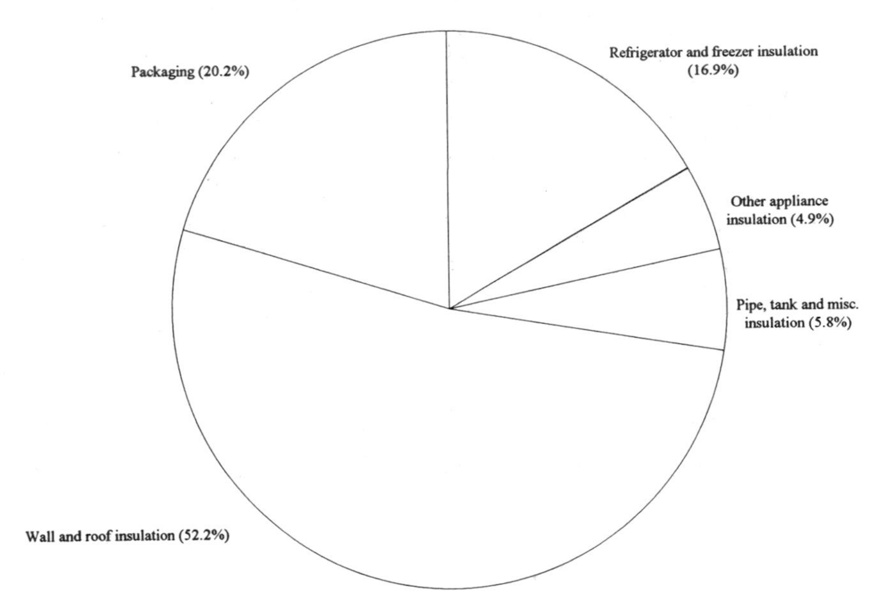

Figure 6.1
Global ODC use by rigid foam application, 1990 (derived from data in UNEP 1991b and AFEAS 1993c). (Total amount: 174,450 tonnes.)

and early 1990s, HCFC-141b and HCFC-142b went into commercial use as foam-blowing agents.

The consumption decline since 1985 has been mostly due to reducing the use of ODCs in the production of packaging materials and some insulating foams. In many instances, CFC consumption in packaging production has been replaced by HCFC-22 and some non-ozone-depleting processes.

The majority of the blowing agents used to produce rigid foams are trapped within the foam structure of the product. The emissions that

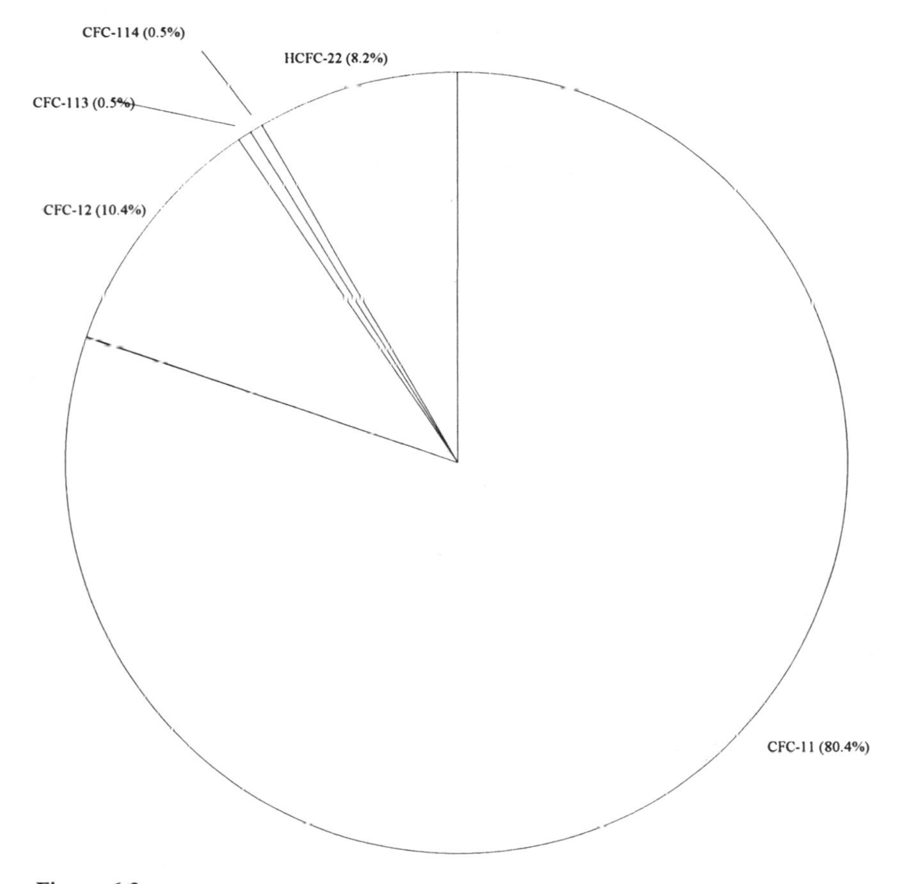

Figure 6.2
Global rigid foam use by compound, 1990 (derived from data in UNEP 1991b and AFEAS 1993c). (Total amount: 174,450 tonnes.)

occur during the manufacture of rigid foams range from approximately 5 to 20 percent of the total amount of ODCs used. The total bank of ODCs in rigid foam in the world at present is roughly 1.3 million tonnes (AFEAS 1993a; AFEAS 1993c). This is approximately equal to the entire worldwide production of all the regulated ODCs in 1985.

Insulating Foams

Because of the low thermal conductivity of many ODCs, these compounds are commonly used to produce insulating foam. The ODC becomes trapped in the closed cells of the foam after injection and formation of the cell structure. This lightweight, flexible insulation has found many applications. It is particularly useful where space is at a premium, as in refrigerator walls, in wood-frame houses, and in commercial buildings.

Closed-cell insulating foams release the trapped blowing agent very slowly (0.5–1 percent per year is a common estimate), maintaining their insulating ability over the lifetime of the product or structure they are associated with (Gramlen et al. 1986). Thus, most of the ozone-depleting compounds that have been used to make ODC-blown insulating foam up until now are still trapped in the foam.

Insulating foams can be produced in sheets or boards for future cutting and installation or sprayed directly onto an object requiring insulation. The latter insulating method has been particularly useful for small spaces and for large or complex objects. Coatings or surface material are often attached to the outer surfaces of insulating foam products to slow the escape of the blowing agents and protect against moisture or impact damage.

The production of insulating foams is the largest use of ODCs within the rigid-foam category, accounting for approximately 80 percent of the ODC used. CFC-11 is the most common compound used. The varieties of insulating foam and the corresponding amounts of ODCs used are listed in table 6.1.

Of the 38,000 tonnes of CFC-11 used in the production of insulating foam for appliances, 75–80 percent was used in the manufacture of refrigerator and freezer insulation. The remainder was used in the production of insulating foams for miscellaneous products, such as portable coolers and vending machines (UNEP 1991b, p. III-1).

Table 6.1
Global use of insulating foam, 1990 (derived from data in UNEP 1991b).

Type of foam	Consumption (tonnes)	ODC used
Polyurethane		
Appliance	38,000	CFC-11
Boardstock/flexible-faced lamination	38,300	CFC-11 (HCFC-22, 141b)[a]
Sandwich panels	24,100	CFC-11 (HCFC-22, 141b)
Spray	12,800	CFC-11
Other	11,300	CFC-11, 12 (HCFC-22, 141b)
Polystyrene board (extruded)	12,000	CFC-12 (HCFC-22, 142b)
Phenolic	2,700	CFC-11, 113, 114 (HCFC-22, 142b)
Total	139,200	

a. ODCs in parentheses were first used as foam blowing agents in the early 1990s. Consumption of these ODCs is not included in the consumption estimates.

Of the approximately 521,000 tonnes of polyurethane rigid foam used in North American, European, and Japanese building insulation in 1989, approximately 78 percent was used in commercial buildings; the remainder was used in the residential sector (Fischer et al. 1991, p. 9.2, table 9.1).

Packaging Foams

ODC-blown rigid foams are also used in a variety of packaging applications. Typical uses include disposable containers and tableware for food packaging and food service, protective packaging for furniture and electronic goods, and semi-rigid fill for automotive bumpers, flotation devices, and shoe soles. The two most important attributes of rigid packaging foams are their ability to insulate products from stresses such as impact and vibration and their ability to resist moisture and thermal influences. Much of the blowing agent within a packaging foam escapes within hours or weeks, the time depending on the foam's thickness and the outer coating.

Table 6.2
Global packaging foam use in 1990 (derived from data in UNEP 1991b and AFEAS 1993c).

Type of foam	Consumption (tonnes)	ODC used[a]
Polystyrene sheet	14,300	HCFC-22 (HCFC-141b)
Polypropylene and polyethylene (extruded and molded)	12,350	CFC-11, 12, 114 (HCFC-22, 142b)
Polyurethane integral skin and miscellaneous (open-celled)	8,600	CFC-11 (HCFC-22)
Total	35,250	

a. ODCs in parentheses were first used as foam-blowing agents in the early 1990s. These ODCs are not included in the consumption estimates.

Manufacturers of polystyrene packaging foam relied predominantly on CFC-12 until the end of the 1980s. For example, approximately 20,000 tonnes of CFC-11 and CFC-12 (more of the latter than of the former) were used in this foam application in 1986 (UNEP 1991b, p. V-3). Polystyrene foam packaging was most commonly used in the fast-food industry for cups and food containers. The considerable decline in CFC use for this application since 1988 is due to both the use of other packaging materials and the use of HCFC-22 in place of CFC-12. Though these levels of use are far below those in the category of non-packaging rigid foam, annual ODC emissions from packaging production are greater. This is because most of the ODCs used in packaging applications are released within weeks, often before the material reaches the final consumer.

6.2 Alternatives to Foam for Insulation

As can be seen from figure 6.1, the preponderance of rigid foam insulation is used in residential and commercial structures. Although plastic insulating foams have the best insulating capacity per unit thickness, they are more costly than other materials; thus, they command only about one-third to one-half of the market for wall and roofing insulation (McElroy and Scofield 1990, pp. 25–26).

In many building applications, fiberglass batt can be substituted for insulating foam to provide equal insulating capacity for about the same cost. Generally, fiberglass batt of twice the thickness achieves approximately the same insulating capacity as ODC-blown building insulation (ibid., p. 5, table 1.3). Fiberglass accounted for approximately 19 percent of the 1986 industrial and commercial roof insulation market in the United States and about 60 percent of the total 1984 insulation market in Germany (EPA-RIA 1987, vol. III, part 1B, p. 3.3-15). The greater insulating-space requirements of fiberglass can be included in new building plans. According to a study performed by the U.S. Environmental Protection Agency, the switch to different insulating materials will, in many cases, produce a net reduction in cost for the same level of insulation (ibid., vol. II, part 2, appendix J).

Other traditional materials applicable to building insulation are cellular glass board, perlite board, and gypsum board. Perlite board accounted for approximately 13 percent of the 1986 industrial and commercial roof insulation in the United States (ibid., vol. III, part 1B, p. 3.3-15). Using greater thicknesses of these materials can compensate for their lower insulating ability per unit thickness. Among the alternative insulating materials, cellulose made from pulverized recycled newspapers is particularly effective and environmentally sound. While nearly equivalent to fiberglass in cost per unit thickness, cellulose has approximately 20 percent more insulating ability (Rowse 1992). After grinding, the pulverized recycled newspaper is coated with boric acid or borax as a fire retardant. Cellulose insulation is not affected by high humidity or even some condensation. However, serious moisture problems such as roof or wall leaks will usually necessitate replacement. Cellulose may serve as a substitute for foam insulation in the retrofitting of existing houses.[4]

Another important application of insulating foam is in the walls of refrigerators and freezers. Eliminating ODC-blown foam will require either increasing the thickness of refrigerator walls and using a different insulating material or making use of new insulation techniques. An important example of the latter, and an insulation technique potentially more efficient than the use of foam, is **vacuum insulation**. Probably the most familiar use of this technology is in Thermos bottles.

Contemporary vacuum-insulation panels have up to four times as much insulating ability per unit thickness as ODC-blown foams. According to preliminary estimates, a refrigerator using double vacuum walls for insulation would be considerably more energy efficient than one using foam insulation (Potter and Benson 1991).

A variety of vacuum-insulation panel designs are being developed. Silica-powder-filled and silica-aerogel-filled panels containing low-thermal-conductivity materials are being pursued and may cost as little as ODC-blown foams per unit of insulating ability (ibid., p. 9, figure 4). Testing of powder-filled vacuum panels in refrigeration units has begun.[5]

Combining vacuum insulation and non-ODC blown foams to provide structural rigidity is another option under investigation (Benson and Potter 1992). Preliminary research indicates that non-ODC blown open-cell foam in combination with vacuum-insulation panels may be one-third less costly than silica-powder-filled insulation panels per unit insulating ability (de Vos and Rosbotham 1993). In addition, this technology has been tested with water-blown foam in refrigerators and has achieved an insuiating ability close to that of conventional ODC-blown foam insulation.[6] The need for an impermeable casing material that will last the life of the appliance is one barrier to commercial development.[7] These options need to be further developed and tested.

In 1992 the Owens-Corning Corporation reported having developed a cost-competitive vacuum/fiberglass panel that could be in production in refrigeration units within approximately two years if a partner company in the refrigeration industry could be located.[8] According to an Owens-Corning press release, this new panel has six times the insulating ability of ODC-blown foam insulation panels, has a projected lifetime of over 20 years, and has been successfully tested in a refrigerator (Owens-Corning 1992).

Insulating panels filled with gases known to have low thermal conductivity, such as argon and krypton, have been developed and show promise in appliance and building insulation applications. "Gas-filled panels" (GFPs) are composed of three components: a plastic outer envelope, a compartmentalized interior made of aluminum-coated plastic, and the insulating gas. Because of the plastic-film construction, GFPs can be molded into a variety of shapes and sizes to suit many insulating configurations. On the basis of preliminary calculations, both air-filled and argon-filled GFPs appear to be cost competitive with ODC-blown foam insulation (Griffith et al. 1991). Krypton-filled GFPs are somewhat more costly, but R values per inch of approximately 15 (versus 7.3 for CFC-blown foam) have been achieved. Those GFPs may be particularly useful for applications where insulating space is at a premium.[9]

Intensive work has been underway to find a chemical replacement for the use of ODCs in the production of insulating foam. Much of this effort has involved HCFCs, especially in the United States, since preliminary investigation revealed problems with HFCs. HFC-152a is somewhat flammable, and HFC-134a seemed to yield unsatisfactory products. The chemicals most commonly evaluated were HCFC-141b and HCFC-22, which have been considered suitable as replacement blowing agents. HCFC-22 and (to a lesser degree) HCFC-141b and HCFC-142b are already being used as alternatives to CFC-11 and CFC-12. U.S. industry seems intent on using these partially halogenated compounds in rigid insulating foams for refrigerators. Frank Dwyer and Kenneth Thrun of Allied Signal have concluded that "HCFC-141b and HCFC-22 are viable alternates for CFC-11 and CFC-12, respectively, in froth-type rigid foam formulations" and that "the commercial availability of Allied Signal's HCFC-141b will allow for a smooth transition of the froth foam industry into the 21st century" (Dwyer and Thrun 1991, p. 737). However, it appears that the manufacturers' rejection of HFC-134a as a principal candidate for insulating foam was premature. The phaseout schedules for ozone-depleting compounds have been far more lax tin the United States than in Sweden. Further, in some countries the prospect of a phaseout of HCFCs is far more immediate than it is in the United States. Such pressures induced a Danish firm to make an intensive and careful effort to develop the manufacturing technology needed to use HFC-134a as a blowing agent. This effort appears to have yielded a satisfactory product, though the conversion costs may not be insignificant. The technical manufacturing problems associated with the use of HFC-134a in insulating foams have been resolved, according to the manufacturer. Further, the energy consumption of a test refrigerator using foam blown with HFC-134a was about 3 percent lower than that of a unit containing CFC-11-blown foam (Klausen and Larsen 1991).[10]

HFC-365 ($CF_3CH_2CH_2CF_2H$), alone and in mixtures with HFC-134a, appears to be a promising non-ozone-depleting blowing agent. In initial laboratory tests, polyurethane foams blown with HFC-365 exhibited lower insulating ability than CFC-11 but somewhat higher insulating ability ($\sim$3 percent) than HCFC-141b when used in polyisocyanurate foams (Murphy et al. 1993).

The fluorinated ethers E245 ($CF_3CH_2OCHF_2$) and E263 ($CF_3CH_2OCH_3$) are also being considered as near-drop-in replacements for CFC and HCFC foam blowing. E245 is nonflammable, appears to have limited toxicity, and exhibits lower thermal conductivity than CFC-11 foams (Blevins et al., n.d.). However, further testing must be carried out to conclusively determine the toxicity and stability of these compounds. In a similar state of development, fluoroiodocarbons (FICs) have been proposed as drop-in replacements in foam production (Nimitz and Lankford 1994).

Flexible-faced rigid foam panels for building insulation have been produced using isopropylchloride (IPC) as a blowing agent (Creyf 1991).[11] This blowing agent has the potential to be used in producing other types of rigid foam,such as block foams, steel sandwich panels, and appliance insulation. IPC-based rigid foams have exhibited, both initially and after aging, slightly better insulating ability than CFC-11-blown foams.

A number of hydrocarbons, including butane and pentane, have also been successfully utilized as foam-blowing agents. The lower insulating ability of hydrocarbon-blown foams (approximately 5–10 percent lower, in the case of pentane) can be overcome by increasing the thickness (UNEP 1991b; Wenning 1993). The use of hydrocarbons in this application is being pursued, particularly in Europe. There are more than a million refrigerator-freezers insulated with pentane-blown foam in Europe (Wenning 1993, p. 318). Flammability and the potential to contribute to photochemical smog are limitations of these compounds.

Carbon dioxide (CO_2) is another alternative blowing agent for the production of insulating foams. Foams made with carbon dioxide generated from the use of water during the foam-forming process are sometimes referred to as "water-blown foams." CO_2-blown foam can be used in the production of foam boards and in spray applications (UNEP 1991b). Because CO_2 tends to diffuse more rapidly from the foam product, reducing the insulating capacity over time, the use of cover materials on the exposed sides of CO_2-blown foam will likely be required. Foil, metal, and plastic can be used as coatings. This measure, combined with increased thickness, can partially compensate for the lower insulating ability per unit thickness relative to ODC-blown foams. The German Greenfreeze refrigerator, insulated with CO_2-blown expanded polystyrene foam, is initially as efficient as equivalent units blown with ODCs (Greenpeace 1992b).[12] Work is underway to lower the diffusion of CO_2 from the foam interior to maintain the initial insulating ability over the life of the product. This refrigerator also uses non-ozone-depleting coolant.

In some instances, carbon dioxide can be blended with hydrocarbons to increase extrudability in extruded polystyrene foams. The addition of a small amount (0.5–2 percent by weight) of perfluoroalkanes contribute to the creation of smaller foam cells, which can achieve an insulating ability per unit thickness equal to that of CFC-11-blown foams (Fischer et al. 1991, p. 4.9).

With currently available options, the use of ODCs as blowing agents in the production of rigid insulating foams can be eliminated by January 1, 1996. In many applications the phaseout can occur before that date. For instance, the use of carbon dioxide as a blowing agent in the production of insulation for district heating pipes enabled the Swedish government to accelerate the phaseout date for this application from an already ambitious target of 1995 to mid-1991 (SEPA 1991, p. 6). Exceptions may be needed where excessive energy-use penalties might arise from a switch to non-ODC-blown insulating foams.

6.3 Alternatives to Foam in Packaging

Paper and cardboard have already replaced many disposable foam containers and foam trays in fast-food restaurants. An added benefit of these materials is their recycling potential.

Polystyrene packaging foam can be blown with non-ozone-depleting blowing agents. In many instances foams can be blown with carbon dioxide, either directly or as a result of increasing the water content in the foam-forming mixture. Dow Chemical, one of the companies marketing this technology, not only claims success with it but has found significant advantages over sheets blown with CFC-12 (Welsh 1991). In particular, the use of CO_2 in the foam-production process allows for easier recycling of the foam product, since at the end of the foam product's life only air remains within the foam structure (ibid., p. 765). Packaging foam can also be blown with a decomposable solid chemical agent, such as azodicarbonamide.

Hydrocarbons such as pentane and butane can be used, alone or in combination with carbon dioxide, as blowing agents in the production of polystyrene packaging foams. On a cost-per-pound basis, hydrocarbons are the least expensive blowing agents for this foam application. However, flammability and hydrocarbon emissions may present barriers to widespread use—particularly in the United States, where there is regional regulation of volatile organic compounds. Installation of emission-control equipment may be an option. For example, it has been estimated that manufacturers utilizing the cheaper hydrocarbon blowing agents can recover the higher initial costs of meeting flammability and pollution criteria within 12–18 months (UNEP 1991b, p. V-5).

Much of the rigid polyurethane used in packaging serves the purpose of physical protection or rigid cushioning more than that of insulation. In these applications, foam products blown with water and hydrocarbons (such as isopentane and butane) have been accepted as alternatives (ibid., pp. III-50–III-51).

In sum, the use of ODCs in the production of foam packaging can be completely eliminated almost immediately through either substitution of alternative products or alternative blowing agents.

6.4 Recovery of ODCs from Existing Foams

The most important fact relevant to the reduction of emissions of ODCs from rigid foams is the large stock of ODCs present in the insulation of existing buildings and appliances—approximately 1.3 million tonnes. Thus, recovering ODC-containing materials from buildings when they are torn down or remodeled is important.

Current emissions from buildings and appliances are small because the majority of buildings and appliances containing ODC-blown foams are still in use. We can expect emissions to increase rapidly in the next two decades. By 1995 the emissions from disposal of appliances and the retirement of buildings exceeded those from manufacturing and diffusion.

In the early 1990s the technology for recovering ODC-blown foams (mostly CFC-11) from appliances such as refrigerators and freezers was developed in Europe. A German company, Adelmann GmbH, developed a process with the following steps (Blessing 1991):

removing the refrigerant (typically CFC-12) by suction
dismantling the refrigerator or freezer by hand or with specially developed tools (rather than shredding it, as in other systems)
shredding the foam panels with a rotary cutter
compressing the foam under high pressure to extract the ODC blowing agent (typically CFC-11).

According to marketing information provided by the company, essentially all of the enclosed blowing agent is recovered. There are over a dozen units based on these principles in operation in Germany, Austria, Switzerland, and the Netherlands (ibid.). The shredded foam, free of ODCs, can be recycled into new foam products.

Refrigerators are generally collected separately from municipal solid waste, so that recovery of refrigerant foam is generally possible. Recovery of insulating foams may also be possible when buildings containing such foams are torn down, but this will generally be much more difficult.

Recovered ODCs may either be reused (after appropriate reconditioning) or destroyed. Reuse allows for the eventual release of these compounds, thus adding to the atmospheric burden. Thus, recycling and reuse should be restricted to the shortest possible period.

6.5 Recommended Phaseout

We recommend eliminating the use of chlorinated ODCs in the production of rigid foams by the beginning of 1996. Every effort must be made to capture and destroy ODCs from the bank of halogenated foam.

7 Solvents

The principal solvents involved in stratospheric ozone depletion are CFC-113, carbon tetrachloride, and methyl chloroform. CFC-113 has mainly been used as a solvent in the electronics industry, and methyl chloroform has been widely used to degrease metal. These compounds are also used as solvents in adhesives, coatings, and inks, in addition to being used to dry clean clothes and other fabrics. The use of carbon tetrachloride as a general solvent has largely been eliminated in the United States (because of its toxicity) but is still prevalent in the Third World, the formerly socialist countries of Central Europe, and the former Soviet Union. Carbon tetrachloride is also used as a feedstock for CFC production, though most emissions appear to be due to solvent applications. All three of these compounds are now regulated by the Montreal Protocol and are scheduled to be phased out by 1996 in industrialized countries and 2006 in the Third World. Considerable progress has been made toward a complete phaseout of ODCs in such applications.

7.1 Historical Use Patterns

CFC-113 and Methyl Chloroform

CFC-113 solvent applications grew partly in response to stricter environmental controls on chlorinated hydrocarbons such as methylene chloride due to both toxicity and regional air pollution problems. Both CFC-113 and methyl chloroform are used to clean electronic circuit boards and a wide variety of metal parts and components. In 1988, CFC-113 and methyl chloroform accounted for approximately 20 and 49 percent, respectively, of chlorinated hydrocarbons used as solvents in

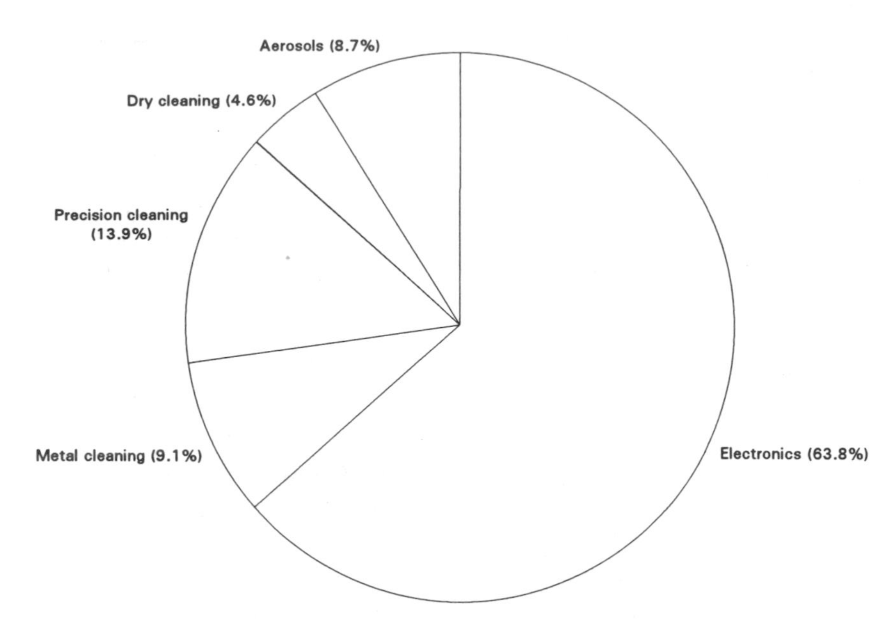

Figure 7.1
Use of CFC-113 in solvent applications, 1989 (derived from data in UNEP 1991f, p. 2-90) (total: 230.5 kilotonnes).

electronics and metal-cleaning applications in the United States, though their share has likely dropped since that time (Fischer et al. 1991, p. 11.4, table 11.2). Other halogenated solvents, including perchloroethylene, trichloroethylene, and methylene chloride, account for the remainder. Small amounts of CFC-113 and methyl chloroform are also used in dry cleaning.

The concentration of CFC-113 in the atmosphere grew rapidly with CFC-113 use, increasing by about 9 percent per year; methyl chloroform grew by about 4 percent per year (WMO 1991, p. 1.4, table 1-1). The worldwide consumption of CFC-113 in 1992 was approximately 126,500 tonnes—down from a peak of about 276,700 tonnes in 1988 (AFEAS 1993b; AFEAS 1992b).[1] In 1988, approximately 36 percent of the global total was consumed in the United States.[2] Figure 7.1 shows the global solvent uses of CFC-113 in 1989.

Global methyl chloroform use amounted to approximately 627,000 in 1992—down from approximately 725,700 in 1990.[3] In 1992, the United

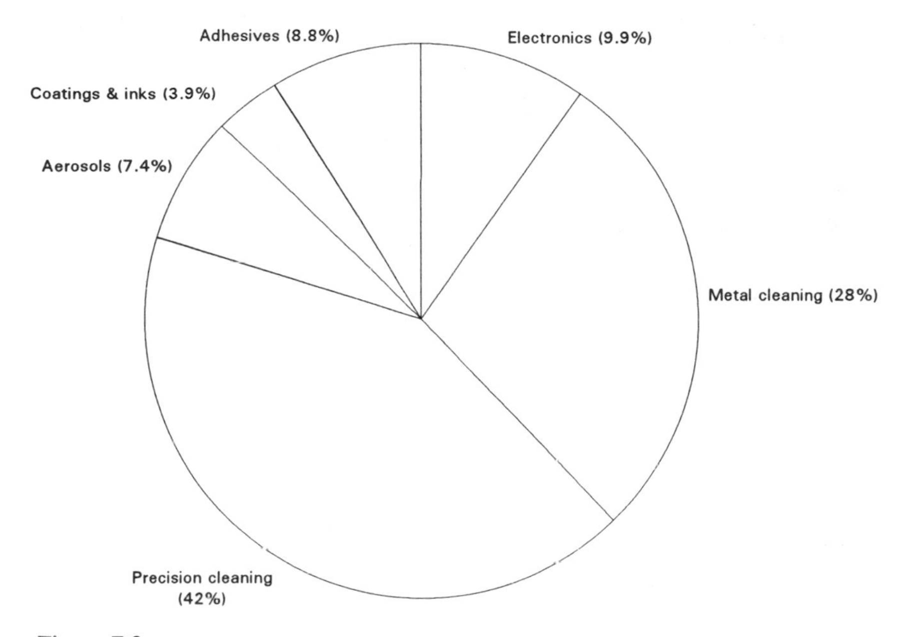

Figure 7.2
Use of methyl chloroform in solvent applications, 1989 (percentage shares derived from data in UNEP 1991f, p. 2–9; total [693 kilotonnes] from Pauline Midgley, written communication, December 8, 1993).

States, Canada, and Western Europe accounted for approximately 56 percent of the worldwide consumption of methyl chloroform.[4] Far East countries accounted for approximately 31 percent. Regional sales data for the early 1990s indicate that U.S. consumption increased between 1991 and 1992 and that much of this increase may have been stockpiled in anticipation of regulatory controls.[5] One of the principal applications of methyl chloroform is sheet-metal degreasing in the automobile industry and elsewhere. According to the World Meteorological Organization (WMO 1985, vol. I, p. 72), methyl chloroform "is also used as a solvent in adhesives, varnishes and paints, where low flammability and low toxicity are important. Sales of methyl chloroform grew rapidly in the 1960s and early 1970s when it replaced tri- and perchloroethylene and CCl_4 [carbon tetrachloride] in many industrial applications." Figure 7.2 presents a breakdown of global methyl chloroform consumption (nearly equal to production), emphasizing the role of this compound in metal cleaning applications. Electronics, metal, and precision cleaning constitute approximately 80 and 87 percent of methyl chloroform and CFC-113 use as solvents, respectively.

In addition to the large-scale uses of CFC-113 and methyl chloroform as a solvent, small amounts of CFC-11 and CFC-12 are used in aerosol spray cans as solvents or cleaning agents.

Carbon Tetrachloride

Worldwide production of carbon tetrachloride amounted to between 800,000 and 1,100,000 tonnes in 1988, of which approximately 90–95 percent was consumed in the production of CFC-11 and CFC-12 (WMO 1985, pp. 72-73; UNEP 1991e, pp. I-1–I-8). Small amounts of carbon tetrachloride are used in the production of a number of chlorinated solvents, such as vinyl chloride and tetrachloroethylene. Carbon tetrachloride is destroyed during the production of ODCs, with the exception of small fugitive emissions which are estimated to be approximately 3 percent of feedstock throughput (Simmonds et al. 1988). Therefore, some reduction in emissions will likely occur as CFC-11 and CFC-12 are phased out.

Though large amounts are used as feedstock, the main source of atmospheric emissions of carbon tetrachloride is solvent use. The World Meteorological Organization estimates that more than 90,000 tonnes per year must be emitted to explain the measured concentration of carbon tetrachloride in the atmosphere (1980 figures); emissions from reporting countries can account for only half that amount (WMO 1985, pp. 72–73). The remaining emissions are likely from developing countries, Eastern European countries, and the former Soviet Union, where this compound may still be used as a solvent. Data from the late 1980s indicate that carbon tetrachloride concentrations continue to grow at a rate of approximately 1.2 percent per year (WMO 1991, p. 1.4, table 1-1).

It appears that carbon tetrachloride has also been used as a fire extinguisher in some developing countries. For instance, the increase in the use of halons in India reflects not only the demands of protecting high-tech equipment but also the need to replace carbon tetrachloride, which has been banned (Puroshothama 1991, p. 693).

Emissions resulting from the use of carbon tetrachloride as a solvent must be reduced because of its long lifetime, its high toxicity relative to other solvents, and its substantial atmospheric concentration, which is larger than that of CFC-113 and almost as large as that of methyl chloroform (WMO 1991, p. 1.4, table 1-1).

7.2 Alternative Solvents

It is now possible to rapidly phase out large-scale uses of CFC-113 and methyl chloroform. Sweden has already effectively phased out the use of CFC-113 in metal, electronics, and precision cleaning. Germany had targeted December 1992 as the phaseout date for the consumption of both CFC-113 and methyl chloroform (UNEP 1991e, appendix F). Austria, Denmark, Finland, and Norway planned methyl chloroform phaseouts for 1994 and 1995 (GECR 1993).

In addition to the accelerated phaseout schedules instituted by many countries, many large ODC solvent users within the electronics and telecommunications industries have established aggressive, independent phaseout schedules. Intel, 3M, Northern Telecomm, and Raytheon pledged a phaseout of methyl chloroform by 1993. IBM and the Rocketdyne division of Rockwell International completely eliminated CFCs from their manufacturing process in October 1993. Northern Telecomm has eliminated the use of CFC-113 entirely, saving money in the process. Comments by Northern Telecomm's assistant vice-president for environmental affairs emphasize this point: "As it turned out, in a three year program we had spent something on the order of $1 million, including some R&D time and some equipment purchases, while in the course of that period we saved $4 million. The savings are in terms of no solvent or less solvent to be purchased, no waste or less waste to dispose of, and doing away with equipment. You don't have a need for maintenance, you don't have a need for the electricity, and you free up some floor space that you can use for other purposes." (GECR 1991, p. 2)

Other CFC-113 phaseout commitments were made by Motorola (end of 1992), Hitachi (1993), IBM (1993), and Nissan (1993) (CBE 1992). These phaseout dates for CFC-113 and especially methyl chloroform in response to regulation are remarkable and noteworthy in view of the vigorous protests of consuming and producing industries in the late 1980s that it would be costly, difficult, or even impossible to phase out these solvents rapidly.

Solvents in Electronics Assemblies

The principle use of CFC-113 and methyl chloroform in the electronics industry is for defluxing. Flux is material applied to a printed circuit board before soldering to improve the surface bonding characteristics of the solder material. Defluxing is the process whereby the remaining flux and solder residues are removed from printed circuit board assemblies. The primary reason for defluxing is to minimize the possibility of corrosion due the remaining surface contaminants.

One important factor in the growth of ODC use in this application has been the adoption by industrial producers of specifications discouraging the use of non-solvent-based cleaning systems. The updating of military specifications in the civilian electronics industry and the cost-effective alternative processes and compounds have already stimulated independent phaseouts of these compounds by some companies and can be applied globally.

The use of military specifications in the electronics industry was described in a recent UNEP assessment: "Much ordinary electronics production, particularly in the smaller firms, was cleaned to military approved requirements using CFC-113 and 1,1,1-trichloroethane [methyl chloroform], even though it was not strictly necessary. To avoid duplication of processes and equipment, the military standards were used for non-military work, probably with large 'overkill' and certainly with a large resultant emission of CFC vapors." (UNEP 1991e, p. 24) The same study pointed out that "an estimated 50 percent of current CFC-113 use results from United States military specifications" (ibid., p. 110). Fortunately, recent revisions to these military specifications are expected to aid in the reduction of ODC use in the electronics industry.

Many alternatives are available for defluxing and cleaning electronic assemblies. One approach has been to use soldering fluxes that require less cleaning afterward because of their low contamination levels. Traditional fluxes contain between 15 and 35 percent solids. Fluxes containing only 1–10 percent solids reduce or eliminate the post-soldering cleaning. Such "no-clean" assembly procedures for printed circuit boards have considerable environmental and economic benefits. Development of a citric-acid-based, water-soluble flux called HF1189 (short for "Hughes flux, November 1989") is also expected to aid in the reduction of ODC defluxing (GECR 1992a).[6] Experience with this system has halved processing time and has reduced operating costs by 90 percent.[7]

Aqueous cleaning (the use of water-based solutions as cleaning agents) has also become far more widespread. Generally, manufacturers have reduced both pollution and cost in the changeover. Alcohols, saponifiers (soaps), and weak water-soluble detergents have been used to modify aqueous methods to suit particular cleaning requirements. Aqueous cleaning is most effective when combined with high-energy sprays and deionized-water rinsing (UNEP 1991e, pp. 69–80).

There is a proliferation of technologies and vendors offering alternative non-ODC solvents. Among the large variety of alternative solvents are organic compounds containing terpenes, alcohols, and esters. Some of these alternatives, derived from natural substances such as wood pulp and orange rinds, are not only non-ozone-depleting but also biodegradable (Hayes 1991; Nigrey and Arzigian 1991). AT&T announced in 1992 that it had developed a new etching process for microchips based on the solvent n-butyl butyrate, found naturally in fruits such as cantaloupe (GECR 1992f). Some of the proposed alternative solvents are hydrocarbon-based compounds and are therefore flammable and possibly toxic. In some places, regulations on the emission of hydrocarbons instituted because of their ability to form toxic ozone in the lower atmosphere may limit their application. These and other solvents can also be combined with water rinsing into what are often referred to as "semi-aqueous" cleaning processes.

Finally, new technological alternatives for precision cleaning have arisen, such as plasma, ultraviolet/ozone, and ice-particle cleaning (UNEP 1991e, p. 182-187). Plasma cleaning is based on passing radiofrequency energy through a process gas such as oxygen, argon, or helium to energize ions and electrons. These charged particles then electrochemically remove surface contamination before being removed by the flow of process gas. UV/ozone cleaning relies on the ability of UV radiation to electrically excite or break up surface contaminants, which then react with atomic oxygen to form simpler gaseous constituents. Both of these processes, though inexpensive to operate, require significant initial investment. Ice-particle cleaning uses a spray of fine ice particles to physically remove surface contaminants. AT&T has developed this concept further, employing particles of CO_2 ("dry ice") (EIN 1994). An important advantage of this technique is that the particles change to a gaseous state after the contaminants have been removed.

Metal Cleaning

CFC-113 and (more commonly) methyl chloroform are used in metal cleaning, including the cleaning of precision components. There have been a considerable number of developments in this field in the past few years. In general, it has been found that aqueous cleaning techniques similar in principle to those in the electronics industry are applicable to metal cleaning. One of IBM's production facilities reduced its use of CFC-113 by switching its disk-drive-cleaning process to a water-based ultrasonic vibration method (CBE 1992). Ultrasonic cleaning equipment creates sub-micrometer bubbles at the surface of the product to be cleaned through the use of high-frequency vibration. These bubbles exert a scrubbing action as they break up and collapse.

In some applications, semi-aqueous techniques are also suitable (however, existing regulations regarding releases of volatile organic compounds must be met in these instances). At AT&T Bell Laboratories it was found that an aliphatic ester mixture was "an excellent cleaner" that met all requirements and could replace CFC-113 (Gillum and Jackson 1991). Similarly, Dow Chemical found that its semi-aqueous formulation met the requirements for a large variety of degreasing applications (Donate and Papajesk 1991). Azeotropic blends of fluoroiodocarbons (FICs) and conventional solvents such as esters, hydrocarbons, and ketones may be "drop-in" substitutes for CFC-113 and methyl chloroform in existing vapor degreasers (Nimitz and Lankford 1993b).

A variety of organic non chlorinated solvents are also potential alternatives in many metal-cleaning applications. Some alcohols (ethanol, isopropanol, glycol ethers) and some ketones (methyl ethyl ketone, acetone) possess favorable solvent properties. Flammability can be reduced by using these compounds in conjunction with inert, nonflammable additives.

A promising approach to metal cleaning that would eliminate the need for hydrocarbons in a semi-aqueous process is the use of "supercritical" carbon dioxide as a solvent. The supercritical state of a chemical is one in which the gas and liquid phases are indistinguishable. There is a particular combination of pressure and temperature at which any compound becomes "supercritical." Carbon dioxide in this state is "an excellent solvent for dissolving nearly every oil that is associated with the manufacturing or machining of metal parts" (Gallagher and Krukonis 1991, p. 262). Carbon dioxide becomes supercritical at a temperature of 31.1°C (about 70°F) and a pressure of 72.8 atmospheres (ibid.).

In its supercritical state, carbon dioxide has a density close to that of a liquid, making for small volumes (compared to the usual gaseous state); at the same time, it can flow like a gas to achieve good cleaning of tight spaces. As it is more expensive than aqueous and semi-aqueous cleaning processes, its use may be advantageous only in precision applications (ibid., p. 264). However, one of the main concerns expressed by industry and by the U.S. military regarding a phaseout of chlorinated solvents has been the cleaning of precision parts. The demonstrated effectiveness of supercritical carbon dioxide in the cleaning of precision equipment removes one of the main obstacles to a rapid phaseout of both methyl chloroform and CFC-113. Hughes Aircraft recently installed a supercritical carbon dioxide system and reduced CFC use by 30 percent (CBE 1992).

Adhesives, Coatings, and Dry Cleaning

The use of methyl chloroform in adhesives and coatings is another area in which industry has expressed particular concern over regulations. However, water-based coatings are now well developed for some applications and should soon be available for other applications (Mertens 1991). Use of methyl chloroform as a solvent in aerosols can be eliminated by reformulating the active ingredient as a solid or a gel, thus avoiding the need for a liquefying solvent (NRDC 1990).

Experience with ODC regulation has shown that, when promising technologies are clearly on the horizon but need some development, a rapid phaseout date can be specified, along with some provisions for exceptions in cases of demonstrated need. The regulation serves to increase demand for alternatives, which in turn stimulates technological innovation. That approach can be taken with methyl chloroform.

Dry cleaning using chemical solvents was originally meant only for particularly sensitive fabrics. Many garments currently dry cleaned can be laundered using gentle aqueous techniques. Spot cleaning using ammonia, hydrogen peroxide, and talcum powder followed by brushing and steaming be can substituted for solvent-based dry cleaning in most cases. Furthermore, some garments labeled "dry clean only" can, in fact, be laundered by water-based methods (Ryan 1993). Where absolutely necessary, perchloroethylene (currently used in the dry cleaning industry) can meet the demand.

7.3 Recommended Phaseouts

The use of methyl chloroform in the electronics industry and all uses of CFC-113 can be phased out globally by 1996. Many specific solvent applications have already been phased out. We have noted some prominent examples.

The only area in which a global phaseout date for methyl chloroform later than 1996 might be desirable is in certain solvent applications in the Third World. The use of carbon tetrachloride as a very cheap solvent (20–60 cents per kilogram) is common in the Third World. It will be very difficult to introduce new cleaning techniques and solvents in time to meet a 1996 phaseout date. Moreover, the fact that in the Third World auto parts such as spark plugs are cleaned rather than thrown away (as is common in the United States) implies that the financial burden of a phaseout may incur an unacceptably high cost, necessitating a slower transition to cover the costs of new equipment and retraining.

Since carbon tetrachloride is a highly ozone-depleting and toxic substance, an immediate phaseout with partial substitution by methyl chloroform for two or three years beyond that time in Third World countries has considerable merit. It is probably the only practical way to accelerate the phaseout of carbon tetrachloride in the Third World.

8

Fire Extinguishers

The ozone-depleting compounds known as halons have been traditionally used for firefighting. The primary difference between halons and other ODCs is that halons contain bromine, and in some cases chlorine as well. Molecule for molecule, bromine is far more effective at depleting ozone than chlorine, mainly because of bromine's slower removal from the stratosphere. Bromine is also involved in nonlinear reactions with chlorine, so the combination of chlorine and bromine can be far more ozone depleting than the same number of chlorine atoms alone. Thus, even though the quantity of halons emitted to the atmosphere is small relative to the quantities of other ODCs, the equivalent chlorine of present concentrations is substantial. Under the Copenhagen Amendments to the Montreal Protocol, new halon production and consumption was to have been completely phased out in complying industrialized countries on January 1, 1994. Third World countries have another 10 years to comply.

8.1 Current Use

Halon fire-protection systems are widely used in computer and process and production control applications, in nuclear facilities, in banks, on airplanes, and "in all electronic computer management and data storage functions" (Mossel 1988). The military is one of the main consumers of halon fire-protection systems. Halons are used on airplanes because of their minimal weight relative to their effectiveness and on ships because of their relatively low volume. They are also used to prevent chemicals from attaining explosive concentrations (Catchpole 1991).

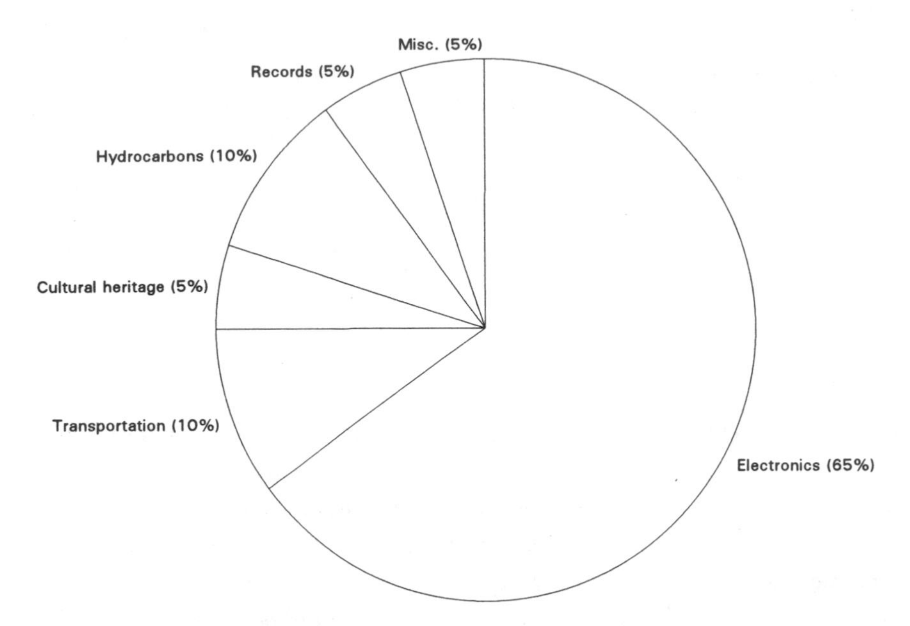

Figure 8.1
Halon 1301 use, 1990 (UNEP 1991c, p. 13) (total: 9.1 kilotonnes).

The speed with which halons extinguish fires has made them attractive firefighting tools. It is thought that halons interrupt fires both chemically (by stopping the reactions that characterize combustion) and physically (by dousing the flames and depriving the chemical reactions of the oxygen necessary to sustain combustion). Halons have been marketed as being particularly well suited to firefighting applications where water might cause widespread damage to electronic equipment and magnetic storage devices (e.g. computer tapes and diskettes).

Three halons are regulated by the Montreal Protocol: halon-1211, halon-1301, and halon-2402. The first two are the most commonly used; the last is produced in very small amounts. Figures 8.1 and 8.2 present the most recent estimates of 1990 halon-1301 and halon-1211 use by application.

Fire protection for electronic equipment is the dominant use of halon systems. Halon-1301 is used primarily in automatic fire-extinguishing systems. Halon-1211 is used in portable fire extinguishers, including some available for household use.

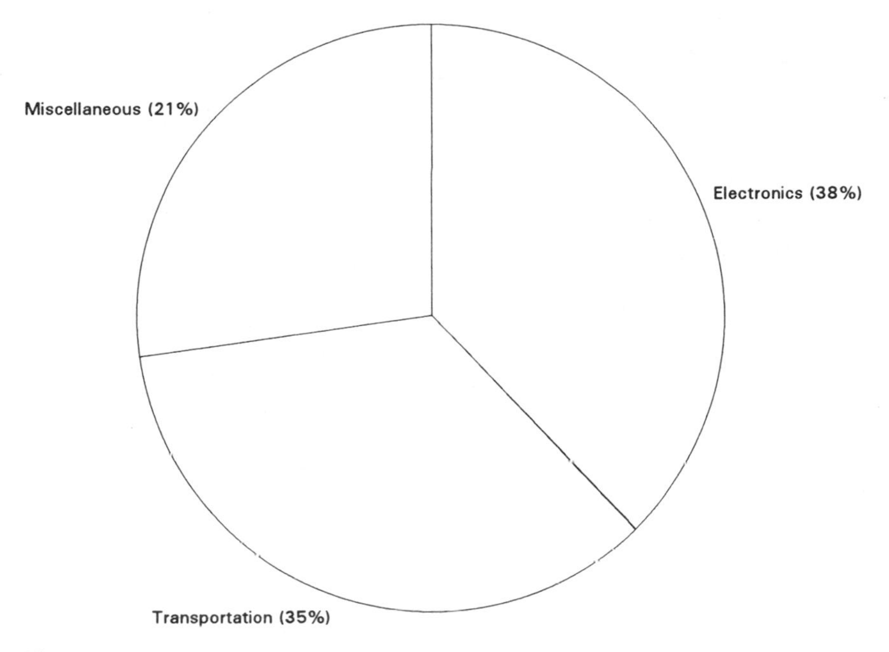

Figure 8.2
Halon 1211 use, 1990 (UNEP 1991c, p. 12) (total: 14.9 kilotonnes).

Global production of halon-1211 and halon-1301 peaked in the year 1988 at 20,200 and 12,600 tonnes, respectively (McCulloch 1992). The most recent published production estimates for the year 1991 show a considerable decline, down to 9037 and 9617 tonnes, respectively.[1] Production is expected to have declined dramatically as a result of the phaseout in the industrial world on January 1, 1994.

Only a small percentage of the annual production is actually used in firefighting. Of the 13,700 tonnes of halon-1211 produced in 1986, approximately 10 percent was used in firefighting situations, 10 percent was used for testing, servicing and accidental discharge, and 80 percent was placed into devices for future use (UNEP 1989a, p. 10). The corresponding estimates for halon-1301 for 1986 are that 7 percent was used in firefighting, 23 percent discharged, and 70 percent banked. The share of halon released via servicing, testing, and accidental discharge has most likely declined since that time.

Most halons are stored in fire-protection equipment or systems before use. We estimate the banks of halon-1211 and halon-1301 in 1991 to have been approximately 62,400 and 86,000 tonnes, respectively.

8.2 Alternatives to Halon

The 1994 Phaseout

As recently as 1988, it was generally accepted that halons were irreplaceable in many applications and that they could not be phased out with other ODCs. In fact, the 1987 Montreal Protocol only put a cap on production of halons by 1992, without making any definite provisions for their eventual phaseout. The 1992 adjustments to the Montreal Protocol greatly accelerated this schedule. Industrialized countries that have complied with the provisions of the Montreal Protocol eliminated the production and consumption of halons January 1, 1994. Third World countries have an additional 10 years to eliminate halon production and consumption.

The success of regulations in stimulating user industries and institutions (such as fire-protection officials in government) to innovate rapidly and to reevaluate the use of chemicals that were to be restricted has been remarkable. This work has precipitated a complete change in the attitudes toward future halon production and consumption. The availability of alternative chemicals today, the promise of many alternatives in the near future, and a reevaluation of the utility of technologies that predated the use of halons (such as those involving water and carbon dioxide) should accelerate the phaseout of halons.

Though production and consumption of halons have been eliminated in many countries, a number of issues remain. Existing halon systems must be retrofitted to non-ozone-depleting alternatives wherever possible so as to avoid the release of halons from these systems in the future. Furthermore, efforts should be made to accelerate the elimination of new halon consumption in Third World countries, to replace halons in existing systems with alternative chemicals, and to replace whole systems where alternative chemicals are undesirable with nonchemical systems.

Survey of Alternatives

It appears that the ability of halons to protect research institutions was somewhat overrated. For example, the firefighting equipment at the Goddard Space flight Center is being converted to water from halon-1301, and a NASA spokesman told a 1991 conference workshop that there will not be any reduction of firefighting capabilities.[2]

Water is particularly good at reducing reignition of fires because it lowers the temperature of combustible objects. However, water fire-extinguishing systems may not be appropriate for equipment rooms where non-encased high-voltage devices are present. A combination of automatic water sprinklers, carbon dioxide, and early warning is the most promising system. Of course, the most obvious advantages of water are lack of toxicity, low cost, and lack of environmental impact.

Water systems in which very fine droplets (less than 200 micrometers) are released have shown particular promise. They are potentially superior to sprinklers in a variety of ways, such as their ability to suppress flammable liquid pool and spray fires and their utilization of much less water than traditional sprinkler systems. Water-use reductions of up to two orders of magnitude relative to conventional water fire-suppression systems have been estimated (UNEP 1993b). These systems depend to a great deal on the characteristics of the nozzle design and, therefore, vary considerably in effectiveness. Like many of the alternative systems, this one may be applicable to specific fire-suppression situations.

Other nonhalon systems have considerable potential to replace present halon systems. Combinations of early-warning systems and carbon dioxide or foam flooding have exhibited promising comparative results in applications where water could cause costly damage. Both carbon dioxide and foam are particularly well suited for liquid fires and the protection of electrical equipment. For portable extinguishers, carbon dioxide and dry chemical systems are already in wide use.

Similar to dry chemical systems, fine solid particulate and aerosol technology is currently being pursued by several companies (UNEP 1993b). Though the technological details are proprietary, these systems are based on the ability of fine solid and liquid particulates to suppress fires by gas-phase cooling. These systems promise to cause less collateral damage to equipment than dry-powder systems

In addition, a large number of non-ozone-depleting compounds are being investigated as potential chemical substitutes for halons in both fixed and portable systems. Among the most promising are perfluorinated alkanes (in particular, perfluorobutane—C_4F_{10}), fluoroiodocarbons (FICs), HFC-125, HFC-23, and HFC-227ea (UNEP 1991c). The role that perfluorinated compounds might play in global warming should be considered before any decision to use these compounds as alternative extinguishing agents. Because of the lower extinguishing efficiency of the HFCs, 2–2.5 times as much of these compounds (by weight) would be required to replace halon-1301. The decomposition products are relatively acidic, unlike those of halons.[3] FICs have shown particularly promising results in preliminary testing. A mixture of 87 percent CF_3I and 13 percent HFC-134a may have the same firefighting effectiveness by volume as halon-1301; if it does, it could be a drop-in replacement in total-flooding systems (Nimitz and Lankford 1993a). $CF_3CF_2CF_2I$ and F_3CF_2I mixed with HFCs appear to be candidates for streaming applications that traditionally used halon-1211.

Mixtures of inert gases—primarily nitrogen and argon—appear to be promising as a firefighting alternative to halon-1301 total-flooding systems (*New Scientist* 1992a; UNEP 1993b). Small amounts of carbon dioxide are often included in the mixture to stimulate breathing in individuals present during the gas release. Mixtures of inert gases displace available oxygen, effectively smothering combustion. Such systems have been approved for use in the Netherlands and Denmark and are being tested in Germany and the U.K. These mixtures can be used in systems currently using halon-1301, although release nozzles and storage tanks will have to be replaced. Inert gases do not thermally decompose and hence do not generate by-products as do other chemical agents.

Since one goal of most halon fire-protection systems is to reduce the risk of damage to equipment (halon systems are generally not designed exclusively for life protection), other fire-prevention measures can compensate for any increased risk arising from the switch to a nonhalon fire-protection system. Better protection of electrical wiring, greater use of fire-resistant materials in equipment rooms, and use of water-resistant materials (where a water-based system is substituted for a halon-based system) may ensure lower damage in case of a fire and its curtailment.

Such measures, combined with existing and emerging technologies, convince us that almost all halon consumption can be eliminated immediately. For those few applications where halon systems are still considered, recovery and stockpiling can meet future needs. It appears that such exceptions will be minimal. Though a number of "halon essential use" applications have been submitted to the Halon Technical Options Committee of UNEP, none have been approved.

8.3 Recovery and Reuse

Recovery of halons after fires is not feasible, nor is their reuse. Fortunately, actual firefighting is not responsible for most halon emissions. Most occur during testing equipment , during training of personnel, and, to a lesser extent, during accidental discharges.

Both minimization and recovery of halons released during equipment testing and personnel training are feasible under many circumstances. ICI, a major manufacturer of halons, is establishing a program of recovery, purification, and reuse (Mossel 1988). Firefighting professionals and associations are instituting programs to reduce false discharges and unnecessary releases of halons by automatic equipment. The current National Fire Protection Association standards require that total-flooding systems be torn down every 6 years. During this process, an estimated 23 percent of the halons are released. Reducing the frequency of system teardown to every 12 years for systems in climate-controlled locations could have a substantial impact on total halon demand (DOD 1991, appendix D). A number of countries have instituted halon-bank-management programs in which recycled halon is distributed or sold on the open market to users of existing halon systems requiring recharging.

Alternative compounds can also be used in situations where testing and training are considered vital. The U.S. military has announced that some of its fire-protection testing and training can be carried out using sulfur hexafluoride (SF_6) rather than halons and that 90 percent of the halon can be replaced with an alternative gas (DOD 1991, appendix D).

In order to facilitate a phaseout of halon production and consumption in the Third World and those countries not adhering to the Montreal

Protocol, present needs can be met by recycling compounds in the existing halon bank. UNEP estimates suggest that, with adequate management, the needs for halon-1301 can be met by the existing banks "well into the next century" (UNEP 1991c, p. 73).

8.4 Recommended Phaseout

We recommend an immediate global phaseout of all halon production. Where halons are required, the existing bank will suffice. We further recommend the immediate collection and destruction of the remaining halon bank.

9

Other Uses

Ozone-depleting compounds are used for a large number of applications that do not fit into the broad industrial classifications examined up to this point, including the use of halocarbons in aerosol devices, the use of CFCs in food processing, emissions of chlorine from solid-fuel rockets, and the use of CFCs in uranium enrichment.

9.1 Aerosols

Probably the best-known use of ODCs is as a propellant in spray cans. The first patent application for the aerosol container came from a Norwegian, Eric Rotheim, in 1926 (Umweltbundesamt 1989, pp. 121–122). The first significant use of aerosol containers occurred during World War II when pressurized carbon dioxide was used to propel insect repellent in heavy steel cans called "bug bombs" (Roan 1990, pp. 33–34). It was found in the 1950s that a mixture of CFC-11 and CFC-12 could be used as an effective propellant at pressures much lower than those required for CO_2. The lower pressures and a newly developed application valve meant that lighter materials, such as aluminum, could be used for containers. This combination ushered in the use of ODCs as propellants in aerosol containers, which by the early 1970s was the largest application of ODCs. These aerosol products were used in dispensing a large variety of products, including hair spray, paints, and medications.

The most common role of a propellant in an aerosol container is to ensure a constant release of an active ingredient (hair spray, paint, etc.) in the form of small particles. ODCs also act as solvents in aerosol containers when the active ingredient requires dissolution. Although constituting a much smaller share of aerosol use, ODCs are also used as the active ingredient in applications such as surface chilling, displacement cleaning for dust removal, and sound horns.

Worldwide use of ODCs in aerosol containers was quite large as recently as 1990, when the estimated consumption was approximately 115,000 tonnes, down from approximately 180,000 tonnes in 1989 (UNEP 1991a).[1] For comparison, use of CFC-11 and CFC-12 in aerosol containers was approximately 430,000 tonnes in 1976. Since 1990 the use of ODCs as propellants appears to have declined rapidly; one 1992 estimate was approximately 31,000 tonnes (CFC-11 and CFC-12 only) (AFEAS 1993a).[2]

The first regulations passed curtailing the consumption of ODCs targeted their use as propellants. In the late 1970s, the United States, Canada (for cosmetic and hygienic products only), Sweden, and Norway banned this use. Many countries have taken similar action since the late 1980s (UNEP 1991a, pp. 19–21).

All of the CFCs, HCFC-22, carbon tetrachloride, and methyl chloroform have been used in aerosol products to varying degrees. CFC-11 and CFC-12 were, and continue to be, the most common. Unfortunately, one of the principal replacements for CFCs in aerosols has been HCFC-22, often misleadingly advertised as "ozone-friendly." Approximately 20,000 tonnes of HCFCs were used in aerosol products in 1990 (ibid., p. 31). We estimate that the decline in ODC use in this application category has accounted for a significant portion of the early 1990s decline in total ODC production.

Special Aerosol Applications

Because of the early regulation of this particular application of ODCs, a large variety of alternatives have been thoroughly explored and utilized. The only use for which possible exemptions might have to be made is for metered-dose inhalers (MDIs), which deliver specific quantities of medication orally in the form of an aerosol spray. This is considered effective for the treatment of bronchial asthma, chronic bronchitis, and other respiratory diseases. It is estimated that approximately 5000–6000 tonnes of ODCs were used in inhalant drug products in 1990 (ibid., p. 33). It is estimated that in the United States alone 25 million people use MDIs

(Zurer 1992). CFC-12 and CFC-114 are used as propellants, with some CFC-11 used as a slurrying agent and cosolvent.

Two ODC-free alternatives to CFC MDIs are dry-powder inhalers and nebulizers. Approximately 10 percent of patients requiring aerosol-delivered medication use dry-powder inhalers (UNEP 1991a, p. 34). In Sweden and Holland approximately 61 and 65 percent of asthma patients, respectively, used this alternative in 1990 (ibid.). For many patients, dry-powder inhalers are easier to use than MDIs. Nebulizers are commonly used in hospital settings, because of the bulk of the compressors they require. There are hand-held nebulizers with pocketbook-size compressors, but the latter are expensive ($200–$300). MDIs are the only effective devices currently available with spacers, which reduce systemic side effects for steroid patients and which helps patients with weak inhalation capabilities administer doses properly (Crompton 1991, p. 154). However, according to a specialist in respiratory ailments, "there is no definite medical basis for the continued use of MDIs. If MDIs were banned, the main problem would be the treatment of children between the ages of 3 and 6. I do not think it would be difficult for the pharmaceutical industry to produce dry powder spacer systems for this group of children who are too young to generate high inspiratory flows necessary for the conventional dry-powder inhalers but too old to have to use nebulizers. The provision of spacer systems for dry powder inhalers would fill this gap." (Dr. G. K. Crompton, written communication, October 1992)

Although dry-powder inhalers and hand-held nebulizers hold promise for eliminating ODC use, this remains an area where continuation of the existing exemption is likely warranted until viable, cost-effective alternatives are found. Existing banks or recycling of CFCs can meet this need if contaminants can be reduced to meet the standards required by this application. Should this not prove possible, small amounts of new CFC production may be warranted.

Exemptions have also been granted to a number of small industrial specialty uses. The most common of these is the use of aerosol products to clean, lubricate, de-dust, and fault-check electronic equipment. In the United States this use accounts for approximately 3500 tonnes of CFCs, predominantly CFC-12 and CFC-113. A variety of options are now available for these uses, including alcohol, compressed air, hydrocarbons, liquid nitrogen, non-aerosols, and (where completely necessary) HFC-134a.

Legislation and consumer action have led to the replacement of many of the ODCs once used in aerosol products with non-ozone-depleting alternatives. A large variety of options continue to be available.

Reformulation of Dispensed Products

Reformulation of products to be dispensed has been used extensively to eliminate the need for aerosols. Common examples are roll-ons, solid-stick applicators, and the use of gels, creams, and pastes in open or squeeze containers. In many instances, manual application with a brush or a cloth can be as effective as application by aerosol, with the added advantages of lower cost and reduced solid waste.

Alternative Dispensing Methods

A number of alternative product-delivery systems have also been effective in replacing propellant-driven aerosols. For example, finger and trigger pumps, commonly used to dispense window and household cleaner in the United States, have taken over large portions of the aerosol market.

A number of two-compartment or "barrier-pack" aerosol products are also available. In these devices, the product and the propellant are separated by a piston or a bag. One advantage of this approach is that the spray can be continuous (rather than intermittent, as with a finger pump). Barrier packs are generally better suited to applications where larger particles are acceptable.

One of the more interesting alternative dispensing methods is a reusable aerosol can (Frutin 1991) with two compartments. The upper compartment contains the product; the lower compartment contains absorbent polymer pellets at the bottom and a compressed gas which can be absorbed into the polymer. The gas in the lower compartment of the can is dissolved in a liquid solvent carrier. The currently proposed formulation uses carbon dioxide as the gas and acetone as the solvent. The compressed gas exerts pressure on the upper chamber via a piston which partitions the two portions of the can. This allows for the discharge of the product via the nozzle. As the piston moves upward, the pressure in the lower compartment momentarily drops, causing more g... leased into the lower compartment from the polymer pellets, thereby maintaining a constant pressure on the upper chamber. Since the propellant is not emitted, it can be reused. The gas is absorbed back into the polymer as the can is refilled with the material to be delivered. Reusability of the container will also lower solid waste. Early performance tests indicate that the pressure exerted on the dispensed product is higher than in conventional aerosols, allowing for top maximum spray performance throughout the life of the product and improved spray pattern and droplet size (*C&EN* 1992).

Alternative Propellants

There are a number of alternative propellant chemicals that can be used in aerosol containers. If relatively course sprays are acceptable, compressed carbon dioxide, nitrogen, and air can be used. These gases are also nonflammable. Currently, these compressed gases are used in approximately 7–9 percent of the aerosol products being produced.

Though flammable, various hydrocarbons (primarily butane, isobutane, and propane), dimethyl ether (DME), and perfluorinated alkanes have been widely used as alternative propellants. For example, an Australian company has introduced pesticides propelled by tetrafluoroethane (CHF_2CHF_2) for use in airplanes (*New Scientist* 1992c). In addition to being flammable, hydrocarbons and DME are volatile organic compounds, which are regulated in the United States because they contribute to the formation of photochemical smog. It is estimated that 85–90 percent of the U.S. propellant market in the early 1990s was for hydrocarbons.[3]

HCFC-22 and a number of other HCFCs have been proposed as alternative propellants; however, the large variety and the ready availability of non-ozone-depleting alternatives should preclude their use in this application.

9.2 Sterilants

CFC-12 is used to stabilize ethylene oxide (EO or EtO), an explosive, toxic, and flammable chemical commonly employed in hospitals, in sterilization facilities, and as a sterilant for spices and other dry foods. The ability of ethylene oxide mixtures to penetrate a variety of packaging materials and rapidly diffuse from the product surface when sterilization is terminated allows sterilized equipment to be transported and handled. In addition, products sensitive to heat or moisture (e.g., plastics that melt or deform at high temperatures) can be sterilized with EtO mixtures rather than steam.

Ethylene oxide is sometimes used by itself and sometimes mixed with CO_2 or CFC-12 to facilitate its delivery and reduce the dangers involved with storage and handling. A mixture of 12 percent EtO and 88 percent CFC-12 (commonly referred to as "12/88") or 10 percent carbon dioxide and 90 percent EtO ("10/90") is nonflammable.

Approximately 18,000–20,000 tonnes of CFC-12 was used for sterilization in 1990, about half of it in the United States. The amount of CFC-12 used in this application was reduced by approximately 5000 tonnes between 1989 and 1991 (UNEP 1991a, p. 48).

The use of ODCs for sterilization has been completely phased out in Sweden (SEPA 1991, p. 3) and in some other countries.

A number of alternatives to the CFC-12/EtO mixture are available and should be able to meet all of the needs of 12/88 users. First, greater efforts must be made to maximize the number of medical products capable of undergoing steam sterilization. This process is safe, well established, and relatively inexpensive. Many products that may be sterilized by steam are not separated thoroughly from those undergoing sterilization by 12/88. Outside the United States, formaldehyde and pure EtO are commonly used. In addition, existing sterilization technologies such as 10/90, dry heat, peracetic acid mixtures, and gluteraldehyde can be utilized to a greater extent. Emerging options such as vapor-phase hydrogen peroxide, ionized gas plasma, and ozone are also useful in many situations (UNEP 1991a, pp. 65–66). HFC-227ea, which entered commercial production in early 1993, has also been proposed as an alternative sterilization agent (Thornton 1993). Although its sterilizing properties appear favorable, its estimated lifetime of 42 years and the fact that it is a greenhouse gas warrant some concern. It may be best suited to retrofitting existing 12/88 systems where no other alternative exists.

The suitability of these alternatives to the use of ODCs in sterilization will vary considerably, depending on such factors as material compatibility, temperature constraints, and worker safety regulations. In addition, on-site use at hospitals (as opposed to central off-site or new-product sterilization) presents significantly different requirements and constraints. Nevertheless, the number and variety of options presently existing for sterilization can meet all ODC sterilization needs.

9.3 Flexible Foam

Flexible polyurethane foam, first introduced in the 1940s, is widely used as cushioning in car seats, bedding, and furniture. ODCs are used as an auxiliary agent in the formation of the fine holes or "open cells" that give soft foams their malleability and their cushioning characteristics. The main functions of ODCs in this process are to cool the exothermic (heat-generating) reactions, to increase the flowability of the molten foam, and to lower the density of the final foam product. The cells themselves are formed by the release of CO_2 as water injected into the molten foam mixture reacts with isocyanate. Since these cells are open (as distinct from the closed cells of rigid polyurethane foam), nearly all of the ODCs used in the manufacture of flexible foam are emitted at the time of production.

Flexible foams are either produced in large non-molded quantities (called "slabstock") or molded (primarily for use in automobiles). CFC-11 is the most common ODC used in the production of these foams; methyl chloroform is also used.[4] Global use of CFC-11 in this application amounted to about 14,000 tons in 1990, down from approximately 60,000 tonnes in 1986 (UNEP 1991b, p. III-36). In 1990, approximately 12,500 tonnes were used in the production of "slabstock." The remaining 1500 tonnes are thought to have been used in the production of molded foam in Japan. The use of ODCs in the production of molded foams has been banned in Sweden since January 1, 1991 (SEPA 1991, p. 6). It has been reduced in other countries, though precise data are not available since the situation is changing rapidly.

A number of process modifications, such as the use of different plastic formulations in the production of flexible foams, have eliminated the need for ODC blowing agents for all but the lowest-density foam applications. Alternative auxiliary blowing agents such as carbon monoxide, acetone, and (where absolutely necessary) methylene chloride can meet the remaining needs for low-density flexible foam.

Greater use of alternative materials (such as polyester batting, fiberfill, and natural latex foams) can reduce the use of halocarbons by reducing the overall demand for flexible polyurethane foam.

9.4 Other Miscellaneous Uses

ODCs have a large number of small miscellaneous uses, ranging from food freezing to warning horns for boats to propellants for whipped cream.

There are non-ozone-depleting alternatives for all of these applications. For example, liquid foods can be frozen with liquid nitrogen, tobacco can be expanded with liquid CO_2, and cleaners and blowers can utilize compressed air or hydrocarbons.

Solid-fuel rockets can deposit ozone-depleting chlorine directly into the stratosphere. Because the potential amount is small relative to industrial emissions, we are not advancing recommendations concerning solid-fuel

Table 9.1
Miscellaneous uses of ODCs (EPA-RIA 1987, vol. III, part 5, p. 1-2; UNEP 1991a).

Application	Tonnes of ODC	ODC used
Freezing of liquid foods[a]	1,300	CFC-12
Tobacco expansion[a]	1,800	CFC-11
Heat detectors[b]	900	CFC-12
Warning devices[b]	600	CFC-12
Cleaners and blowers[b]	900	CFC-12, 114
Mold release agent	675	CFCs, CH_3CCl_3
Skin chillers/cleaners[b]	52	CFC-11, 12, 113
Stabilizer in whipped toppings[b]	14	CFC-115
Total	6,241	

a. Worldwide consumption, 1991.
b. U.S. consumption, 1985.

rocket emissions at this time. It is important to note that the potential ozone depletion from this source is tied to the frequency of launches and to atmospheric circulation. The current launch rates give rise to relatively small amounts of stratospheric ozone depletion, but the implications of greater launch frequency should be considered. Furthermore, much of the work now done by astronauts on the Space Shuttle (which requires chlorinated fuel) could be done with unmanned rockets powered by nonchlorinated liquid fuels.

9.5 Recommended Phaseout

We recommend immediate elimination of the production of ODCs for use in aerosols, flexible foams, and miscellaneous applications. Exemptions may be necessary for the use of CFC-12 in metered-dose inhalers for children. The CFC-12 required for this application should come from the existing uncontaminated bank or some limited new production. The production of ODCs for sterilization should be eliminated by the beginning of 1996.

10

Ozone-Depleting Compounds with Natural and Anthropogenic Origins

A few of the chemicals responsible for stratospheric ozone depletion have both natural and anthropogenic origins. The most prominent of these are methyl bromide and methyl chloride. In this chapter, we examine the relative magnitude of natural versus anthropogenic methyl chloride and methyl bromide emissions and the associated policy issues.

10.1 Natural Inorganic Chlorine Sources

The largest source of chlorine emissions into the atmosphere is sodium chloride (NaCl)—or common salt—from sea spray. Some chlorine from the oceans may also be released as a gas, mostly in the form of hydrogen chloride (HCl). In general, this inorganic chlorine from the oceans does not affect stratospheric ozone, since both sodium chloride and tropospheric HCl dissolve in water and are removed by rain within days. Sodium chloride also returns to the ocean by dry deposition. This is thought to be the main mechanism by which sea salt returns to the ocean surface (WMO 1989, vol. II, chapter XI).

Episodic emissions of HCl from volcanic eruptions are another potentially large natural source of inorganic chlorine. In order to affect the abundance of chlorine in the stratosphere, volcanic eruptions must have sufficient energy to send their primary chlorine-containing compound, HCl, through the tropopause into the stratosphere.[1] Few volcanoes erupt with sufficient force to directly inject HCl into the stratosphere. Furthermore, those that do will not have a lasting effect; they will alter stratospheric chlorine concentrations for only about two or three years. HCl deposited in the troposphere is removed quickly by precipitation.

While non-anthropogenic inorganic chlorine sources such as large volcanoes can add to stratospheric chlorine, one organic compound with both a natural and an anthropogenic emission source, methyl chloride, supplies far more.

10.2 Methyl Chloride

The formulation of public policy to protect the ozone layer has focused exclusively on the emission of industrial chemicals. These chemicals are indeed primarily responsible for stratospheric ozone depletion; however, policy makers have virtually ignored methyl chloride (CH_3Cl), a significant ozone-depleting compound with both natural and human sources.

The present concentration of methyl chloride in the atmosphere is approximately 600 parts per trillion by volume. This represents approximately 17 percent of the current atmospheric chlorine concentration.[2] The oceans are the primary natural source of methyl chloride emissions; there are small contributions from spontaneous (natural) forest fires and from terrestrial fungi (Harper 1985). Biomass burning, the primary anthropogenic source, encompasses the combustion of fuelwood, the burning of savannas in Africa, and the burning of forests (particularly in the Amazon region). Industrial processes also contribute to the anthropogenic component of methyl chloride emissions.

Owing to the wide dispersal of both the natural and the anthropogenic sources of methyl chloride and the lack of good measurements, emission levels are difficult to quantify. To calculate emissions requires knowledge of both the concentration in the atmosphere and the atmospheric lifetime. Unfortunately, there is considerable uncertainty associated with the lifetime of methyl chloride, which is generally estimated at 1.5 years (WMO 1991, p. 8.8). The few direct measurements of methyl chloride emissions are spatially limited and must be extrapolated globally. As a result, the concentration of 600 pptv corresponds to emissions ranging from 2 million to 5 million tonnes per year (WMO 1985, vol. I,

p. 73). There is even considerable uncertainty in the estimate of the average concentration, although 600 pptv is widely used in the research literature. Some attempts have been made to quantify the anthropogenic sources on the basis of laboratory data and a few isolated field measurements.

Industrial Emissions

The worldwide industrial production of methyl chloride is estimated to be about 0.5 million tonnes per year (WMO 1985, vol. I, p. 73). The U.S. production in 1988 was approximately 0.24 to 0.27 million tonnes (USITC 1989; CMR 1989). According to industry estimates, approximately one-third of the methyl chloride produced is used to produce other chlorofluoromethanes (CMR 1989). The remainder is used to manufacture a variety of chemical products, including silicones and agricultural chemicals. Thus, most emissions of methyl chloride must come from the stacks of the factories where it is used as a feedstock.

The exact amount of methyl chloride used as a chemical feedstock is not known. Generally, emissions of chemicals used as feedstocks are a small fraction of total use. If industrial emissions of methyl chloride represent less than 10 percent of its use, this yields an upper-bound estimate of 0.05 million tonnes per year for methyl chloride emissions. The actual emissions from feedstock uses are likely to be lower. For instance, about 3 percent of emissions of carbon tetrachloride are estimated to be from feedstock uses (Simmonds et al. 1988). A similar percentage for methyl chloride would yield a release estimate of about 0.01 million tonnes. These estimates indicate that industrial emissions of methyl chloride are likely to be small relative to the 2–5 million tonnes of total emissions.

Emissions from Biomass Burning

Most biomass burning is associated with the clearing of land for agricultural use, the burning of savannas, and the use of biomass as a fuel. There is also a contribution from natural forest fires. Biomass generates methyl chloride emissions only when it is burned in smoldering low-temperature fires. In high-temperature fires, such as combustion in modern boilers, little or no methyl chloride is emitted.

Biomass burning in the Amazon region is a large proportion of the total. Furthermore, dramatic increases have occurred in this region, while biomass combustion elsewhere has remained relatively constant (Skole and Tucker 1993).

Biomass Burning Outside the Amazon Region

To determine the anthropogenic contribution to biomass burning, it is useful to account for the amount of forest lost to natural fires. Unfortunately, accurate worldwide data on such fires are not available. Moreover, some forest fires are not natural, but are due to accidents arising from human activity. It has been estimated that temperate and boreal wildfires account for approximately 4 percent of the biomass burned each year (Crutzen et al. 1979, p. 254, table 1). Thus, it appears that natural wildfires are likely to account for only a small proportion of the biomass burned annually, especially in view of the lower incidence of wildfire in the humid tropics. Because the data are limited and because our aim here is to produce a first approximation of the anthropogenic contribution to methyl chloride emissions, we will ignore the contribution of natural wildfires.

The primary difficulties in estimating anthropogenic methyl chloride emissions from biomass burning stem from the sheer diversity of biomass that is burned and the lack of reliable data regarding each kind of biomass. The laboratory and *in situ* measurements that have been performed indicate that the methyl chloride emissions per unit of carbon burned in various kinds of biomass vary greatly with the type of biomass and with the conditions of combustion.[3]

In order to generate an estimate of the annual emissions, we selected a range of 30–80 parts of methyl chloride to a million of carbon dioxide on a volume basis (or about 34–90 parts of methyl chloride to a million of carbon dioxide on a weight basis) in order to represent the fact that the biomass burned is some mixture of wood, agricultural residues, dead leaves, etc.[4]

Biomass combustion estimates typical of the late 1970s and the 1980s, representing a variety of biomass types (other than Amazon forest biomass), range from approximately 4.5 billion to 7.0 billion tonnes of biomass burned with a mean of 5.75 billion tonnes (Andreae 1991, p. 8, tables 1.2 and 1.3). The mean estimate corresponds to carbon emissions of about 2.6 billion tonnes, equivalent to about 9.5 billion tonnes of carbon dioxide.

Using the above data on biomass burning and the ratios of methyl chloride to carbon dioxide yields a range of approximately 0.25–1.04 million tonnes per year of methyl chloride emissions due to biomass burning outside of the Amazon region.

We do not expect methyl chloride emissions to increase as fast as population, since the urban population in the Third World is increasing faster than rural population and since there has been a trend away from wood and straw as fuels. We will assume as a first approximation that changes in methyl chloride with population have been small enough to be ignored in the context of the order-of-magnitude calculations that we are making here. Therefore, we assume that these emissions have been constant.

Methyl Chloride Emissions from Amazon Forest Burning

The burning of the Amazon forest on a large scale is thought to have arisen during the late 1970s and the 1980s. It is believed that 28,600 square kilometers (km^2) of forest were lost in 1975, and more than 77,000 km^2 in 1978 (Henderson-Sellers 1987, p. 472). Deforestation at the latter rate appears to have continued through much of the 1980s; the estimate for 1987 was approximately 75,000 km^2.[5]

As before, it is necessary to first estimate the amount of carbon dioxide being released from biomass burning in the Amazon. Tropical forests are estimated to have a stock of about 200 tonnes of carbon per hectare (Revelle 1987, p. 56). Combining this with the 1987 estimate of Amazon forest burned (75,000 km^2) implies that approximately 4.4 billion tonnes of carbon dioxide would be released, allowing for partial combustion.[6] This is about 2.7 billion tonnes of biomass, placing our total at 7.2–9.7 billion tonnes. We use the 1987 estimate to give an idea of the peak contribution that burning of this scale may be making.

In the absence of data on ratios of methyl chloride to carbon dioxide specific to Amazon forest burning, the range (34–90 ppm by weight) used for all other biomass produces an estimate of the contribution of the Amazon forest burning to methyl chloride of approximately 0.15–0.4 million tonnes per year in 1987.

Summary: Anthropogenic Emissions of Methyl Chloride

Table 10.1 summarizes our estimates of the anthropogenic emissions of methyl chloride relative to total emissions of 2–5 million tonnes per year. To convert annual emissions into a mixing ratio, we equate the figure for total emissions to 600 pptv, assuming that annual emissions are in a steady state. This is a reasonable assumption since emissions are probably changing slowly relative to the lifetime of methyl chloride.

It is evident from table 10.1 that the uncertainties associated with these estimates are very large. Nevertheless, the total anthropogenic contribution to methyl chloride emissions is significant. The main contribution to anthropogenic emissions is the highly dispersed activities in the Third World that involve the burning of biomass. One-sixth to one-third of this is due to Amazon forest burning; almost all the rest is due to biomass burning outside this region.

Table 10.1
Estimated contemporary methyl chloride emissions.

Source	Emission (million tonnes)	Concentration (pptv)	Percentage of total CH_3Cl	Percentage of total Cl
Industrial	0.01	1–3	0.2–0.5	0.03–0.09
Biomass, except Amazon	0.25–1.04	30–310	5–52	0.86–8.9
Amazon	0.15–0.4	18–120	3–20	0.5–3.0
Total anthropogenic (rounded)	0.4–1.44	48–430	8–72	1.4–12
Total emissions	2.0–5.0	600	100	17
Natural emissions (rounded, item 5 less item 4)	0.56–4.6	170–550	28–92	5–16

If the midpoint of the ranges associated with the anthropogenic and total methyl chloride emission estimates is considered, the amount of anthropogenic emissions is approximately 26 percent of total methyl chloride emissions worldwide. This proportion corresponds approximately to published estimates ranging from 22 to 46 percent (Andreae 1991, p. 13, table 1.6; Lobert et al. 1991, p. 301, table 36.6; Maño and Andreae 1994).

This calculated share of total methyl chloride emissions means that anthropogenic methyl chloride due to the combustion of biomass is an important component of anthropogenic chlorine levels in the atmosphere, amounting to about 160 pptv out of a 1989 total of 3500 pptv.

Natural Emissions

As we have discussed, natural emissions are thought to come primarily from natural forest fires and the oceans. The contribution of natural forest fires is probably small. This leaves the oceans as the predominant natural source. Although there are measurements of methyl chloride emissions from the oceans, there is considerable uncertainty as to how the scattered measurements that have been made can be used to derive the entire emissions of the oceans.

Measurements of methyl chloride emissions in the Eastern Pacific region, along the coasts of South and North America, have resulted in an estimate of total natural oceanic emissions of 5 million tonnes per year (Singh et al. 1983). This figure is at the upper limit of the total methyl chloride emission range presented here. Though this may represent an overestimate of the oceanic emissions, it indicates that the upper end of the anthropogenic emission range estimated in table 10.1 is probably too high.

Assessing the Relative Importance of Methyl Chloride

Table 10.2 presents the relative contributions to total atmospheric chlorine from methyl chloride and the other ODCs. As is evident, the most important sources of anthropogenic emissions are industrial compounds, and among these CFC-11, CFC-12, carbon tetrachloride, and methyl chloroform are the most important, in descending order. Thus, the international process that focused on the phasing out of these and related

Table 10.2
Contributions of various compounds to total atmospheric chlorine, 1989 (derived from WMO 1991, p. 1.4, table 1-1).

Chemical name	Chemical formula	Contribution to Cl (pptv)	Percentage of total Cl
CFC-11	CCl_3F	785	22.6
CFC-12	CCl_2F_2	906	26.1
CFC-113	CCl_2FCClF_2	192	5.5
CFC-114	$CClF_2CClF_2$	35	1.0
CFC-115	$CClF_2CF_3$	5	0.1
HCFC-22	$CHClF_2$	110	3.2
Carbon tetrachloride	CCl_4	428	12.3
Methyl chloroform	CH_3CCl_3	405	11.7
Halon-1211	CF_2ClBr	2	0.1
Methyl chloride[a]	CH_3Cl	~160	~4.6
Total anthropogenic		3028	87.3
Natural methyl chloride		~440	~12.7
Total chlorine (rounded)		3470	100

a. Anthropogenic only.

compounds did address the most important sources from a global perspective.

However, table 10.2 shows that the present contribution of anthropogenic methyl chloride is as important as that of the most important CFC solvent, CFC-113, and that it is much more important than some of the currently regulated compounds. (We should note, however, that the growth rates and lifetimes of these other chemicals are large.) Thus, we must include anthropogenic methyl chloride in order to address the totality of the problem of ozone-depleting chlorine emissions.

Although anthropogenic methyl chloride emissions are only a few percent of total ozone-depleting chlorine emissions, their relative importance within the Third World is much greater, both because per-person use of halogenated industrial compounds is far lower in the Third World than in the OECD countries and because biomass burning is widespread in the Third World.

Overall, methyl chloride was about 26 percent of the total concentration of ozone-depleting compounds attributable to emissions from the Third World in 1987, somewhat lower than the concentration resulting from CFC-11 and CFC-12 together and comparable to the contribution from carbon tetrachloride (Makhijani et al. 1988). However, the use of ozone-depleting compounds has been increasing rapidly in many countries of the Third World. Thus, the relative share of methyl chloride in these emissions is declining, quite rapidly in some cases.

The role of methyl chloride is probably larger in China, India, Brazil, and the countries of tropical Africa than the average for the Third World. In China and India, the most populous countries in the world, biomass is a common fuel. Besides large use of fuelwood for residential, commercial, and industrial energy applications, considerable savanna burning occurs in tropical Africa. This is a principal contributor to the estimates of biomass burning. Finally, a large proportion of Amazon forest burning occurs in Brazil. There is relatively little burning of biomass as a fuel in Latin America (on a scale that matters for calculations of global totals), since modern fuels are used for most applications even in rural areas.

No resources are currently being allocated to reduce methyl chloride emissions, nor is it a priority among the various investment strategies to reduce emissions of ozone-depleting compounds from the Third World. Policies that ignore the Third World's contribution of methyl chloride from biomass burning to atmospheric chlorine levels are missing one of the most important components of emissions from many of these countries.

We recognize that biomass burning, especially use of biomass as a fuel, is a very difficult and complex issue affecting the lives of hundreds of millions of people and the jobs and livelihoods of millions. Yet we should note that reducing biomass burning in the Third World in order to protect stratospheric ozone is compatible with attempts to improve the overall health and welfare of the poor within the Third World. For instance, smoldering fires used for cooking are responsible for increased morbidity and mortality among women in the Third World. Burning of the Amazon forest reduces biodiversity and adds considerably to global carbon dioxide emissions. In other words, ozone-layer depletion adds a strong reason for a more urgent policy on the matter of biomass burning, which is needed for a large number of pre-existing environmental and human health reasons.

10.3 Methyl Bromide

Current Use

Like methyl chloride, methyl bromide has both a natural and an anthropogenic component. It is generated naturally by oceanic biological processes. This natural component, like its chlorinated counterpart, participates in the natural bromine cycle and thereby helps maintain the natural balance of the ozone layer. The current concentration of methyl bromide in the troposphere is approximately 9–13 pptv, making it the most important source of organic bromine, with considerably greater concentrations than any of halons (UNEP 1992, section II, p. 5). It is estimated that there is about 1.3 times more methyl bromide in the northern hemisphere than in the southern hemisphere.

A considerable portion of the methyl bromide in the atmosphere may be due to emissions from industrial and agricultural applications. The main application of methyl bromide is as a pesticide in soil fumigation. It is also used as a space and structural fumigant, a chemical feedstock, and a commodity fumigant for grains, fruits, and flowers. By volume, methyl bromide is the second largest pesticide used in the world (CATs 1992).

Figure 10.1 gives a breakdown of the industrial uses of methyl bromide worldwide in 1990. As shown in the figure, soil fumigation accounts for the greatest share. It is estimated that global methyl bromide production for fumigant use in 1990 was approximately 63 kilotonnes, of which about 30 kilotonnes were released (UNEP 1992). The amount not released is destroyed by chemical reactions during use, such as those that occur within soils.

Like methyl chloride, methyl bromide is emitted during the combustion of biomass. Measurements have been made of the amount of methyl

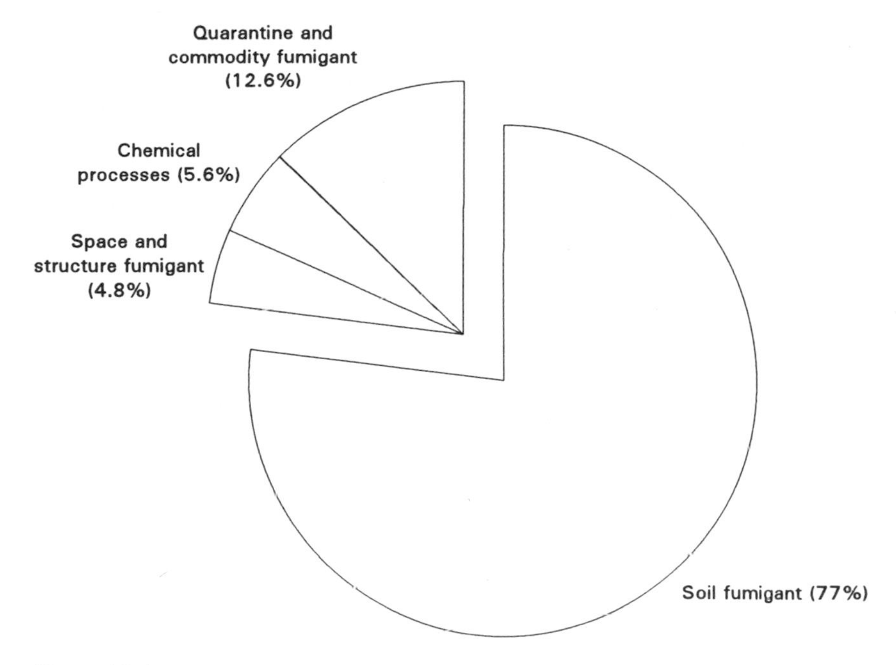

Figure 10.1
Global industrial use of methyl bromide, 1990 (derived from data in UNEP 1992) (total: 647 kilotonnes).

bromide released during the combustion of material such as forest and savanna grass relative to the amounts of carbon dioxide and methyl chloride (Maño and Andreae 1994). Using estimates of global carbon dioxide release from biomass combustion, the amount of methyl bromide emitted from this source has been estimated to be 10–50 kilotonnes per year (ibid.).[7]

Methyl bromide is also a component of automobile exhaust where ethylene dibromide is added to gasoline, as it still is in many parts of the world (Bruno 1991). Methyl bromide emissions from automobile exhaust are estimated to be between 5632 and 16,236 tonnes per year (UNEP 1993b, p. 3-26).

Figure 10.2 shows the worldwide emissions of methyl bromide by source. Assuming that the concentration of methyl bromide in the atmosphere is in a steady state and that the globally averaged lifetime is

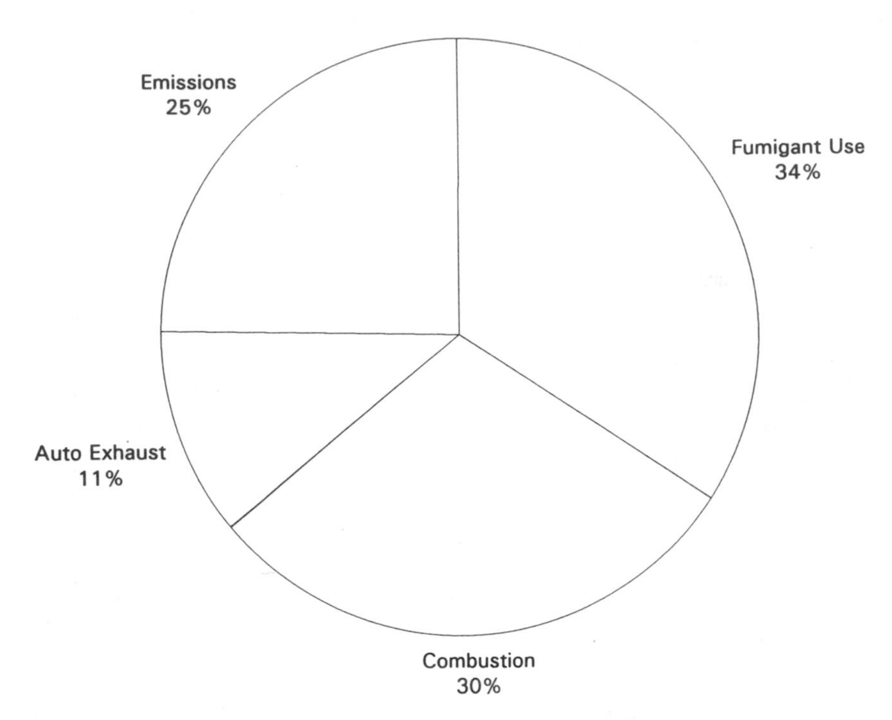

Figure 10.2
Global methyl bromide emissions, 1990 (derived from data in UNEP 1992, UNEP 1993b, and Maño and Andrae 1994) (total: 100 kilotonnes).

approximately 2 years, emissions would have to total approximately 100 kilotonnes per year (Khalil et al. 1993; Reeves and Penkett 1994; Singh and Kanakidou 1994). There is considerable debate over the amount of global anthropogenic emissions of methyl bromide and its apportionment among the individual emission sources. The above estimates therefore, should be viewed as very approximate.

Alternatives

In view of the greater ozone-depleting efficiency of bromine and the toxicity of methyl bromide, alternatives to the industrial uses of methyl bromide must be fully explored. For example, methyl bromide can lead to respiratory failure and is a suspected human carcinogen. In the state of California, methyl bromide has caused more hospitalization and death than any other pesticide (JPR 1991; PAN 1992). The Netherlands phased out the use of methyl bromide completely in 1992 because of its toxicity (MBr Workshop 1992).

Alternatives to methyl bromide are somewhat limited. Many of the short-term chemical alternatives are toxic chemical pesticides and may present a risk to the public. The biological controls typically used by organic farmers may offer suitable alternatives. In the Netherlands, the elimination of methyl bromide use in glasshouse nurseries was accomplished by changing to artificial growth substrates (e.g. hydroponics) and steam sterilization (UNEP 1992, section III, p. 7). However, many pesticide alternatives are pest-specific, whereas methyl bromide eliminates essentially all pests. A combination of reduced methyl bromide use, some chemical substitution, steam and solar sterilization, and organically based pest-management methods can reduce methyl bromide use in the near term. Estimates made by the United Nations suggest that up to 90 percent of the existing uses of methyl bromide can be replaced by 1997 with such alternatives (ibid.).

Alternatives to the use of methyl bromide as a commodity fumigant are also somewhat limited at this time. Nonperishable and durable goods can be fumigated by controlled-atmosphere, biological, irradiation, and temperature-control procedures. Irradiation is controversial. Worker, food, and public safety will have to be carefully examined should irradiation be used. Only relatively short-lived isotopes should be used, owing to the problems of disposing of long-lived radionuclides. Improved sanitation and inspection may lessen the need for fumigation of many products. Methyl bromide fumigation of fruits and vegetables may be difficult to eliminate in the near term. Many Third World countries use methyl bromide fumigation to facilitate export of income-generating agricultural products. Alternative technologies and chemicals must be made available to such countries.

Recovery or emission control is an option when using methyl bromide as a commodity fumigant. A pilot project in New Zealand recovered between 95 and 97 percent of the methyl bromide used in a fumigation chamber. However, the cost of such a recovery method may be a barrier for Third World countries.[8]

A variety of nonchemical, structural pest-control options exist. Among them are high- and low-temperature (liquid nitrogen) treatments, physical barriers (e.g. sand around the perimeter of a foundation) and electrical shock for spot treatment of pests. Microwave irradiation and material replacement are additional options. Alternative chemical treatments such as sulfuryl fluoride and aluminum phosphide are available; however, like methyl bromide, they are highly toxic. In view of the small amounts of methyl bromide used and the variety of alternatives, use of methyl bromide in this application could be eliminated within a few years.

Recommended Phaseout

Given the limited alternatives, we tentatively recommend a phaseout of the production of methyl bromide for use as a fumigant by 1996 (except in the Third World where no viable alternatives exist—in such instances, an extension of the phaseout date to 1997 may be necessary). We further recommend that efforts at recovery and emission control, where possible, be instituted immediately.

The consumption of ethylene dibromide (EDB) in the United States, used as a lead scavenger in leaded gasoline, was reduced substantially during the 1970s and essentially eliminated during the 1980s, mainly because of its acute toxicity. Leaded gasoline is still used in parts of Europe and the Third World. The cost of refinery modernization and octane additives in the absence of regulation have presented a barrier to the elimination of EDB. Because the emissions of methyl bromide from this source are not insignificant and because leaded gasoline poses a serious threat to public health, we recommend the elimination of leaded gasoline by the year 2000 or as soon as it is feasible.

In view of the realization that methyl bromide emissions arise from biomass burning as well as from emissions of methyl chloride, this source of ozone-depleting substances must be examined much more closely. Even though the quantity of methyl bromide emitted is far smaller than the quantity of methyl chloride, its role in ozone depletion is greater because bromine is a more powerful ozone-depleting substance. As we have discussed, there are strong health and environmental reasons other than ozone depletion to greatly reduce smoldering low-temperature fires.

In recognition of the considerable uncertainties associated with biomass burning and its importance to both public and environmental health, the next conference to revise the Montreal Protocol, to be held in Vienna in late 1995, should commission a large, multi-faceted effort toward research into its scientific, technological, social, economic, health, and environmental aspects. Only such an effort can be the basis for concerted action that will benefit the poor of the Third World and protect the environment.

III

Policies and Recommendations

11

The Existing Regulatory Framework

Efforts to regulate chemicals suspected of causing ozone depletion began very soon after the U.S. National Academy of Sciences completed its first review of the ozone-depletion theory in the mid 1970s. For the most part, initial concern gave way to governmental and corporate complacency in the early to mid 1980s. Since the publication in 1985 of the discovery of the Antarctic ozone hole and the subsequent steady stream of adverse news about stratospheric ozone depletion, regulations have continually tightened at all levels, from local to global, and corporate response has gone from moribund to energetic in many cases.

11.1 Early Efforts

In 1976 the National Academy of Sciences upheld the basic science of the Molina Rowland hypothesis, which claimed that the emission of chlorofluorocarbons into the atmosphere would lead to serious ozone depletion (NAS 1976). The first legislative action was taken in the United States at the state level. In June 1975 the Oregon state legislature passed a law banning nonessential uses of CFCs in aerosol devices. Numerous states and municipalities followed this action with a variety of chemical production and consumption controls and product bans in the decade that followed.

Action at the national level began when the Environmental Protection Agency, the Food and Drug Administration, and the Consumer Product Safety Commission announced a timetable, to begin in late 1978, banning the use of chlorofluorocarbons in nonessential aerosol products. By August 1977, Congress had amended the Clean Air Act to give the EPA official regulatory power in this matter. This primarily affected the aerosol industry. However, by that time, numerous alternatives had been developed to aerosol propellants in applications ranging from hair sprays to pesticides. Though aerosol producers had initially claimed that regulation would ruin their industry, the shift away from CFC propellants caused little, if any, disruption. A spokesman for Union Carbide stated, concerning regulatory controls: "Any impact would be very minimal since we've been working on the problem for some two and a half years." (Cagin and Dray 1993, p. 213) Positive action in the face of actual or near-certain impending regulation was to become a pattern in industry's response to the crisis of ozone depletion.

In other parts of the world, legislative action was also limited. Four countries enacted nearly total bans on the use of CFCs in aerosols in the late 1970s: the United States, Sweden, Canada, and Norway. In 1980, the European Community (EC) adopted a 30 percent reduction in CFC aerosol consumption from 1976 levels. In 1982, they adopted a production capacity freeze of these compounds at 1980 levels. The EC's action, however, was considered a hollow effort by many. European sales of CFCs for aerosols had already declined by approximately 28 percent since 1976, and the production-capacity freeze was such that output could increase by more than 60 percent over 1980 levels—approximately 480,000 tonnes—before running into the legal limit (Benedick 1991, p. 25; Haas 1992, p. 201). Action by Japan and the rest of the world at that time was minimal or nonexistent.

Little regulatory action concerning ozone-depleting compounds occurred in the early 1980s. At the same time, producers of ODCs continued to vigorously pursue new markets and expand existing ones for their products, resulting in the development of thousands of applications, major and minor, by the time ODC regulation was put back on the agenda. By 1983, production of ODCs was greater than ever, despite the considerable reduction in production caused by the ban on most aerosol applications in the United States and a few other countries.

11.2 International Efforts

The Vienna Convention

The United Nations Environment Programme (UNEP) was one agency that did not slacken its efforts in the early 1980s. As early as 1977, UNEP had convened the International Conference on the Ozone Layer. Attended by 32 countries, this conference released the World Plan of Action on the Ozone Layer, which called for a treaty on ozone-layer protection. In 1981, UNEP began work on such a treaty through formation of the Ad Hoc Working Group of Legal and Technical Experts for the Elaboration of a Global Framework Convention for the Protection of the Ozone Layer.

The outcome of these international efforts was the 1985 agreement known as the Vienna Convention for the Protection of the Ozone Layer.[1] Although this agreement did not mandate controls on production, consumption, or emissions, it created a process by which such regulations could be created by the governments of countries meeting under the auspices of the convention. As described by Richard Benedick, the leader of the U.S. delegation to the UNEP negotiations,

> The convention created a general obligation for nations to take "appropriate measures" to protect the ozone layer (although it made no effort to define such measures). It also established a mechanism for international cooperation in research, monitoring, and exchange of data on the state of the stratospheric ozone layer and on emissions and concentrations of CFCs and other relevant chemicals. These provisions were significant because, before Vienna, the Soviet Union and some other countries had declined to provide data on CFC production. Most important, the Vienna Convention established the framework for a future protocol to control ozone-modifying substances. (Benedick 1991, p. 45)

In addition, the Vienna Convention specified that parties to the convention would meet at regular intervals. This mechanism was especially important in that it allowed for routine consideration of specific ODC production and consumption curbs for the future.

The Montreal Protocol

The publication of the discovery of the Antarctic ozone hole in 1985 led to fundamental revision of theories about how fast and to what extent ODCs could damage the ozone layer (Farman et al. 1985). This provided the scientific background for the first international agreement setting forth a specific program to curb the use of these compounds. This agreement, known as the Montreal Protocol on Substances that Deplete the Stratospheric Ozone Layer, was signed by 24 countries in September 1987 and took effect on January 1, 1989.[2]

The Montreal Protocol specified a timetable for reducing the production and consumption of CFCs and halons. It also set restrictions on trade in these compounds and in products containing them or made with them. The protocol had been ratified by 148 countries as of March 1995.[3]

The control provisions of the Montreal Protocol were as follows: After a freeze at 1986 levels in mid 1989 and a 20 percent cut in 1993, production and consumption (excluding recycled materials) of CFC-11, CFC-12, CFC-113, CFC-114, and CFC-115 were to be reduced to 50 percent of the 1986 levels by 1998. Production of halon-1211, halon-1301, and halon-2402 was to be brought back to 1986 levels by 1992.[4]

Developing countries with a consumption level below 0.3 kilogram per capita, called "Article 5 countries" (and referred to as "Third World countries" in this book), were given an additional 10 years to comply with the restrictions applied to the industrialized ("non-Article 5") countries and were allowed to increase their consumption to 0.3 kg per capita in the interim.[5] Parties in the industrialized countries were allowed to exceed the stated control provisions by anywhere from 10 to 15 percent in order to meet the basic domestic needs of the Third World countries during the additional time the latter have to comply. Furthermore, a pledge of technical assistance and financial support was made to Third World parties to encourage participation.

The Montreal Protocol required parties to report production and consumption of regulated substances annually. This was an important practical provision because in the mid 1980s there was very little knowledge of the actual use of ODCs in many countries. Even production data were highly uncertain. This situation has improved somewhat as a result of these reporting requirements. However, there continue to be significant gaps in the reporting and the availability of data. Finally, the protocol specified trade sanctions and restrictions for parties who violated the treaty's provisions and obligated signatories to impose sanctions and restrictions on non-parties.

The Montreal Protocol did not include all the ODCs now considered threats to the stratospheric ozone layer. In particular, carbon tetrachloride and methyl chloroform were absent, as were HCFCs. Methyl chloride and methyl bromide were not considered either. This last omission has turned out to be a considerable problem, since emissions of methyl bromide are thought to have grown considerably in the late 1980s. Furthermore, it turns out that the ozone-depleting ability of methyl bromide is substantial. The 20-year ozone-depletion potential of methyl bromide relative to CFC-11 is about 2.8, approximately four times its steady-state ODP.

It was soon apparent that, even with regard to the compounds it did regulate, the Montreal Protocol was seriously deficient. Less than six months after the protocol was signed, and well before it was ratified by enough parties to put it into legal effect, a new scientific assessment showed that ozone depletion was far worse than models had projected.

In March 1988, the Ozone Trends Panel, a group of distinguished scientists who had been studying ozone measurements, announced that significant ozone depletion had already occurred over populated areas in the northern and southern middle latitudes. It also reached a firm conclusion that stratospheric ozone depletion was being caused by ODCs (WMO 1988). The controversies and uncertainties regarding the causal relationship between ODCs and ozone depletion were definitively resolved by the assessment and by measurements that showed a close anticorrelation between levels of chlorine monoxide and spatial variations in stratospheric ozone levels.

Soon after the Ozone Trends Panel released its summary report, DuPont, the world's largest CFC producer, announced that it would stop producing CFCs by the year 2000 (Shabecoff 1988). A new date for reconvening the international conference to amend the Montreal Protocol was set for June 1990. Research on alternatives by producing and consuming companies and by new entrants into the market was accelerated greatly.

In the summer of 1988, Sweden passed a pioneering and historic law that required the phasing out of CFCs by the end of 1994 and of halons "as soon as possible."[6] This law was prepared in consultation with industry.

The London Amendment

In June 1990, the parties to the Montreal Protocol met to reassess the provisions of the treaty. Various proposals for a complete phaseout of

ODCs had been made by environmental organizations, portions of the chemical industry, and national governments since the signing of the Montreal Protocol in 1987. Given the increasing stratospheric ozone loss, the fact that alternatives for most applications were rapidly becoming available, and the pressure from the environmental movement, the London meeting resulted in significant amendments and adjustments to the original Montreal Protocol.[7]

The changes made to the Montreal Protocol as a result of the London meeting made it the first international treaty in which governments agreed to eliminate a group of chemicals in order to protect the environment. The core of these changes was the phasing out of CFCs and halons by the year 2000. Carbon tetrachloride and methyl chloroform were added to the list of controlled chemicals and scheduled to be phased out in 2000 and 2005, respectively. As in the original protocol, Third World parties were granted an additional 10 years and production in the industrialized countries was allowed to exceed the control limits by 10–15 percent in order to meet the basic domestic needs of Third World parties.[8] The phaseout schedule of the Montreal Protocol, with the adjustments and amendments agreed to in London, is given in table 11.1.

Among the other significant accomplishments of the London meeting was the creation of a multilateral fund to be used to assist the Third World in the transition away from the ODCs presently in use. Aside from meeting the incremental costs to Third World countries, the fund was also established to finance country studies, technical assistance, and training programs.

The fund was formally established in 1991 with $160 million for the first 3 years. In addition, there was a provision to increase the fund from 160 million U.S. dollars by $40 million each upon accession to the treaty of China and India. Though those countries became parties (in 1991 and 1992, respectively), they argued that the industrialized countries had created the problem of ozone depletion and should pay the entire cost of phasing out ODCs. India estimated that the cost of a phaseout would be far larger than the $40 million allocated to it from the $240 million fund (Billimoria & Co. 1990). As of September 1993, approximately $125

Table 11.1
The Montreal Protocol as amended and adjusted in London in 1990. (All provisions are for production and consumption.)

Industrialized countries	Third World
CFCs	
Freeze in mid 1989	Freeze in mid 1999
50% reduction by 1995	50% reduction by 2005
85% reduction by 1997	85% reduction by 2007
100% reduction by 2000	100% reduction by 2010
Halons	
Freeze in 1992	Freeze in 2002
50% reduction by 1995	50% reduction by 2005
100% reduction by 2000	100% reduction by 2010
CCl_4	
85% reduction by 1995	85% reduction by 2005
100% reduction by 2000	100% reduction by 2010
CH_3CCl_3	
Freeze in 1993	Freeze in 2003
30% reduction by 1995	30% reduction by 2005
70% reduction by 2000	70% reduction by 2010
100% reduction by 2005	100% reduction by 2015

million had been contributed out of the $240 million pledged for the entire interim fund (OzonAction 1993a, p. 5). Out of the $127 million pledged for the years 1991 and 1992 combined, $24 million had not been received (UNEP 1993a, p. 19).

HCFCs, notably HCFC-22, HCFC-141b, and HCFC-142b, were not regulated, though the London Amendments recognized the problem and set a non-binding target of eliminating HCFCs by the year 2040. Unfortunately, methyl chloride and methyl bromide were not addressed at all in the London Amendments. The combined anthropogenic component of these compounds contributes approximately 15 percent of the total chlorine and bromine from human origins currently in the stratosphere in terms of weighted equivalent chlorine.[9]

The London Amendments entered into force on August 10, 1992, with the accession of Chile. By March 1995, 102 countries had ratified these amendments to the Montreal Protocol.[10]

The Copenhagen Amendment

The next significant meeting of the parties to the Montreal Protocol occurred in late 1992 in Copenhagen (Ozone Secretariat 1993). Ozone levels declined dramatically in the early 1990s, and heterogeneous depletion was strongly suspected to have occurred over the Arctic. As with the period leading up to the London meeting, the two years preceding the Copenhagen negotiations witnessed further development of ODC alternatives. Furthermore, it was realized that since the ozone-depleting ability of methyl bromide was of the same order of magnitude as that of CFCs, methyl bromide had to be included in the control provisions. With these developments as a backdrop, further adjustments and amendments to the Montreal Protocol were made. The control schedule is given in table 11.2.

The phaseout date for all the compounds originally scheduled for elimination in the year 2000 was moved up to 1996, and controls on both HCFCs and methyl bromide were established for industrialized countries. The additional 10 years for Third World countries and the limited production allowance in the industrialized countries to meet the basic domestic requirements of the Third World remained.

There was considerable debate over the control of HCFCs and methyl bromide at the Copenhagen meeting. The large chemical manufacturers, present in considerable numbers, argued that premature elimination of HCFCs would remove commercial incentives for producers to make the chemicals and for manufacturers to use them in their products and processes.[11] Without these compounds, they argued, companies would continue to use CFCs and other fully halogenated ODCs. However, this overlooked the variety of non-ozone-depleting alternatives that had been developed, many of which are listed in this book. Because many of these alternatives were based on inexpensive, widely available chemicals or were not chemically based, they were absent from the industry perspective. In the end, a schedule that extended production to 2030 was agreed upon, with interim reductions and a consumption cap in the year 1996.

The amount of HCFC consumption allowed by the cap is equivalent to 3.1 percent of the 1989 CFC consumption plus the total 1989 HCFC

Table 11.2
The Montreal Protocol as amended and adjusted in Copenhagen in 1992. Except where specifically stated otherwise, all provisions are for production and consumption and base-year levels are the same as in table 11.1.

Industrialized countries	Third World
CFCs	
Freeze in mid 1989	Freeze in mid 1999
75% reduction by 1994	75% reduction by 2004
100% reduction by 1996	100% reduction by 2006
Halons	
Freeze in 1992	Freeze in 2002
100% reduction by 1994	100% reduction by 2004
CCl_4	
85% reduction by 1995	85% reduction by 2005
100% reduction by 1996	100% reduction by 2006
CH_3CCl_3	
Freeze in 1993	Freeze in 2003
50% reduction by 1994	50% reduction by 2004
100% reduction by 1996	100% reduction by 2006
HCFC (consumption)	
Production cap (3.1%) in 1996	pending
35% reduction by 2004	
65% reduction by 2010	
90% reduction by 2015	
99.5% reduction by 2020	
100% reduction by 2030	
CH_3Br	
Fumigant uses to be frozen in 1995 at 1991 levels[a]	pending

a. Post-harvesting uses exempted.

consumption, weighted by the steady-state ozone-depleting potential (S-S ODP) of each compound. This can be expressed as

$$\sum C^{1996}_{HCFC_i} \cdot ODP_{HCFC_i}$$

$$= 0.031 \cdot \left[\sum C^{1989}_{CFC_i} \cdot ODP_{CFC_i}\right] + C^{1989}_{HCFC\text{-}22} \cdot ODP_{HCFC\text{-}22},$$

where C is an amount of ODC consumption, the subscript denotes the specific ODC, the superscript is the year in which the production amount is being considered, and ODP is the steady-state ozone-depleting potential.[12]

The specific phaseout schedule for the last 10 years of HCFC consumption was lobbied for by industry representatives and the U.S. delegation to the negotiations. Because the United States is a particularly large user of commercial-scale air conditioning systems, which have lifetimes of roughly 40 years, the continued (though small) production scheduled between 2020 and 2030 was to guarantee the availability of HCFCs to meet the servicing requirements of these devices.

Methyl bromide used in fumigation, except in post-commodity applications, was included in the list of regulated chemicals at Copenhagen only after considerable debate. In this instance, the industrialized countries, led by the United States, were pushing for a production and consumption freeze in 1995 and a 25 percent reduction in 2000. Israel (a large producer of methyl bromide) and a number of developing countries objected to any regulation, citing the use of methyl bromide as essential in the Third World. In the end, a freeze in 1995 at 1991 levels was agreed upon. This did not include post-commodity uses, and, like the HCFC regulations, it applies only to industrialized countries. The precise nature of methyl bromide and HCFC controls to be adhered to by Third World parties will be determined at the seventh meeting of the parties, to be held in Vienna in late 1995.

The multilateral fund was made a permanent institution, and the funding level was established at $510 million for the period 1994–1996 (OzonAction 1993b, p. 5). Since its inception, the executive committee of the multilateral fund has involved the World Bank, UNEP, the United Nations Development Agency (UNDP), and the United Nations Industrial Development Organization (UNIDO) in implementing the activities of the fund.

By March of 1995, the 44th instrument of ratification of the Copenhagen Amendments to the Montreal Protocol had been received. The Copenhagen Amendments entered into force on June 14, 1994.[13]

Though the amendments and adjustments agreed to in Copenhagen represent considerable progress, there is much within the protocol that can be improved. It has now been suggested that, aside from its use as a fumigant, methyl bromide may be emitted from biomass burning and from the combustion of leaded gasoline in automobiles. This has effectively enlarged the total amount of methyl bromide emissions thought to arise from controllable human activities. Because the ozone-depleting ability of bromine is much larger than that of chlorine, a global phaseout of all industrial uses of methyl bromide must be enacted. Furthermore, methyl chloride, also a by-product of biomass burning, has yet to be discussed within the context of the Montreal Protocol.

Another crucial issue that must be considered for the next revision of the protocol (due in 1995) is minimizing releases from the existing banks of ODCs. These releases are large and will likely continue growing until ODCs are phased out completely. These banks are a large source of current emissions and will continue to leak to the atmosphere.

Finally, the technical potential exists for a faster phasing out of ODCs in most applications. With a real commitment to technology transfer and international cooperation, the longer phaseout period allowed for Third World countries can be eliminated in most cases, thereby putting these countries on essentially the same phaseout schedule as the industrialized countries. Similarly, the phaseout schedule for HCFCs can be accelerated in all parts of the world.

Because both the number of parties to the Montreal Protocol and the number of substances regulated are growing, the financial mechanisms will not be adequate to meet the needs of the Third World. Some estimates have placed the required funding at approximately $2 billion between 1997 and 2010, far greater than the funding levels agreed to thus far (Markandya 1992, p. 6).

11.3 The U.S. Clean Air Act and Regulatory Actions in Other Countries

In 1990 the U.S. government amended the U.S. Clean Air Act to include provisions for the regulation of ODCs (US CAA 1990). Specific rule-making for this portion of the Clean Air Act, labeled Title VI, was handed over to the U.S. Environmental Protection Agency. In comparison to the Montreal Protocol, Title VI is more comprehensive, dealing with issues such as the labeling, recycling, and disposal of products and the market allocation of production credits. Like the Montreal Protocol process, it allows for the promulgation of stricter regulations should evidence show that the legislated phaseout dates are not sufficient to protect the ozone layer.

A key difference between the U.S. control schedule and the Montreal Protocol concerns the HCFCs and methyl bromide. Methyl bromide production and consumption was to have been frozen at 1991 levels in the year 1994, with a complete phaseout scheduled for the year 2001 (FR 1993b). For most HCFCs, production and consumption are to be frozen in 2015 and eliminated by 2030.[14] All uses of HCFCs in new equipment will be banned by 2015 (2010 for HCFC-22 and 142b) except uses as feedstock or as a domestic refrigerant. Particular compounds are to be phased out sooner. HCFC-141b is to be eliminated by the year 2003, and both HCFC-22 and HCFC-142b are to be frozen in 2010 and phased out by 2020. Furthermore, after the freeze takes effect for HCFC-22 and HCFC-142b, these compounds can be used only to service equipment manufactured before 2010.

In addition to somewhat stricter regulation of HCFCs and methyl bromide, the U.S. Clean Air Act contains a number of other provisions.[15] Beginning in 1992, it became necessary for persons servicing automobile air conditioners to be trained and authorized to use recycling equipment during service, maintenance, and repair (FR 1992). A one-year delay was allowed for small-volume automobile service shops. In mid 1992 it became illegal to knowingly dispose of or vent refrigerants during servicing, maintenance, or repair of appliances or industrial-process refrigeration (EPA 1993). Beginning in early 1993, all products made with or containing ODCs had to be labeled as such (FR 1993a). All ODC containers must also be labeled.

Many other countries have also taken actions that put them on a phaseout schedule faster than that set by the Copenhagen Amendments. For example, HCFC-22 use in Germany will be banned by 2000 except for use in equipment made before that year (GECR 1993). Denmark will phase out HCFCs by 2002 and methyl bromide between 1995 and 1998 (EW 1993). Generally, experience in Germany, Sweden, the United States, and elsewhere has shown that realistic, tough regulation encourages innovation, making the conduct of business more compatible with ozone-layer protection. In 1995 Sweden will ban the importing of refrigeration equipment containing CFCs from all countries, not just those that are not parties to the Montreal Protocol (GECR 1992e). The Swedish government also intended to ban all uses of HCFCs except those used in refrigeration and the production of insulating foams by January 1, 1995. Canada will ban HCFC production and importing by 2020 (GECR 1992b). Nonrecoverable uses of HCFCs (such as aerosols and foams) will be banned by 2010.

Many local governments in the United States have also taken action, spurred by the demands of nongovernment organizations, scientists, and popular organizing efforts. The first of many such local laws was enacted in 1989 by the town of Irvine, California (Irvine 1989). It was far more stringent in many respects than international efforts. For instance, this law was the first in the United States to include provisions for recycling of refrigerants. A number of other local and state ordinances and laws have been passed taking action on specific applications or approaching regulation more comprehensively.

Local legislation has two beneficial aspects. First, it can spur international and national efforts toward better and more comprehensive ozone-layer protection. Second, local legislation is often an essential complement to national laws and international treaties. This is particularly true of the adoption of regulations to recover and recycle ODCs from existing equipment, such as automobile air conditioners and refrigerators. Even when there are national laws in this regard, enforcement can only be effective with the wholehearted cooperation of local governments.

11.4 Taxes

In our discussion of the technologies that are available to phase out ODCs, we noted that government regulations and broad disincentives for the use of these compounds have played a very effective role in the development of effective and economical alternative technologies in the last few years.

Taxes are needed on the use of ODCs in the industrialized countries over the next several years, even as they are being phased out. There are several reasons for this.

The first reason relates to minimizing the amounts of ozone-depleting compounds used over the next few years. This will minimize future emissions. It will also reduce future costs of recovery of ODCs from banks. Such recovery and destruction is far more expensive than the costs of using these compounds in the equipment. Therefore, overall social cost minimization requires that taxes be imposed immediately on ODCs.

Secondly, taxes will discourage nonessential uses and encourage the more rapid development of alternative technologies. The rapid development of such technologies is important to economic success for corporations and countries. Companies that have lagged behind or insisted on the use of chemicals that may be banned with the next round of bad news will tend, in the long run, to incur higher costs and have smaller markets at home and abroad.

Finally, taxes can help raise revenues for assisting domestic industries that will suffer financial hardship in the rapid phaseout of ODCs. The revenues can also be used to enhance the multilateral fund to assist Third World countries in phasing out ODCs on essentially the same schedule as the industrialized countries.

12

Projections of Atmospheric Chlorine and Bromine Concentrations

The potential for ozone depletion beyond what has already been ensured by past releases of ozone-depleting compounds is intimately tied to the amount and pattern of future emissions. Because stratospheric ozone depletion is exhibiting a nonlinear response to the present chlorine and bromine burden in the atmosphere, any future emissions of chemical compounds that would contribute to this problem must be minimized.

In this chapter we examine the potential magnitude and timing of future atmospheric chlorine and bromine levels by constructing a model of ODC emissions under various control strategies. The reasoning behind the use of this approach was discussed in chapter 4. We start with a variety of future ODC-emission scenarios, translate these emissions into ODC concentrations in the atmosphere, and finally calculate the amount of chlorine and bromine contributed by each compound to total atmospheric chlorine and bromine abundance. By examining the trends in atmospheric chlorine and bromine under different policy constraints, the relative efficacy of policy measures aimed at limiting future ozone loss can be judged.

Future levels of ODC emission are difficult to predict. To do so requires a projection of future population growth patterns, ODC demand, a correct accounting of ODC banks and their emissions levels, and, of course, future regulatory constraints on production and consumption. These variables depend on, among other things, the rigor and universality with which regulations and treaties are implemented and the changing relationship among production, consumption, and emission. The ODC emission projections and the resulting atmospheric chlorine and bromine concentrations presented here are not intended to be predictive. Rather, they highlight the effects that various policy decisions can have on stratospheric ozone and the timing of its recovery.

12.1 Emission Scenarios, Constraints, and Assumptions

To examine the effects that various policies might have on ODC emissions and the atmospheric chlorine and bromine levels that result, we have constructed a model containing three emission scenarios. We refer to these as the **Copenhagen Amendments scenario**, the **Accelerated Phaseout scenario**, and the **Saving Our Skins scenario**. The primary differences among these ODC emission scenarios concern five regulatory issues:

the phaseout schedule of ozone-depleting compound production and consumption
the ODC production and consumption phaseout schedule followed by Third World countries[1]
the extent of future HCFC production
the future control or elimination of emissions from ODC banks
the future control of methyl chloride and methyl bromide emissions due to low-temperature biomass burning.

All of the scenarios are global in scope, representing all production and all emissions. Similarly, it is assumed that the policy measures advanced are adopted internationally and strictly adhered to. There are three reasons for this. First, annual, country-specific data are limited, making country-specific projections of ODC production and consumption difficult and prone to error. Second, as was shown in chapter 1, ozone depletion has been increasing in severity. Continued ozone loss and the initial signs of ultraviolet radiation's impacts will likely result in greater international participation and compliance. Third, the availability and reliability of ozone-safe technologies, chemicals, and processes have increased significantly in recent years and, in general, are proving much less costly than was originally thought. As the barriers of cost and accessibility are lessened and the cost of noncompliance (i.e., the direct cost of controlled chemicals due to taxation and supply limitations) is increased, interest in and compliance with international regulatory efforts are likely to increase.

Admittedly, comprehensive international participation has not been achieved for the most recent round of ODC control amendments to the Montreal Protocol, negotiated in Copenhagen. As we noted in chapter 11, nearly all the countries producing or consuming ODCs have ratified the original Montreal Protocol and the London Amendments; however, this is not yet the case for the more stringent Copenhagen Amendments. As with the London Amendments, ratification by many Third World countries may be slowed by the lack of adequate financial and technical assistance (Rosencranz and Milligan 1990). Compliance may be limited in many of the formerly socialist countries of Europe, since many are not currently slated to receive financial assistance to meet the incremental costs of ODC phaseouts. Thus, considerable efforts will have to be made to get the universal compliance that we have assumed in all three scenarios.

The time domain of the model we have constructed extends from 1985 to 2090. Annual production (considered equivalent to consumption) and emissions data from a variety of sources are used for the years 1985–1991; data on most compounds are available up to 1992.[2] After 1992, production and emissions are computed for each year.[3]

In all three of the scenarios, we have taken into account banked releases and prompt releases. Because of the magnitude of emissions from ODC banks, differences in how these compounds are used or disposed of in existing equipment will prove critical to future levels of ozone-depleting chlorine and bromine. Allowing the existing ODC banks to leak out over the next few decades can cause substantial emissions to continue despite production and consumption bans. We attempt to address this by explicitly including the growth of and the resulting emissions from the larger ODC banks for which data is available. These include CFC-11 banked in foams, CFC-12 banked in hermetically and non-hermetically-sealed cooling systems, CFC-113 banked in cooling systems, all banked HCFC-22, and both halon-1211 and halon-1301 banks. For all other processes and compounds, we assume that the production in a particular year is followed by consumption and releases within the same year. This corresponds reasonably well to the actual use patterns of these compounds. The calculations used to derive banked production and emissions are supplied in the appendix.

We have not modeled a few compounds with short atmospheric lifetimes that together may be contributing to the increase in atmospheric chlorine. These include chloroform ($CHCl_3$), dichloromethane (CH_2Cl_2, commonly called methylene chloride), tetrachloroethene (CCl_2CCl_2, commonly called perchloroethylene), 1,2-dichloroethane (CH_2ClCH_2Cl), and others. Many countries, including the United States, already regulate some of these compounds as toxic substances within provisions that do not involve ozone-layer protection.[4] Should these compounds be shown to contribute to ozone depletion, we recommend policies to minimize emissions.

Copenhagen Amendments Scenario

The base-line scenario we have constructed, the Copenhagen Amendments scenario, adopts all the control measures up to and including those specified in the November 1992 Copenhagen Amendments to the Montreal Protocol.[5] The Accelerated Phaseout and Saving Our Skins scenarios build upon this base line, incorporating more stringent controls. These two scenarios will be described later.

In order to apply the controls specified in the Copenhagen Amendments to our base-line scenario, the industrialized countries and the Third World are separated into two distinct ODC-producing and ODC-consuming entities. It is assumed that production is equivalent to consumption within each geographic block. In reality, there is a net transfer of ODCs from the industrialized countries to the Third World. However, this is taken into account in our model by both maintaining global totals for past years and assuming that Third World consumption will be met in the future by a combination of domestic production and the production allowed in the industrialized countries to meet the needs of the Third World.

It is difficult to predict the extent to which consumption in Third World countries will grow before the turn of the century—their first compliance date under the amended Montreal Protocol. Many of the applications for which ODCs are used are in growing demand in the Third World countries. Data from a few of these countries suggest that consumption may be growing at a significant rate and that domestic production capacity combined with allowed imports will enable continued growth in the near term (UNEP 1991f, p. 6-3). For example, total ODC consumption in India and Thailand tripled between 1985 and 1991 (Billimoria & Co. 1990; EIN 1991, p. 15). Consumption of ODCs in China, the Third World's largest producer and consumer, has been growing at approximately 11 percent per year (World Bank 1993). Particular compounds are experiencing even greater growth. For example, methyl chloroform consumption in India increased by a factor of 8 during the same period. Halon-1211 and halon-1301 consumption increased by a factor of 6.

Some analyses estimate that consumption of ODCs in the Third World may decrease in the late 1990s (UNEP/ExCom 1992). These analyses rely on the stated intentions of some Third World countries and transnational companies to reduce consumption or eliminate production capacity. Though such declines are possible, they appear unlikely in view of the trends evident in actual consumption since the Montreal Protocol came into being.

A published estimate of 1985 ODC production by geographic region is used to initialize the base-year production in the Third World.[6] This production is increased beyond the base year until the first compliance date for Third World countries is reached in the year 2000. The projected growth in production in the Third World countries is calculated by assuming that a small portion of the population in these countries (a relatively wealthy, high-ODC-using class) will increase its per-capita consumption of the regulated ODCs to the present average per-capita consumption in the industrialized countries, approximately 2 kg.[7] This consumptive demand is reduced by 20 percent due to the expected substitution of non-ozone-depleting HFCs and not-in-kind technologies. Even with this growth, Third World ODC consumption remains within the amended Montreal Protocol limits. Table 12.1 presents our assumptions regarding the projected population in the year 2000 and the portion of the population consuming at 2 kg per capita.

The total consumption calculated for the year 2000 (613,000 tonnes) is distributed among specific chemicals according to the 1985 Third World compound distribution with some adjustment for HCFC displacement.[8]

Table 12.1
Third World ODC consumption in 2000.

Region	Projected population[a] (millions)	Percent of population at 2 kg/person[b]	Total consumption[c] (1000 tonnes)
China	1300	5%	104
India	1000	5%	80
Rest of Asia	1300	10%	208
Africa	900	2%	29
Latin America and Caribbean	800	15%	192
Total	5300		613

a. Based on data in WRI 1992, table 16.1.
b. The proportions of people using ODCs at 2 kg/person are based on Gross Domestic Product (GDP) and GDP growth rate. A 5% population portion was assumed if GDP growth rate was 5% or more during the 1980s and GDP was under $500 per person in 1989. A 2% portion was assumed if growth rate was under 5% and GDP was under $500 per person in 1989. A 15% portion was assumed for Latin America, which has a GDP greater than $1500. An intermediate rate (between India and China and Latin America) of 10% was assumed for the rest of Asia, since this area includes countries with relatively high, intermediate, and low GDPs. Source for GDP data: WRI, table 15.1.
c. HFC and not-in-kind displacement have been accounted for.

Between the base year and the year 1992, the industrialized countries are allotted the remainder of the annual global ODC production. After 1992, production in the industrial countries is eliminated according to the provisions of the Copenhagen Amendments.

As was described in chapter 3, a number of alternative chemicals have been developed to serve as substitutes for the ODCs scheduled for elimination in 1996. Some of these alternatives, the hydrochlorofluorocarbons (HCFCs), are ozone depleters; others, the hydrofluorocarbons (HFCs), contain no chlorine and do not contribute to ozone depletion. The extent to which HCFCs deplete ozone varies from compound to compound, depending on the number of chlorine atoms per molecule and the rate at which particular HCFCs are destroyed in the lower atmosphere.

We make no assumptions in this work about the elimination of HFCs, since they do not affect the policies that we have discussed. However, we recognize that they are greenhouse gases, and we believe that we must make strong efforts to limit and reverse the buildup of greenhouse gases in order to avoid a crisis in this area similar to that which we are facing with ozone depletion. However, we also recognize that the immediacy and the severity of ozone depletion can be lessened by replacing ozone-depleting substances in existing equipment, where that is possible. HFCs could play an important role in this replacement in the next few years. They will also alleviate economic disruption in the refrigeration and air conditioning industry as we make the transition to technologies that are environmentally sound with respect to greenhouse-gas accumulation and ozone depletion. We believe that, with appropriate efforts at recovery and reuse of these compounds, emissions can be kept to a minimum. In addition, it is likely that overall gains in energy efficiency in the industrialized countries and elsewhere could dramatically reduce greenhouse gas emissions at significantly lower cost than short-term elimination of HFCs. This is a matter that must be addressed in more detail in the context of limiting greenhouse-gas concentrations.

Since the exact mix of HCFC production in the future is uncertain because of incomplete toxicity testing and price determination, three HCFCs are included in our analysis. One of these, HCFC-22, has been commercially produced for over 40 years and thus is treated somewhat differently from the other HCFCs included in the model scenarios. These other two HCFCs represent the as-yet-unknown mix of new HCFC production, which was begun in the early 1990s.[9] They are HCFC-123 (CF_3CHCl_2, with a 2-year lifetime and one chlorine atom) and HCFC-141b (CCl_2FCH_3, with a 10.8-year lifetime and two chlorine atoms). While chemical manufacturers are marketing a large variety of HCFC chemicals, HCFC-123 and HCFC-141b provide an adequate representation of the many compounds planned, considering the range of their atmospheric lifetimes and chlorine content.[10] These are the two factors of concern when calculating their contribution to atmospheric chlorine and bromine loading.

International regulations now contain HCFC production controls for the industrialized countries. The Copenhagen Amendments to the Montreal Protocol have specified a production limitation or "cap" in 1996, followed by a number of interim reductions and an ultimate elimination of production in 2030.[11]

An HCFC-control schedule for Third World countries has not yet been formally identified within the international negotiating framework. This control schedule will be determined at the 1995 meeting of the parties to the Montreal Protocol. However, to facilitate the construction of our base-line scenario, we assume that the production of each HCFC increases from a base-year amount to the year-2000 consumption target described previously. Production of HCFC-22 is begun in 1985, production of HCFC-123 and HCFC-141b in 1993. The production of these ODCs increases up to the year 2006, after which a production-elimination schedule identical to that specified for the industrialized countries is followed, though delayed by 10 years.[12]

HCFC-22 production between the model base year and the year 1992 in the industrialized countries accounts for the rest of annual global HCFC-22 production. After this, annual production increases at the historic rate of 7 percent per year up to 1996, the year of the HCFC production cap.[13] After this time, the Copenhagen Amendments' phaseout schedule is strictly adhered to in the scenario. Because HCFC-22 is used widely and much of its future production will be devoted to servicing existing equipment, the cap constraint is met by production controls on new HCFCs.

After new HCFC production in the Third World is accounted for, the industrial countries are allotted the remainder of the global HCFC production cap in 1996. This is divided evenly between HCFC-123 and HCFC-141b. After 1996, the reduction schedule specified in the Copenhagen Amendments is adhered to.

The two remaining ODCs included in the Copenhagen Amendments base-line scenario are methyl bromide and methyl chloride. Because of data limitations, the production of these compounds are not divided between the industrialized countries and the Third World but are left as global estimates. Furthermore, we account for both the anthropogenic and natural emissions of these compounds.

The methyl bromide emissions arising from its use as a fumigant, which we estimate at approximately 37,000 tonnes in 1991, are frozen in 1995 at 1991 levels.[14] As specified in the Copenhagen Amendments to the Montreal Protocol, quarantine and preshipment applications are exempted from this control measure.[15] This is an uncontrolled, and hence unpredictable, portion of methyl bromide emissions. For modeling purposes, we assume that it is frozen in 2005 at 2001 levels. This corresponds roughly to the emission pattern that might result were there a 10-year delay for Third World methyl bromide use.[16] Actual emissions may be higher, which would not contravene the Copenhagen regulations.

The emissions of methyl bromide arising from the use of ethylene dibromide (EDB) in leaded gasoline are held constant at current levels (approximately 11,000 tonnes) until 2000, after which a 5-year phaseout is instituted. This is an attempt to reflect the possible future decline in the use of leaded gasoline in the Third World.

Of the three primary uses of methyl bromide (preharvest fumigation, post-harvest fumigation, and structural fumigation), approximately 50 percent of the production of methyl bromide for soil fumigation and 80 percent of that for post-harvest and structural fumigation are ultimately emitted. All three scenarios assume that there is no significant future change in the percent of production emitted by each application. In view of the generally large uncertainty associated with the anthropogenic versus nonanthropogenic apportionment of methyl bromide production, this will not introduce significant additional error.

No controls have been placed on the emissions of methyl chloride or methyl bromide resulting from low-temperature biomass combustion in any regulations to date. However, in order to reflect the probable decline in low-temperature biomass burning in the future, we have assumed that half of the anthropogenic component is eliminated between 2010 and 2040 in the Copenhagen Amendments base-line scenario.

The Copenhagen Amendments base-line scenario contains no provisions to control or eliminate emissions of banked ODCs. Banks of ODCs continue to increase until the production-phaseout dates. After production ceases, the banks continue to emit for decades until all their contents have escaped.

The basic features of all three scenarios are presented in table 12.2.

Accelerated Phaseout Scenario

Some of the larger ODC-producing and ODC-consuming countries have adopted domestic regulations that are more stringent than the 1992 amendments and adjustments to the Montreal Protocol. The Accelerated

Table 12.2
Emission scenarios.

Copenhagen Amendments	
Industrialized countries	*Third World*
CFCs	
Freeze in 1990	Freeze in 2000
75% reduction by 1994	75% reduction by 2004
100% reduction by 1996	100% reduction by 2006
Halons	
Freeze in 1992	Freeze in 2002
100% reduction by 1994	100% reduction by 2004
Carbon tetrachloride	
85% reduction by 1995	85% reduction by 2005
100% reduction by 1996	100% reduction by 2006
Methyl chloroform	
Freeze in 1993	Freeze in 2003
50% reduction by 1994	50% reduction by 2004
100% reduction by 1996	100% reduction by 2006
HCFCs	
Production cap (3.1%) in 1996	Growth to 2006
35% reduction by 2004	35% reduction by 2014
65% reduction by 2010	65% reduction by 2020
90% reduction by 2015	90% reduction by 2025
99.5% reduction by 2020	99.5% reduction by 2030
100% reduction by 2030	100% reduction by 2040
Methyl bromide	
Fumigant uses	
Freeze in 1995 at 1991 levels[b]	Freeze in 1995 at 1991 levels[b]
Emissions from auto exhaust	
Eliminate between 2000 and 2005	Eliminate between 2000 and 2005
Emissions from biomass burning	
50% reduction between 2010 and 2040	50% reduction between 2010 and 2040
Methyl chloride (biomass only)	
50% reduction between 2010 and 2040	50% reduction between 2010 and 2040

Accelerated Phaseout

Industrialized countries	*Third World*
Methyl bromide (all fumigant use)	
Freeze in 1994 at 1991 levels	Freeze in 1994 at 1991 levels
100% reduction by 2001	100% reduction by 2001
HCFC-141b	
Production cap (2.6%) in 1996	
100% reduction by 2003	
Remaining HCFCs	
Production cap (2.6%) in 1996	
35% reduction by 2004	
60% reduction by 2007	
80% reduction by 2010	
90% reduction by 2013	
100% reduction by 2014	

Saving Our Skins (all countries)[a]

CFCs
- Freeze in 1990
- 75% reduction by 1994
- 100% reduction by 1996

Halons
- Freeze by 1992
- 100% reduction by 1994

Carbon tetrachloride
- 85% reduction by 1995
- 100% reduction by 1996

Methyl chloride (biomass only)
- 50% reduction bet. 2000 and 2020

Methyl chloroform
- Freeze in 1993
- 50% reduction by 1994
- 100% reduction by 1996[c]

HCFCs
- 100% reduction by 1996

Methyl bromide
- Fumigant uses:
  - 100% reduction by 1996
- Emissions from auto exhaust
- Eliminate between 1995 and 2000
- Emissions from biomass burning
- 50% reduction between 2000 and 2020

Other measures
- Bank controls

a. Only post-1994 controls apply to Third World countries.
b. Post-harvesting uses are exempted.
c. Third World elimination date is 1997.

Phaseout scenario combines the more restrictive ODC control provisions of United States and European Community regulations.[17] Relative to the Copenhagen Amendments base-line scenario, this results in adjustments to the methyl bromide and HCFC production elimination schedules only. All other ODCs are unaffected.

The portion of global emissions arising from methyl bromide's use as a fumigant are frozen in 1994 at 1991 levels and completely eliminated in 2001, with no exemption for post-harvest use.

The production cap specified for HCFCs in 1996 is reduced from the 3.1 percent in the base-line scenario to 2.6 percent. HCFC-141b production in the industrialized countries is eliminated in 2003; all other HCFC production in those countries is eliminated by 2014. The HCFCs eliminated in 2014 adhere to a series of interim reductions, listed in table 12.2. The control schedule of the Third World countries is adjusted to maintain a 10-year delay relative to the industrialized countries.

The accelerated phaseout of methyl bromide and HCFC-141b is part of recent rulemaking by the U.S. Environmental Protection Agency; the reduced production cap and the elimination schedule of the remaining HCFCs were established by the European Community.

Saving Our Skins Scenario

The Saving Our Skins scenario represents what we deem to be the most stringent technically feasible ODC-control strategy. In addition to moving the phaseout dates for HCFCs and methyl bromide forward significantly, we assume that emissions from the largest of the ODC banks are reduced and that both industrial and Third World countries adhere to the accelerated phaseout schedule.

Half of the methyl chloride emissions and half of the methyl bromide emissions resulting from biomass burning are eliminated between 2000 and 2020. In addition, the Saving Our Skins scenario assumes a complete phaseout of the fumigant component of methyl bromide production between 1994 and 1996. Though we have recommended that some exceptions be made to the 1996 phaseout where no viable alternatives exist for use in the Third World, we have not modeled these additional emissions, which are likely to be small. Finally, emissions of methyl bromide arising from the use of EDB in leaded gasoline are assumed to be eliminated between 1995 and 2000.

The Saving Our Skins scenario also assumes a complete global production phaseout of HCFCs by January 1, 1996. Since these compounds have an impact on stratospheric ozone, and since the applications now using HCFCs can use HFCs or switch to halocarbon-free technology, they can and should be eliminated, with provision for very limited exceptions (see chapters 5 and 13). Moreover, as we have discussed, the short-term effects of HCFCs are considerably greater than their steady-state ODPs.

Limited production of HCFC-22 may be needed for use in existing equipment presently using this compound. This could probably be avoided by a sufficiently large tax on new HCFC-22 production to encourage recovery and reuse. Elimination of avoidable leaks will also reduce this demand.

The Saving Our Skins scenario includes a number of measures to significantly reduce the emissions of ODCs from service banks.[18] We assume that the emissions of CFC-12 banked in non-hermetically-sealed cooling systems, such as car air conditioners, are reduced by 60 percent beginning in 1996 through more universal and improved recovery during servicing and through leak reduction. Essentially all of the emissions from CFC-12 banked in hermetically sealed cooling systems, such as home refrigerators, are eliminated beginning in 1996 by recovery prior to permanent disposal.

For non-hermetically-sealed systems, recharging with a non-ozone-depleting compound will be required after 1996. As was explained in chapter 5, the recovered CFC-12 can be recycled in the remaining applications for which non-ozone-depleting alternatives are not available. These applications must be sealed systems, and recycling and/or removal of the ODCs must be done when non-ozone-depleting alternatives become available. At that time, the remaining CFC-12 should be destroyed.

The Saving Our Skins scenario assumes removal, or destruction without removal, of 50 percent of the CFC-11 banked in closed-cell foams upon disposal of ODC-containing foams beginning in 1996. In addition, we assume that 50 percent of the HCFC-22 bank emissions will be eliminated beginning in 1996. This reflects recovery and/or replacement with a non-ODC compound during servicing or disposal of central air conditioning units and large refrigeration units. This represents an intensification of a trend that is already underway.

Half of the halon-bank emissions in the Saving Our Skins scenario are also eliminated beginning in 1996. This will require reclaiming halon-containing fire-extinguishing equipment from end users, in addition to the various production stockpiles of halon. As with collected CFC-12, we assume that the recovered halons will be destroyed to prevent their emission to the atmosphere.

With these measures, emissions from the service banks would be reduced significantly in the next few years and essentially eliminated around the turn of the century.

The Saving Our Skins scenario assumes that, with some exceptions, the Third World adopts the phaseout schedule specified for the industrialized countries. This will require an arrangement whereby the Third World can afford to purchase ODC-free technology or be given financial assistance to do so. Third World countries with significant domestic production of ODC-using refrigerators and similar equipment will need the means to either retrofit existing production facilities to alternative compounds or construct new facilities. This will require significant transfer of technology and funds from the industrialized countries to the Third World. As was described in chapter 11, a multilateral fund was initiated to meet the incremental costs to Third World. As of yet, this fund has not met its original intentions in terms of financial support. Creating a single, global ODC-elimination schedule, as recommended here, will likely require a considerable further expansion of the multilateral fund.

We recognize that it will not be possible in practice for all Third World countries to follow precisely the same phaseout schedule adopted in the industrialized countries. Renegotiation of the Montreal Protocol will occur in late 1995, making a 1996 phaseout improbable within the context of international regulations. However, voluntary efforts can and should be taken now to the extent possible. Adopting a goal of Third World acceleration could result in emissions very close to those we have modeled. This is because most of the differences between scenarios, so far as

Third World emissions are concerned, come from post-1995 growth in these compounds.

Since new technologies are being developed rapidly and existing technologies are being refined for a much broader range of applications, it may be possible to increase the pace and the breadth of these controls. As was discussed in chapter 11, some countries have enacted domestic regulation mandating the elimination of ODCs prior to 1996. Since it is difficult to judge the precise dates of commercialization and regulatory approval of new technologies or new applications of existing technologies within different countries, we have not included any phaseouts prior to 1996.

Emission trends by compound for the three scenarios are supplied in the appendix.

12.2 Concentration Model and Summary Results

Once emitted into the atmosphere from the surface of the Earth, pollutants can be removed in a number of ways. For example, they can be chemically altered, adsorbed onto surfaces, or removed by dissolution in rainwater and seawater. Ozone-depleting compounds are primarily removed by either photodissociation or chemical reaction with atmospheric oxidants. The most important of these oxidants is a compound called the hydroxyl radical (OH). Careful calculations of the rate at which these processes proceed, given the quantity and distribution of the hydroxyl radical in the atmosphere and the intensity and distribution of photodissociating radiation, has been performed by atmospheric scientists using multi-dimensional models of the atmosphere. One outcome of such studies is an estimate of the length of time emitted amounts of various ODCs will persist in the atmosphere, a characteristic referred to as the atmospheric lifetime. Table 3.1 provides the list of ODC lifetimes adopted here.

Combining these atmospheric-lifetime estimates with the ODC emissions calculated in our emission scenarios, allows us to estimate the atmospheric concentrations of all the ODCs. We assume that all the compounds are thoroughly mixed in the troposphere regardless of the location of emission. Given the long lifetimes of the ODCs relative to the global mixing times in the troposphere (weeks to months), this assumption is a reasonable approximation. Concentrations, therefore, represent global, annual averages.

It should be mentioned that thorough mixing within a single year does not occur in the real atmosphere. For example, a concentration gradient of approximately 10 percent exists between the northern and the southern hemisphere for both CFC-11 and CFC-12 (Singh et al. 1979). This is the result of larger emissions in the northern hemisphere and the approximate one to two years required for long-lived compounds emitted in the northern hemisphere to cross the equator. For our purposes, this small gradient is inconsequential.

Because of the relatively long lifetimes of most ODCs and the rapid mixing in the troposphere, ODCs emitted at the surface will eventually reach the stratosphere. When these compounds reach altitudes where sufficient amounts of short-wavelength UV radiation is abundant, they are photodissociated. At the time of photodissociation, chlorine or bromine is released. As was explained in chapter 4, the concentration of chlorine and bromine in the troposphere serves as a proxy measure for the status of ozone depletion. The time required for transport from the surface to the stratosphere is not explicitly included in the model used here. This would effectively delay all concentration estimates by 2–5 years.

Because bromine is more efficient at destroying stratospheric ozone, we consider two different methods of characterizing these concentrations. In the first we combine the number of chlorine atoms and bromine atoms to arrive at what will be called the **nonweighted equivalent chlorine** concentration; in the second we weight the bromine released by a factor of 40, a crude measure of the comparative ozone-depleting ability of bromine (WMO 1991, p. 6.14).[19] This measure will be called the **weighted equivalent chlorine** concentration. This highlights the true importance of brominated compounds and is the reason for our recommending the rapid elimination of methyl bromide.

The concentration of equivalent chlorine is achieved by calculating the concentration of each ODC and summing the associated chlorine and bromine atoms in either "weighted" or "nonweighted" form.[20] Details of the concentration calculations for each of the scenarios are supplied in the appendix.

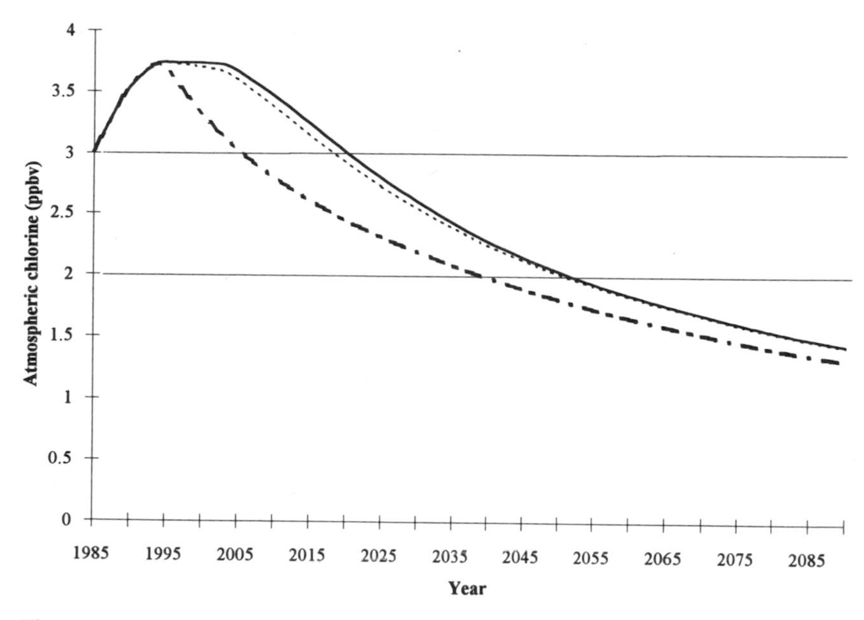

Figure 12.1
Unweighted chlorine loading comparison. Transport time has been estimated at approximately 2–5 years, with a midpoint of 3 years assumed. The timing of all equivalent chlorine values will be delayed by this amount. Scenarios: (solid line) Copenhagen Amendments, (dots) Accelerated Phaseout, (dots and dashes) Saving Our Skins.

Figures 12.1 and 12.2 graphically depict the weighted and non-weighted equivalent chlorine concentration over time, respectively, for the three scenarios. The concentration that is thought to have initiated the dramatic stratospheric ozone loss over the Antarctic (approximately 2 ppbv nonweighted equivalent chlorine or 2.4 ppbv weighted equivalent chlorine) is represented in the figures as a horizontal line. The figures serve to highlight the different ways in which ozone depletion can be reduced. The first and perhaps the most important is control over the peak of equivalent chlorine concentration expected to occur within the next decade or so. Equally important is the rapidity with which the threshold between 2.0 ppbv of chlorine and 2.4 ppbv of *equivalent* chlorine threshold is crossed. Finally, the area between each of the curves and the 2.0/2.4-ppbv threshold can be thought of as the integrated equivalent chlorine burden, the magnitude of which reflects the peak of equivalent

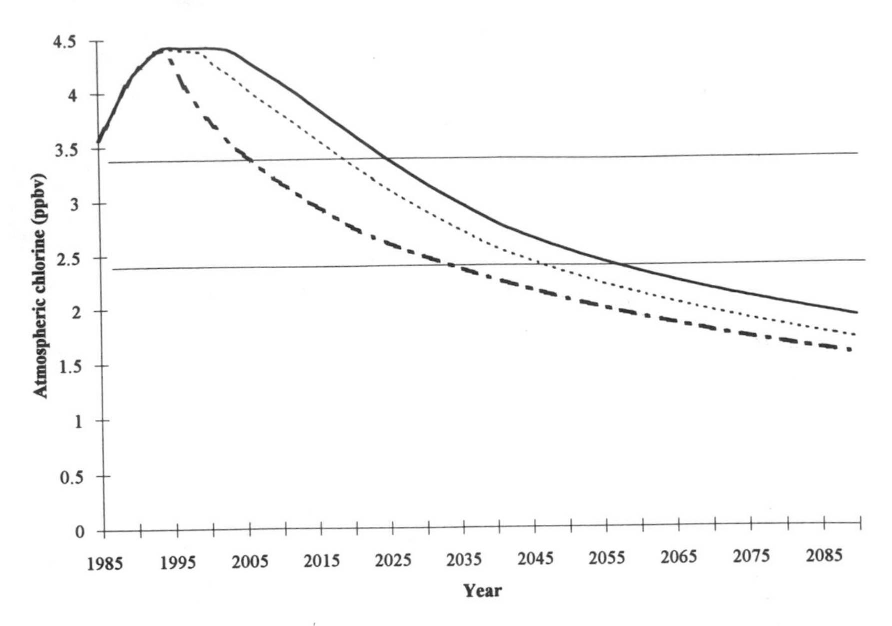

Figure 12.2
Weighted chlorine loading comparison. Transport time has been estimated at approximately 2–5 years, with a midpoint of 3 years assumed. The timing of all equivalent chlorine values will be delayed by this amount. Scenarios: (solid line) Copenhagen Amendments, (dots) Accelerated Phaseout, (dots and dashes) Saving Our Skins.

chlorine, the time at which the aforementioned threshold is crossed, and the rapidity with which the policy measures outlined above are applied.

Tables 12.3–12.5 contain the trends of the compound-specific contribution to weighted equivalent chlorine for the scenarios in addition to the peak, the time at which the 2.4-ppbv threshold is crossed, and the integrated burden.

If we assume that severe ozone depletion will be eliminated when equivalent chlorine concentrations drop below the level that triggered the Antarctic ozone depletion in the late 1970s, we see that the Saving Our Skins scenario would result in an elimination of serious ozone depletion between one to three decades earlier than the other cases. The maximum weighted equivalent chlorine concentration will be reached by about 1997 (including a transport delay of 3 years) in the Saving Our Skins

Table 12.3
Weighted equivalent chlorine concentrations (ppbv) for Copenhagen Amendments scenario. Peak equivalent chlorine: magnitude 4.4, year[a] 2002. 2.4 ppbv year[a]: 2057. Integrated burden: 80 ppbv-years.

Year	CFC-11	CFC-12	CFC-113	CFC-114	CFC-115	Halon-1211	Halon-1301	HCFC-22	CCl_4	MCF	HCFC-123	HCFC-141b	CH_3Cl	CH_3Br	Total
1985	0.666	0.764	0.090	0.010	0.004	0.062	0.068	0.080	0.400	0.390	0.000	0.000	0.600	0.440	3.574
1990	0.797	0.924	0.192	0.015	0.006	0.096	0.091	0.112	0.414	0.464	0.000	0.000	0.600	0.462	4.174
1992	0.821	0.967	0.215	0.015	0.007	0.098	0.096	0.128	0.416	0.468	0.000	0.000	0.600	0.481	4.313
1994	0.838	0.996	0.226	0.016	0.007	0.098	0.101	0.147	0.414	0.452	0.004	0.014	0.600	0.503	4.416
1996	0.848	1.015	0.229	0.016	0.008	0.095	0.107	0.168	0.407	0.376	0.010	0.043	0.600	0.499	4.420
2000	0.856	1.033	0.229	0.017	0.008	0.084	0.115	0.207	0.393	0.267	0.013	0.092	0.600	0.506	4.421
2005	0.847	1.034	0.225	0.018	0.008	0.068	0.124	0.239	0.365	0.176	0.012	0.124	0.600	0.481	4.319
2010	0.811	1.005	0.215	0.018	0.008	0.051	0.130	0.250	0.328	0.078	0.009	0.128	0.600	0.463	4.093
2020	0.710	0.929	0.197	0.017	0.008	0.027	0.135	0.200	0.265	0.015	0.002	0.081	0.577	0.442	3.605
2050	0.427	0.720	0.150	0.015	0.007	0.003	0.122	0.037	0.140	0.000	0.000	0.006	0.522	0.391	2.538
2090	0.207	0.510	0.104	0.012	0.007	0.000	0.086	0.003	0.060	0.000	0.000	0.000	0.522	0.391	1.901

a. Transport time has been estimated at approximately 2–5 years. (Assume a midpoint of 3 years.) The timing of all equivalent chlorine values will be delayed by this amount.

Table 12.4
Weighted equivalent chlorine concentrations (ppbv) for Accelerated Phaseout scenario. Peak equivalent chlorine: magnitude 4.4, year[a] 1995. 2.4 ppbv year[a]: 2047. Integrated burden: 67 ppbv-years. Improvement: 16%.

Year	CFC-11	CFC-12	CFC-113	CFC-114	CFC-115	Halon-1211	Halon-1301	HCFC-22	CCl_4	MCF	HCFC-123	HCFC-141b	CH_3Cl	CH_3Br	Total
1985	0.666	0.764	0.090	0.010	0.004	0.062	0.068	0.080	0.400	0.390	0.000	0.000	0.600	0.440	3.574
1990	0.797	0.924	0.192	0.015	0.006	0.096	0.091	0.112	0.414	0.464	0.000	0.000	0.600	0.462	4.174
1992	0.821	0.967	0.215	0.015	0.007	0.098	0.096	0.128	0.416	0.468	0.000	0.000	0.600	0.481	4.313
1994	0.838	0.996	0.226	0.016	0.007	0.098	0.101	0.147	0.414	0.452	0.004	0.014	0.600	0.489	4.403
1996	0.848	1.015	0.229	0.016	0.008	0.095	0.107	0.168	0.407	0.376	0.009	0.038	0.600	0.487	4.401
2000	0.856	1.033	0.229	0.017	0.008	0.084	0.115	0.207	0.393	0.267	0.011	0.064	0.600	0.449	4.334
2005	0.847	1.034	0.225	0.018	0.008	0.068	0.124	0.239	0.365	0.176	0.010	0.060	0.600	0.285	4.057
2010	0.811	1.005	0.215	0.018	0.008	0.051	0.130	0.244	0.328	0.078	0.006	0.051	0.600	0.261	3.805
2020	0.710	0.929	0.197	0.017	0.008	0.027	0.135	0.180	0.265	0.015	0.001	0.021	0.577	0.239	3.320
2050	0.427	0.720	0.150	0.015	0.007	0.003	0.122	0.029	0.140	0.000	0.000	0.001	0.522	0.188	2.323
2090	0.207	0.510	0.104	0.012	0.007	0.000	0.086	0.002	0.060	0.000	0.000	0.000	0.522	0.188	1.698

a. Transport time has been estimated at approximately 2–5 years. (Assume a midpoint of 3 years.) The timing of all equivalent chlorine values will be delayed by this amount.

Table 12.5
Weighted equivalent chlorine concentrations (ppbv) for Saving Our Skins scenario. Peak equivalent chlorine: magnitude 4.4, year[a] 1994. 2.4 ppbv year[a]: 2034. Integrated burden: 44 ppbv-years. Improvement: 45%.

Year	CFC-11	CFC-12	CFC-113	CFC-114	CFC-115	Halon-1211	Halon-1301	HCFC-22	CCl_4	MCF	HCFC-123	HCFC-141b	CH_3Cl	CH_3Br	Total
1985	0.666	0.764	0.090	0.010	0.004	0.062	0.068	0.080	0.400	0.390	0.000	0.000	0.600	0.440	3.574
1990	0.797	0.924	0.192	0.015	0.006	0.096	0.091	0.112	0.414	0.464	0.000	0.000	0.600	0.462	4.174
1992	0.821	0.967	0.215	0.015	0.007	0.098	0.096	0.128	0.416	0.468	0.000	0.000	0.600	0.481	4.313
1994	0.838	0.996	0.226	0.016	0.007	0.098	0.101	0.147	0.414	0.452	0.004	0.014	0.600	0.489	4.403
1996	0.833	0.998	0.226	0.016	0.008	0.090	0.104	0.145	0.398	0.356	0.002	0.016	0.600	0.392	4.183
2000	0.800	0.978	0.218	0.016	0.007	0.071	0.105	0.121	0.366	0.185	0.000	0.011	0.600	0.286	3.765
2005	0.750	0.943	0.209	0.015	0.007	0.052	0.106	0.091	0.329	0.082	0.000	0.007	0.584	0.249	3.424
2010	0.696	0.905	0.200	0.015	0.007	0.037	0.106	0.067	0.296	0.036	0.000	0.004	0.565	0.229	3.163
2020	0.591	0.831	0.183	0.014	0.007	0.018	0.103	0.036	0.239	0.007	0.000	0.002	0.526	0.194	2.749
2050	0.347	0.642	0.139	0.012	0.007	0.002	0.084	0.005	0.126	0.000	0.000	0.000	0.522	0.188	2.075
2090	0.168	0.455	0.097	0.010	0.006	0.000	0.056	0.000	0.054	0.000	0.000	0.000	0.522	0.188	1.557

a. Transport time has been estimated at approximately 2–5 years. (Assume a midpoint of 3 years.) The timing of all equivalent chlorine values will be delayed by this amount.

scenario, compared to 1998 for the Accelerated Phaseout scenario and 2005 for the Copenhagen Amendments base-line scenario.[21] Most important, the integrated burden is improved by 16 percent in going from the Copenhagen Amendments scenario to the Accelerated Phaseout scenario, and by 45 percent in going from the Copenhagen Amendments scenario to the Saving Our Skins scenario.

The resulting chlorine and bromine trends in the three scenarios arise from the different control measures taken in each. In order to illuminate the effectiveness of the four primary control measures considered here, it is useful to isolate the effect each has on the equivalent chlorine concentration. This is done in figure 12.3, where the various control measures are introduced into the emissions-scenario model successively.

Although there are differences in the way each control measure reduces equivalent chlorine, no single measure stands out as significantly more effective than the others. Table 12.6 lists the effect each of these control measures has on the weighted equivalent atmospheric chlorine peak, the time at which the 2.4-ppbv threshold is crossed, and the integrated burden. Table 12.7 displays the same, except there the individual measures are enacted independently rather than successively. This helps elucidate the gains made by measures where some overlap exists.

The last control step in tables 12.6 and 12.7 reflects the additional control on methyl bromide and methyl chloride. Of these two actions, over 90 percent of the improvement in the integrated burden is due to controls on the emissions of methyl bromide.

A number of conclusions can be drawn from the policy comparison in tables 12.6 and 12.7. One is that although controlling the emission of methyl bromide into the atmosphere is perhaps the most effective of the measures considered, accelerating the schedule of the Third World, reducing the amount of banked ODCs and eliminating the use of HCFCs can also have a sizable impact on the ozone layer's recovery. As we have noted, the gain made by accelerating the Third World countries' ODC-elimination schedule depends to a sizable degree on the future ODC growth rates assumed in these countries. Because we have attempted to be conservative in this respect, the gains made by altering the schedule of Third World countries could, in fact, be much larger.

Before the Copenhagen negotiations, there were no controls on the production and consumption of HCFCs. The phaseout schedule that was

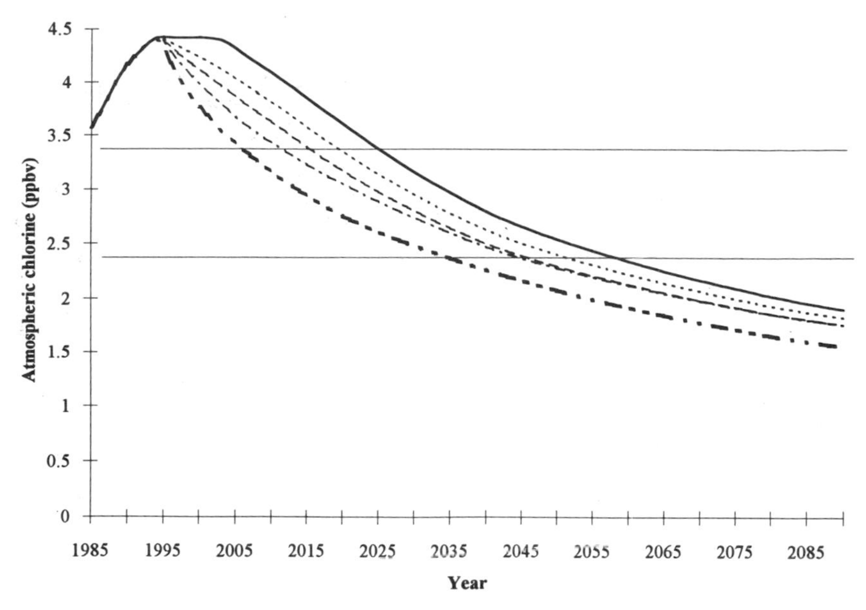

Figure 12.3
Weighted chlorine loading control measure comparison. Transport time has been estimated at approximately 2–5 years, with a midpoint of 3 years assumed. The timing of all equivalent chlorine values will be delayed by this amount. (solid) Copenhagen Amendments scenario, (dots) adjustment of Third World schedule, (dashes) bank elimination, (light dots and dashes) HCFC control, (heavy dots and dashes) Saving Our Skins scenario.

agreed upon in Copenhagen has limited the impact of HCFCs considerably. For this reason, the control of HCFC production and consumption does not figure as prominently in ozone-layer recovery as it did before 1993. Furthermore, our treatment of all ODCs as equally able to contribute to ozone loss overestimates the gain made by HCFC control.

As we mentioned in the introduction to this chapter, the emission scenarios and individual policy measures described in this effort should not be viewed as predictive. Rather, we have shown what relative gains can be expected should this series of technically feasible policy measures be enacted. There are, of course, other impediments that must be overcome if these gains are to be realized. However, significant progress toward accelerating the recovery of the ozone layer could be realized immediately.

Table 12.6
Impact of individual policy measures: successive enactment.

	Copenhagen Amendments	Third World delay removed	Bank removal	HCFC elimination	CH_3Cl and CH_3Br
Weighted equivalent chlorine peak					
Magnitude	4.4	4.4	4.4	4.4	4.4
Year[a]	2002	1994	1994	1994	1994
2.4 ppbv crossing					
Year[a]	2057	2051	2045	2044	2034
Integrated burden	80	68	60	55	44
Improvement		15%	25%	32%	45%

a. Transport time has been estimated at approximately 2–5 years. (Assume a midpoint of 3 years.) The timing of all equivalent chlorine values will be delayed by this amount.

Table 12.7
Impact of individual policy measures: independent enactment.

	Copenhagen Amendments	Third World delay removed	Bank removal	HCFC elimination	CH_3Cl and CH_3Br
Weighted equivalent chlorine peak					
Magnitude	4.4	4.4	4.4	4.4	4.4
Year[a]	2002	1994	1994	1994	1994
2.4 ppbv crossing					
Year[a]	2057	2051	2050	2056	2048
Integrated burden	80	68	68	70	66
Improvement		15%	14%	12%	17%

a. Transport time has been estimated at approximately 2–5 years. (Assume a midpoint of 3 years.) The timing of all equivalent chlorine values will be delayed by this amount.

12.3 Uncertainties

There are a number of uncertainties associated with the calculations performed here beyond those inherent in the assumptions mentioned in the introduction and description of the emission scenarios. Many can be identified and crudely approximated; characterization of others is beyond the scope of this study.

In order to initialize the emissions of ODCs, reliable production and emissions data are required. Because manufacturer-specific information concerning production levels is considered proprietary, reliable data are difficult to assemble. The data that can be assembled are susceptible to underreporting. In those instances where the magnitude of underreporting has been estimated, we have made the appropriate corrections.

A related uncertainty is the future rate of release from ODC banks. Release rates were calculated from existing data and are expressed as a fraction of the release from the bank in a given year. If no discernible trend in the bank release rate exists in recent data, we assume that existing release rates continue at the present rate through the model period. In cases where a trend in the release rates is evident, this is extrapolated to future years. Details of the compound-specific banking methodology are given in the appendix.

The atmospheric lifetimes for the compounds modeled here are taken from a variety of sources. Some rely solely on the results of chemical-dynamical models; some are calculated from observed concentrations of the particular ODCs in the atmosphere. Because of measurement difficulties and the required approximations in model studies, atmospheric lifetimes exhibit significant uncertainty.

The consistently shorter lifetimes of the alternative ODCs are due to the fact that they are partially removed in the troposphere by the hydroxyl radical (OH). In order to estimate the atmospheric lifetimes of these compounds, the amount and distribution of OH in the atmosphere must be specified. As will be explained later, OH is difficult to measure or calculate and exhibits large variations in atmospheric concentration. This gives rise to another significant uncertainty associated with the calculated lifetimes for the ODCs.

Though we do not attempt to calculate the actual loss of ozone due to reactive chlorine and bromine in the stratosphere, atmospheric chlorine and bromine concentrations are a reasonable proxy for the cumulative ozone-depletion potential of all ODCs combined. One problem with such an approach is that not all chlorine and bromine atoms are equally effective in terms of their potential to deplete ozone. A crucial difference is the position in the atmosphere at which the reactive chlorine or bromine associated with a particular ODC is released, which varies from compound to compound. Compounds with long stratospheric lifetimes and/or short tropospheric lifetimes may have quite different ozone-depleting characteristics than compounds that are destroyed relatively quickly once they reach the stratosphere (Weisenstein et al. 1992). We have not attempted to model these properties.

Finally, as was mentioned in chapter 4, the projections of atmospheric chlorine and bromine we have calculated here do not reflect the nonlinear nature of the ozone loss now measured over the southern pole or that which might arise elsewhere.

12.4 The Hydroxyl Radical

A large variety of chemical compounds, of both natural and human-derived origins, are emitted to the atmosphere each year. To maintain the relatively stable chemical composition exhibited by the atmosphere, a number of removal mechanisms (or sinks, as they are commonly called) are constantly at work. Chemical compounds entering the atmosphere are primarily removed by four processes: they can be chemically altered either by reaction with another compound or through interaction with solar radiation, dissolved (into rainwater or seawater), or adsorbed onto surfaces. For many compounds, including the ODCs, the first two mechanisms, chemical reaction and photodissociation, are by far the most important.

As was discussed in chapter 1, photodissociation of ODCs occurs in the upper stratosphere, where ultraviolet radiation of sufficient energy is available to break the molecular bonds of these compounds. Little UV-B and essentially no UV-C radiation penetrates to the troposphere, because of the filtering effect of the ozone layer and the oxygen in the stratosphere. In order for such a compound to undergo photodissociation, it must travel to the level in the stratosphere where such high-energy UV radiation is found.

Compounds will naturally be carried to the stratosphere by general circulatory motions or smaller-scale convective events (such as cloud formation), provided that they are not rapidly removed in the lower atmosphere. In some cases, a molecule can circulate through the stratosphere and the troposphere a number of times before reaching an altitude where UV radiation of the appropriate wavelength is available. Generally, it requires anywhere from 2 to 5 years for a compound to travel to the middle stratosphere, as vertical motion in this portion of the atmosphere is slow. Therefore, compounds for which stratospheric photodissociation is the sole removal mechanism exhibit longer atmospheric lifetimes than those removed in the troposphere. This can be seen in table 12.8, where atmospheric lifetimes and removal mechanisms are given for a number of ODCs and HFCs. The longer-lived compounds (about 30 years or longer) tend to lack any hydrogen atoms in their molecular makeup.

Equally important for many chemical compounds that enter the atmosphere is removal by reaction with another compound. For many of the ODCs, reaction with an oxidizing agent known as the hydroxyl radical (OH) is by far the most common. This process occurs primarily in the troposphere, where the hydroxyl radical is most abundant. Because removal takes place in the troposphere, compounds controlled by this mechanism generally have shorter atmospheric lifetimes than those controlled by photodissociation. This can be seen in table 12.8. This process tends to be the primary removal mechanism for the compounds containing hydrogen, which are referred to as partially halogenated ODCs.

The important atmospheric removal processes for ODCs and related compounds are depicted schematically in figure 12.4.

The reaction of hydrogen-containing ODCs can be depicted as

$$RH + OH \rightarrow R + H_2O, \qquad (12.1)$$

where RH is a partially halogenated compound before reaction with OH and where R is the haloalkyl radical remaining after removal of the hydrogen atom by OH. The remaining haloalkyl radical undergoes a sequence of reactions that lead to soluble products, which are then removed from the atmosphere by rain or seawater.

Table 12.8
Atmospheric removal mechanisms for ODCs and related compounds (WMO 1991, pp. 6.7, 8.7, 8.8; UNEP 1992, section II, p. 5; WMO 1989, vol. II).

Common name	Chemical formula	Lifetime (years)	Removal mechanism
Fully halogenated			
CFC-11	$CFCl_3$	55	photolysis, O'D reaction
CFC-12	CF_2Cl2	116	photolysis
CFC-113	$CFCl_2CF_2Cl$	110	photolysis
CFC-114	CF_2ClCF_2Cl	220	photolysis
CFC-115	CF_2ClCF_3	550	photolysis
Halon 1211	CF_2ClBr	11	photolysis
Halon 1301	CF_3Br	77	photolysis
Halon 2402	CF_2BrCF_2Br	20	photolysis
Carbon tetrachloride	CCl_4	47	photolysis
Partially halogenated			
Methyl chloroform	CH_3CCl_3	6.1	OH reaction
HCFC-22	CHF_2Cl	15.8	OH reaction
Methyl chloride	CH_3Cl	1.5	OH reaction
Methyl bromide	CH_3Br	2	OH reaction
HCFC-123	$CHCl_2CF_3$	1.7	OH reaction, UV
HCFC-124	$CHFClCF_3$	6.9	OH reaction
HCFC-141b	CH_3CFCl_2	10.8	OH reaction, UV
HCFC-142b	CH_3CF_2Cl	22.4	OH reaction
HCFC-225ca	$CF_3CF_2CHCl_2$	2.8	OH reaction
HCFC-225cb	CF_2ClCF_3CHFCl	8	OH reaction
HFC-134a	CH_2FCF_3	15.7	OH reaction
HFC-125	CHF_2CF_3	40.7	OH reaction
HFC-143a	CH_3CF_3	64.6	OH reaction
HFC-152a	CH_3CHF_2	1.8	OH reaction
HFC-32	CH_2F_2	7.3	?

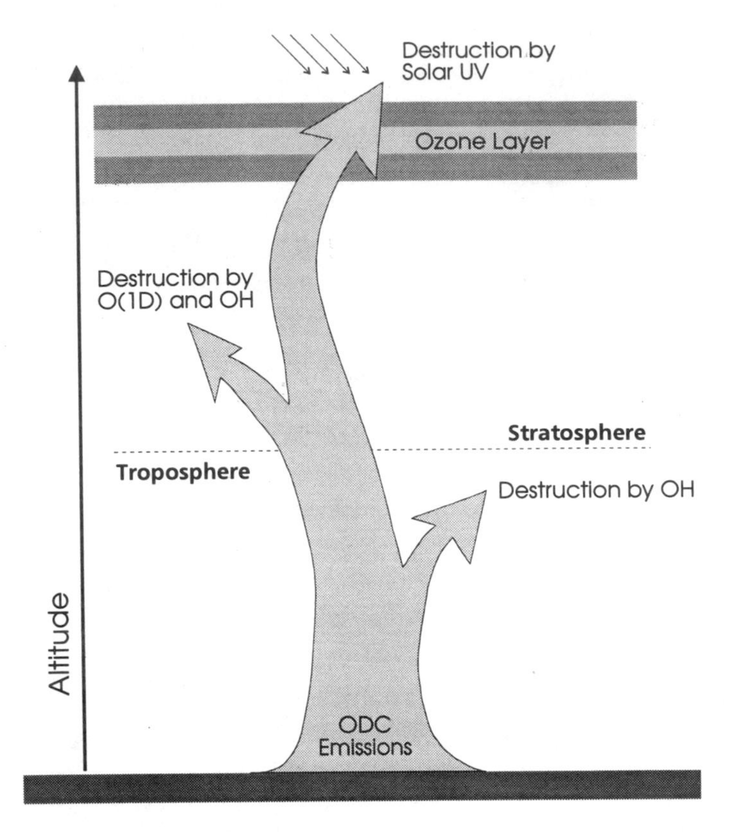

Figure 12.4
Generalized chemical life cycle of an ODC in the atmosphere (WMO 1989, vol. II, p. 311, figure 5).

The hydroxyl radical not only serves to remove many of the partially halogenated ODCs from the atmosphere; it is also the primary oxidizing agent for a large number of gaseous chemical compounds emitted at the surface. Although OH is the most important oxidant in the atmosphere, two other compounds that provide additional oxidizing capacity to the atmosphere are tropospheric ozone (O_3) and hydrogen peroxide (H_2O_2).

Tropospheric ozone controls, to a large extent, the amount of OH in the troposphere. The importance of hydrogen peroxide rests on the fact that it is primarily responsible for conversion of sulfur dioxide (SO_2) to sulfuric acid in cloud and rain droplets. Because of its ability to oxidize so many compounds in the lower atmosphere, OH has been referred to as the "tropospheric vacuum cleaner" (Graedel 1978).

As we have mentioned, the atmosphere's ability to remove the many chemical compounds being introduced into it by anthropogenic activities is crucial in maintaining the energy balance and the natural chemical balance required for atmospheric stability. The hydroxyl radical is the most important atmospheric constituent in maintaining this balance. Therefore, changes in the concentration of OH have significant implications for the overall oxidizing capacity of the atmosphere.

Atmospheric Chemistry of OH

The primary production path for OH is initiated by the photodissociation of ozone given as

$$O_3 + hv \rightarrow O('D) + O_2, \qquad (12.2)$$

where hv represents solar radiation and O('D) is an electronically excited oxygen atom. Wavelengths of 320 nanometers or less are required to photodissociate ozone. OH is then formed when this excited oxygen atom combines with a water vapor molecule:

$$O('D) + H_2O \rightarrow OH + OH. \qquad (12.3)$$

The major sinks or removal pathways of OH in the clean troposphere (polar or remote marine regions) are reaction with carbon monoxide (CO, ~70 percent) and with methane (CH_4, ~30 percent). The reactions are

$$CO + OH \rightarrow CH_3 + H_2O \qquad (12.4)$$

and

$$CH_4 + OH \rightarrow CH_3 + H_2O. \qquad (12.5)$$

In polluted urban air, OH also reacts with nonmethane hydrocarbon pollutants such as gasoline fumes. The chemical reactions resemble those shown in reactions 12.1 and 12.5. In the presence of sufficient amounts of nitric oxide, these reactions lead to the formation of tropospheric ozone, an agent in photochemical smog. All of this is part of the process by which OH helps remove pollutants from the troposphere.

At the global level, the role of OH is no less important. As is shown in reaction 12.5, OH is the primary removal mechanism for methane and many of the ODCs. Methane is an important greenhouse gas. ODCs contribute to global warming and, of course, ozone depletion. In addition, OH plays an important role in the gas-phase conversion of sulfur oxides (SO_X) and nitrogen oxides (NO_X) into sulfuric and nitric acid, which contribute to acid rain and snow. An important question, in view of the crucial function of OH in the atmosphere, is whether the amount of OH is changing over time.

OH: Past Changes and Current Distribution

Because of its short lifetime (about 1 second) and its small concentration in the atmosphere (approximately 10^6 molecules per cubic centimeter, or about 0.037 pptv), OH is difficult to measure directly. Further, because of its short lifetime, spatial and temporal variations can be substantial. As a result, modeling and observation of compounds in the atmosphere that are linked to OH concentrations are generally used, with varying degrees of success, to derive the OH distribution and its trend.

Figure 12.5 shows the results of a two-dimensional modeling effort aimed at simulating the distribution of OH. Globally averaged values range from 5×10^5 to 1×10^6 molecules/cm^3 throughout the year (Isaksen 1987). As can be seen in the figure, OH concentrations vary considerably by latitude, principally as a result of seasonal variation in sunlight. As a result, compounds whose removal depends on OH will also exhibit seasonal fluctuations in concentration.

A recent effort at calculating the concentration of OH has used levels of methyl chloroform in the atmosphere and knowledge of methyl chloroform emission levels to derive a global-average range of $7.7 \times 10^5 \pm 1.4 \times 10^5$ molecules/cm^3 (Prinn et al. 1992). Table 12.9 shows a collection of recent OH-trend calculations emphasizing the net change in OH since preindustrial times.

The calculated trends rely heavily on the assumed emission distribution and on trends associated with methane, carbon monoxide, nitrogen oxides, and nonmethane hydrocarbons. Though the range in model results is quite large, there is general agreement that OH levels have dropped globally since preindustrial times, primarily because of increases in methane and carbon monoxide from human activities.[22] This decline in OH concentration would, in turn, have contributed to the

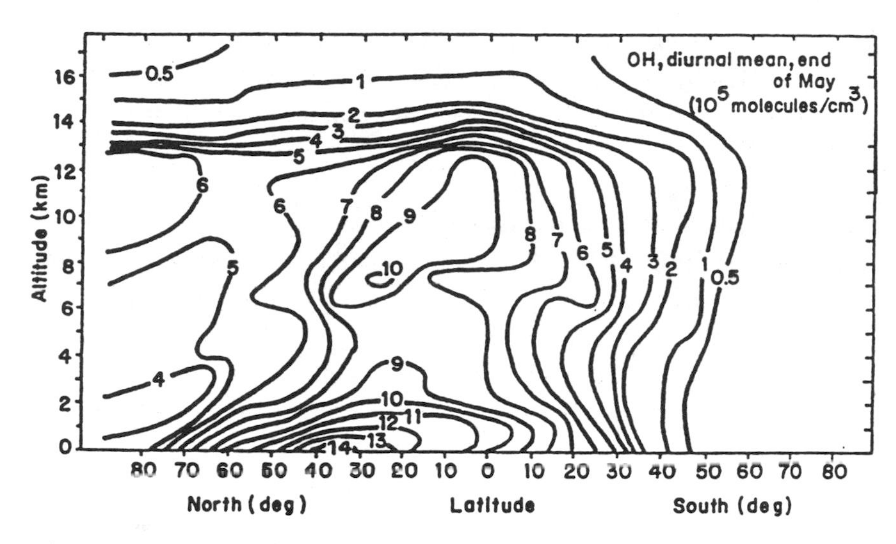

Figure 12.5
Diurnally averaged OH distribution calculated in a two-dimensional model (Isaksen 1988, p. 150, figure 4).

Table 12.9
Model-calculated changes in OH (Thompson 1992). Abbreviations: (NA) not available; (AREAL) Atmospheric Research and Exposure Assessment Laboratory; (OGI) Oregon Graduate Institute; (GSFC) Goddard Space Flight Center; (NH) northern hemisphere; (SH) southern hemisphere.

Model and type	Current global OH ($10^6/cm^3$)	OH change since pre-industrial times
Harvard University (1D)	NA	60% loss
Cambridge University (2D)	0.95	NH: 50% loss SH: 20% gain
Max-Planck-Institut (2D)	0.91	NH: 50% gain to 40% loss SH: 20–40% loss
Max-Planck-Institut (3D)	0.6–0.8	NH: 10–20% gain SH: 10–20% loss
USEPA/AREAL (1D)	0.6	4% loss
OGI (mlti-1D)	0.8	11% loss
GSFC (multi-1D)	0.6	20% loss

buildup of methane and carbon monoxide, because less OH would have been available to remove these compounds from the atmosphere. Consequently, there would have been a greater buildup from the same emission level.[23]

Future Changes in OH and Their Implications

Predictions of future OH levels are complicated by a variety of competing processes. Evidence that tropospheric ozone is increasing would imply that OH levels would also increase, since that is the main process by which OH radicals are created (reactions 12.2 and 12.3). The increase in tropospheric ozone results from the human-derived emissions of nitrogen oxides and nonmethane hydrocarbons. In addition, stratospheric ozone loss allows UV to penetrate the troposphere in greater amounts, leading to the formation of ozone in the troposphere. Finally, a global average temperature increase may lead to greater OH levels through an increase in available water vapor. On the other hand, increasing human-derived emissions of methane, carbon monoxide, and nonmethane hydrocarbons suppress OH concentrations.

There is still considerable uncertainty about future trends in OH concentrations, but there is no doubt about the ability of human-derived emissions of chemical compounds to perturb global OH levels. It is likely that changes in OH levels will occur in the coming decades. Lower OH concentrations in the future imply increased lifetimes of compounds whose removal relies on reaction with OH. Among compounds in this category are methane (a potent greenhouse gas) and many of the ODCs, particularly the HCFCs. The implications for ozone depletion are that the long-term effect of emissions of HCFCs could be more severe than is indicated by their steady-state ODPs. By contrast, increased OH levels would lead to potential decreases in these compounds but may increase the rate of acid formation in the troposphere.

13

Industry's Response to Ozone Depletion

A careful examination of the history of the responses of the chemical manufacturers to the problem of stratospheric ozone loss holds important lessons for future protection of the ozone layer and also for action on other environmental concerns, such as the increase in greenhouse gases in the atmosphere.

The evolution of the relation between ODCs and ozone depletion followed two independent paths. One path was that of a U.S. government agency acting to fulfill the intent of environmental laws. The other was that of the scientific community, which was concerned about the effects of seemingly harmless chemicals on the environment.

The scientific investigation that hypothesized the role of chlorine in stratospheric ozone depletion was undertaken to fulfill the requirements of a U.S. law: the National Environmental Policy Act (NEPA). It prompted the National Aeronautics and Space Administration (NASA) to fund a study in the early 1970s by Richard Stolarski and Ralph Cicerone on the possible problems associated with chlorine released in the exhaust of the Space Shuttle's solid-fuel rocket motors. When the Space Shuttle program was planned, NASA anticipated a large number of flights per year. Hence the inquiry and its unexpected result: frequent shuttle flights might cause increases in the concentration of stratospheric chlorine, which could have a deleterious effect on the ozone layer.

The role of chlorofluorocarbons in depleting the ozone layer was first identified in 1974 by two other scientists working independently of Stolarski and Cicerone: Mario Molina and F. Sherwood Rowland. Their investigation started when Rowland began wondering what happened to CFCs once they were released into the atmosphere.[1] Knowing that they are nonreactive in the troposphere, he suspected they would be transported to the stratosphere (see chapter 1). He hypothesized that the destruction of these compounds was accomplished by ultraviolet radiation when they were stirred slowly across the tropopause into the stratosphere. This theory held that destruction by UV resulted in the release of the ozone-depleting chlorine bound within the CFC molecule.

Before these two investigations, scientific attention concerning the ozone layer had been focused on the effects of nitrogen oxides from supersonic airplanes or from global nuclear war.

These investigations coincided with significant legislative developments concerning environmental protection in the United States. One powerful expression of the movement toward such protection was the passage of the National Environmental Policy Act in 1969. In spite of this, the dominant attitude of corporate leaders toward these activities was one of hostility. In general, they did not view environmental protection as a responsibility or an ethical imperative. Rather, it was regarded mainly as a nuisance and a hindrance to profit.

13.1 The Early Corporate Response

After the publication of the 1974 Molina-Rowland paper linking ozone depletion to CFCs, a burst of scientific and environmental concern led to a 1978 ban on the use of these compounds in most aerosol products in the United States. Similar bans were instituted in Sweden, Norway, and Canada. ODC-manufacturing corporations began searching for alternative chemicals as a hedge against the possibility of total or partial bans on the use of ODCs. Research on alternative chemicals continued until about 1980. At that time, industry abruptly cut back on relevant research and development. It is particularly instructive to examine the reasons why there was a six-year hiatus in developing practical alternatives to ozone-depleting compounds and related technology by the major ODC manufacturers.

In the case of aerosols propellants, several non-ozone-depleting aerosol propellant approaches and chemicals were identified in the 1970s for most uses. These were put into place rapidly in the United States,

Canada, Norway, and Sweden after the aerosol ban was implemented in 1978.

Unfortunately, the use of ozone-depleting compounds for other purposes, such as automobile air conditioning, foam production, and cleaning solvents continued to grow in the United States. In addition, the efforts of the other major industrialized countries to reduce aerosol emissions were weak and ineffective—so much so that aerosols continued to be among the largest sources of ODC emissions until the late 1980s, despite the near-total U.S. ban. Worldwide CFC emissions from aerosols totaled about 2.5 million tonnes from 1978 to 1988, when a rapid reduction in this use began.

The U.S. ban, although nearly unilateral, contributed a great deal to the reduction of ODC emissions. In just 10 years, the U.S. reduction in emissions of CFCs amounted to about a billion kilograms; equal to three years' worth of U.S. CFC consumption in all applications in the mid 1980s.[2] It is safe to say that the U.S. aerosol ban has made a major contribution to the reduction of future ozone depletion. Had the other major ODC-aerosol users followed the United States' lead, cumulative CFC emissions from all sources worldwide during the decade following the U.S. ban would have been halved, considerably improving the prognosis for the ozone layer.

After the legislative action in the late 1970s, regulations on other uses of ODCs seemed possible. Several promising alternative chemicals were identified by the major ODC producers, and considerable toxicological testing was begun. Nonetheless, the scale of expenditures was not large. For example, DuPont, with 25 percent of the world's CFC sales, spent an average of $2.5 million per year between 1974 and 1980 (DuPont 1987). By the time research into alternatives was stopped, major ODC producers had identified several substances that passed acute and some subacute toxicity tests and possessed the necessary physical and chemical properties to replace the ozone-depleting compounds in use at the time. Most notable among these was a non-ODC called HFC-134a, which has come into widespread use in the 1990s.

DuPont, the world's leading manufacturer of ODCs, abruptly suspended toxicological testing and the development of large-scale manufacturing processes for alternatives in late 1980 or in 1981. As far as we can determine, most chemical companies essentially stopped or drastically scaled back research and toxicological testing on alternatives at about the same time, almost in unison. A 1987 statement by DuPont, made shortly after the company had resumed its research and development, claimed that the reason for the suspension was as follows: "At about this time [1980], scientific and regulatory concern over CFCs began to decrease. This, in combination with market surveys which indicated little interest in more expensive alternatives, led to a de-escalation of effort." (DuPont 1987)

Other corporations have given the same explanation for suspending further work. For example, a representative from ICI, another major multinational chemical manufacturer, stated in 1988:

> Following the revelation of the so-called Ozone Depletion Theory in the early 1970s, ICI carried out a program of work on potential substitutes for CFC 11 and 12 in particular.... At the same time we were keeping close to the development of the science which was starting to indicate a lower level of concern with regard to the part that CFCs might play in the alleged depletion of ozone.... Therefore, having developed a process for a potential alternative CFC and conducted a series of toxicological trials, we decided to suspend such work, since the potential costs of these products at approximately five times the price of existing CFCs did not justify a research program based upon the state of the science at that time. (ICI 1988)

Both DuPont and ICI had a point, from a business perspective. After all, it comes as no surprise that a market survey would indicate a preference for a less expensive product that serves an existing function as well as or better than a proposed replacement under existing conditions.

Scientific Concern

The decision to curtail research on alternative chemicals was made even though no scientific work had disproved the link between CFCs and ozone depletion. Indeed, even without taking into account the rapidly growing use and emissions of the solvent CFC-113, or the growth in use of CFC-11 and CFC-12, various computer modeling efforts in the 1979–1981 period projected global-average ozone depletion of about 12 percent (EPA-RIA 1987, vol. I, p. 2-8. exhibit 2-6).[3]

In 1979 the National Academy of Sciences performed a thorough review of the problem. Rather than showing any diminution of concern, the NAS report projected an ozone depletion of 16.5 percent late in the next century if CFC-11 and CFC-12 continued to be emitted at the rate prevailing at that time (NRC 1982, p. 13). This estimate omitted CFC-113, as well as other important sources of stratospheric chlorine such as carbon tetrachloride. As we have discussed, a 20 percent ozone-depletion level could have very serious adverse ecological and health effects. This was appreciated in the 1970s and was the underlying reason for the widespread concern about ozone loss. These model estimates call into question the corporate statements justifying suspension or even substantial reduction of research in 1980 or 1981 on the grounds of decreased scientific concern.

Even after an NAS panel's downward revisions of ozone-depletion projections in 1982, models still indicated depletion levels in the range of 5–9 percent within 100 years (NRC 1982). Again the new projections were made without taking into account either the growth associated with ODC use occurring in the 1980s or the use of the solvent CFC-113.

Even the more optimistic scientists on the NAS panel agreed that there were uncertainties in their conclusions and that ozone depletion could be a factor of 2 higher, or a total of 10–18 percent. Moreover, the scientists on the NAS panel could not agree that a factor of 2 was an appropriate expression of the uncertainty. Some of them felt that, because the models did not correspond to known ozone measurements at 35 kilometers altitude, important factors might be missing from the models. Thus, this group of scientists concluded that there was no basis upon which to limit uncertainty to a factor of 2 (ibid.). Subsequent events have vindicated that judgment.

In 1983 the NAS again revised ozone-depletion projections downward, this time to about 3 percent (NRC 1983). However, it also warned that if CFC use grew at 3 percent per year, and if emissions of carbon dioxide and other gases that might enhance ozone production were reduced, ozone depletion might reach 10 percent. Furthermore, considerable uncertainties still remained. The NAS was direct about one point: CFCs were a cause of ozone depletion (ibid., pp. 9–16).

The 1983 results from the Lawrence Livermore Laboratory ozone-depletion model, and further modeling exercises in 1985, estimated eventual depletion levels between 10 and 12 percent (EPA-RIA 1987, vol. I, p. 2-8, exhibit 2-6).

In sum, DuPont's implication that research into alternative chemicals could safely be suspended because scientific concern had decreased in the early 1980s to the point where ozone depletion could be dismissed as a serious environmental problem is unwarranted. It is also remarkable that, although evidence of the alarming Antarctic ozone hole was published in 1985, DuPont did not resume research and toxicological testng on alternatives until late 1986. This occurred in spite of the fact that, according to its own External Affairs Department, the corporation continued to keep abreast of developments and research on the ozone-depletion question even after testing of alternatives was suspended.[4]

Regulatory Concern

The argument that there was a decrease in regulatory concern is similarly without merit. For example, in April 1980, the EPA announced its intention to issue a rule that would freeze emissions of nonaerosol uses of CFCs. Certainly this was a serious expression of an intent to regulate a rapidly growing market. DuPont representative Kathleen H. Forte was unable to explain why the EPA's announcement in 1980 of its intent to further regulate CFC use did not spur DuPont to continue even minor expenditures on alternatives to ozone-depleting compounds.[5]

There were other regulatory signals. In April 1980, the United Nations Environment Programme (UNEP) urged governments to reduce production of CFCs. In 1984, the Natural Resources Defense Council filed suit against the EPA for failing to move on the 1980 regulatory initiative. In March 1985, twenty countries signed the Vienna Convention, which spelled out the obligation of countries to control activities likely to damage the ozone layer.

It is interesting, in view of these persistent regulatory and scientific concerns, that (so far as we know) all large-scale producers of ODCs in all major manufacturing countries abandoned or scaled back research on alternatives. Ethics and environmental responsibility aside, none apparently wanted to invest even a modest amount of money for the enormous competitive advantage that would be gained should governments decide on new restrictions on CFC use.

Corporate Response to the Regulatory Environment

Our conclusion is that the suspension of research on alternatives by the chemical manufacturers was more closely related to the Reagan-Bush administration than to the state of scientific or regulatory concern in 1980. Ronald Reagan's election led to the widespread anticipation among corporations that existing regulations would be eased and proposals for many new regulations would be reviewed and probably scrapped. As the U.S. government had previously taken the lead in regulatory efforts to protect the ozone layer, European manufacturers also chose to abandon research efforts around this time.

In November 1980, president-elect Reagan appointed a panel to look into deregulation. That panel was headed by vice-president-elect George Bush. It recommended a freeze on all pending environmental regulations. The April 1980 notice of intent to regulate CFCs was among them. The EPA was subsequently stalled for many years, taking no action on ODCs. Then it acted only after being goaded by legal pressure brought to bear by the Natural Resources Defense Council.

Our conclusion that the explanations offered by DuPont's literature in the late 1980s is misleading is confirmed by a DuPont statement offered in June 1980, several months before Ronald Reagan's election in November of that year. At that time, DuPont had explicitly announced its intention to continue research: "In 1974 when the theory of ozone depletion by chlorofluorocarbons (CFCs) was first published, the DuPont Company initiated extensive research programs. First, research was necessary to resolve the uncertainties in the theory. Second, since the outcome of this research could not be predicted, product research was necessary to develop alternative materials which might be substituted for CFCs. In 1980, the science [of CFCs and ozone depletion] is still not resolved and it is prudent to continue both research programs." (DuPont 1980)

Thus, when regulation was imminent, DuPont took the official position that it was "prudent" to continue research on alternatives because of uncertainties in the theory. But later in that year or in 1981, DuPont abandoned research.[6] Clearly, in abandoning research on alternative chemicals, DuPont committed itself to what was, by its own definition of June 1980, an imprudent course. DuPont's claim to have abandoned its ODC-related research due to decreased scientific and regulatory concern seems, in light of their own literature, to be disingenuous, at best.

DuPont did not resume serious work on alternative chemicals until late 1986, more than a year after the Antarctic ozone hole was determined to be a critical new phenomenon associated with ODC releases.[7] The resumption of the research, in fact, coincided with the impending reality that the discovery of the ozone hole would likely lead to restrictions on ODC production and consumption.

It seems clear that corporations did not internalize even minimally the social and economic costs of the harm done by their products, until governmental regulations and citizen pressure forced them to do so. Indeed, in a 1988 interview, the External Affairs Department of DuPont explicitly cited the absence of regulations as a factor in suspending work: "In the absence of regulations there was nothing to drive the search for alternatives, because there was no market demand. Given that nothing was drawing customers to buy them, research was scaled back."[8]

After the findings of the Ozone Trends Panel were released in March 1988, DuPont announced that it was instituting a program to completely phase out production of regulated ozone-depleting compounds by about the year 2000. Other corporations followed suit.

A similar course of events occurred in most of the ODC-consuming industries. Indeed, the situation was often much worse, since in many cases these industries took no action in the 1970s and the early 1980s to reduce emissions even though it would have been economical for them to do so.

In this regard, the consumption of ODCs prior to the Montreal Protocol and the Ozone Trends Panel report was parallel to that of energy consumption before the oil-price rises of 1973–74. Individual, governmental, and corporate consumers could have taken many energy-saving actions prior to that date—actions that would have been economical at then-prevailing prices. But since energy was cheap and availability was not in question, only a few acted while most did nothing.

However, after the signing of the Montreal Protocol (which mandated sharp restrictions on ODC availability) and the subsequent clear indications by mid 1988 that these compounds would be phased out entirely, corporations dependent on ozone-depleting compounds began to pay serious attention.

13.2 Corporate Developments after 1988

In 1988 Sweden, rather than being driven by short-term political compromises with corporate interests, enacted regulations that were environmentally ambitious and based on the best available technological information. This program was developed in consultation with industry. (Unlike many countries, most corporations in Sweden tend to see environmental regulation by government as a normal part of business.) The goal was to reduce CFC use by 50 percent by the end of 1991 and phase it out entirely by January 1, 1995. The actual achievement by 1991 was a reduction of 80 percent (SEPA 1991). A complete phaseout of CFCs had been accomplished by early 1994.[9]

A large part of this achievement was possible because corporations that consumed ozone-depleting compounds began to take a close look at the options in the face of a certain, rapid phaseout of the traditionally used ODCs and probable future restrictions on HCFCs. Many of them found that ODCs were an important and even central part of their processes, though they were only a small fraction of their overall costs (even in the case of ODC-intensive products, such as insulating foam). They also found out in many cases that equally good and economical alternatives were available. In the more difficult areas, such as refrigeration, the options were not always immediately available, but they were developed far more rapidly than was anticipated even by most environmentalists at the time.

For those cases where alternatives were not readily available, Scandinavian manufacturers undertook serious and intense research to find alternatives. It is not a surprise that the technology to adapt the use of HFC-134a to blowing insulating foam for refrigerators and other appliances came from Denmark.

A similar process has been going on in the United States. However, it is hampered by a number of things. First, between 1980 and 1992 the regulatory climate was weak in comparison with many countries in Europe. As a result, American corporations tended to assume that they would have more time to develop alternatives and that HCFCs (which deplete ozone but to lesser degree than the currently used compounds) would remain available for a considerable time. This has tended to slow down innovation and the pace of introduction of new technologies, and also to divert a large portion of the funds available for innovation and conversion into HCFCs rather than into non-ozone-depleting chemicals and processes. Second, the influence of U.S. industry and governmental institutions in the process set up by UNEP to review alternative technologies and chemicals appears to be large, particularly in the area of air conditioning and refrigeration technology.

As we have discussed, the use of waste heat from on-site electricity generation has the potential to replace ozone-depleting compounds in large-scale air conditioning while accomplishing a broad range of goals in the energy sectors. Yet neither U.S. agencies (the EPA or the Department of Energy) nor UNEP have seriously considered the potential of this technology in the context of eliminating ODCs. This commercial technology uses waste heat to drive an absorption air conditioning system in the summer and to provide heating in the winter.[10] Generally, it can reduce building energy use compared to buildings air conditioned with vapor-compression systems and heated using oil- or gas-fired furnaces. Widespread use of cogeneration has the potential to reduce ODC emissions rapidly in nonunitary air conditioning applications while contributing to increasing electricity-generation capacity and reducing summer peak loads. But the UNEP panel on refrigeration ignored this technology, and the members of the panel refused to provide the public an explanation for this omission at the 1991 International Conference on Alternatives to CFCs and Halons.

One workshop at the conference included several members of the UNEP refrigeration panel. A question was raised from the audience about the conspicuous absence of representatives from the cogeneration industry on the panel. A panelist responded that industry was not represented because the technology is already commercial. When Arjun

Makhijani pointed out that a large part of the panel's purpose was to consider commercial technologies to replace ODCs, Dr. Kenneth Hickman (of York International, a maker of ODC-using air conditioning equipment), a member of the UNEP team that wrote a report on refrigeration technologies, simply said that on-site generation was "available" and that "the market will decide."[11] The clear implication was that the UNEP panel was not going to consider this technology, even though it was commercial and could be especially beneficial in immediately reducing ozone depletion and global warming impacts and could produce a number of other benefits, especially in the Third World.

A U.S. Department of Energy representative at the conference was similarly evasive, claiming that there was no representative of the on-site-generation industry on the panel and that none of the hundred or so members had thought to bring it up for inclusion! None of the papers presented at the international conference dealt with this very important technology. Thus, even as other parts of the government might be considering it as an energy conservation step or as a method of increasing electric generation capacity, it has not been integrated into an urgent schedule for ozone-layer protection.

The proceedings of the 1993 CFC and Halon Alternatives International Conference do contain one paper on the use of waste heat in absorption air conditioners. Curiously, this deals with mixtures containing HCFC-22 as possible replacements for commercially available, non-ozone-depleting lithium bromide–water and ammonia-water systems. The research, however, showed that, under circumstances applicable to air conditioning systems (as distinct from applications requiring temperatures below the freezing point of water), existing systems have adequate, and often even superior, efficiency performance (Bhaduri 1993).

Users of ODCs in the United States are obliged by regulations to consider the energy-use consequences of a change to new chemicals and processes. In this context, it is all the more remarkable that absorption air conditioning using waste heat has not been pursued more vigorously either by corporations that could use such equipment or by government.

We have already discussed specific examples of other countries with more ambitious phaseout programs for HCFCs (see chapter 5 and 11). Let us discuss the program of Sweden in the context of a positive model of government-industry interaction. Sweden planned to phase out HCFCs by the beginning of 1995 but will allow for exceptions in some refrigeration and foam-blowing applications. This is partly because many users converted to HCFCs during the rapid CFC phaseout. However, these exceptions will not be of indefinite duration. Rather, companies needing extra time will be issued a permit to use an HCFC. According to the Office of Science and Technology of the Embassy of Sweden in Washington, this permit will probably be for a limited time—"until an appropriate substitute is found."[12] At that time, the user will be obliged to shift to the new, non-HCFC substitute. This effectively places the burden of proof on those who would use an ODC, and obliges them to make an extra investment to phase out the HCFC when a substitute is found. This creates a disincentive for the kind of HCFC use that the Eastman Company decided upon, where it could have used a number of alternatives for air conditioning but decided on HCFC-22 (see chapter 5). Ironically, this is one of the conversions away from CFCs that have been cited as positive by the EPA.[13]

Most ODC-consuming corporations have come a long way since the early resistance to change in the 1980s. Chemical manufacturers have also taken many steps in response to the inevitability of regulation. Today there is not the near-general resistance to regulation that there was in 1980–81, when all of the chemical manufacturers suspended or drastically cut back research on alternatives. Rather, different countries and different corporations are proceeding at widely different rates. Yet there is every reason for governments and corporations to choose the best technologies that are non-ozone-depleting and energy efficient, both to avoid unnecessary costs and to increase the speed of the ODC phaseout.

13.3 Phaseout and Taxation Policy

The history of corporate reaction indicates that regulations based on the technical feasibility and the realistic short-term promise of new technologies are needed to spur corporations to action and innovation. This includes both the corporations that produce ozone-depleting chemicals and those that consume them.

The range of products and processes using ozone-depleting compounds is vast. And, as we have shown, the technologies are available in essentially all major areas to make a phaseout of ozone-depleting compounds by 1996 possible (see chapters 5–9). However, the delay in issuing regulations and the fact that the next revision of the Montreal Protocol will not occur until late 1995 mean that the potential cannot be practically fulfilled for some HCFCs and for methyl bromide.

In order to minimize the amount of new production before 1996 and spur development of substitutes that can be used in existing equipment, without jeopardizing large amounts of valuable equipment, individual nations should impose taxes on the use of ODCs, as the United States has done under the amended Clean Air Act. The United States and other industrialized countries should impose a tax on HCFC-22 consumption as an important complement to a policy of phaseout for new uses. A tax on new production HCFCs would result in an acceleration of recovery of inventories from existing equipment. Some of the proceeds of the taxes could be used to acquire these recovered ODCs for destruction, so as to achieve the goal of minimizing emissions rapidly.

14

Summary

1. Ozone-layer damage is serious. Improvement will not begin until the early years of the 21st century.

The seasonal "ozone hole" over the Antarctic continent exhibits declines of up to 60 percent each austral spring. It is beginning earlier, lasting longer, and widening further each year. Though not yet as serious, seasonal depletion over the northern pole has begun and may be due to mechanisms similar to those responsible for the Antarctic ozone hole. Ozone depletions of up to 20 percent have been observed over high-latitude northern-hemisphere locations for periods lasting approximately a month. The possibility that nonlinear ozone loss will occur outside the Antarctic ozone hole has serious health and environmental implications. Nonlinear ozone depletion implies that each additional amount of chlorine and bromine emitted into the stratosphere depletes ozone to a greater extent than the last. Such nonlinear effects work on the model of "the straw that broke the camel's back." We estimate that atmospheric chlorine and bromine concentrations will not begin to decline until approximately 2005 under present regulations. As a result, the level of ozone depletion will depend to a great extent on weather patterns and other natural variations in the years to come.

2. Ultraviolet radiation is now showing annual increases over populated locations.

Though measurements are limited and preliminary, the amount of UV-B radiation reaching the surface of the Earth at mid-latitude locations appears to be increasing. Increases of approximately 35 percent per year during winter and 7 percent per year during summer have been observed over Toronto between 1989 and 1993.

3. The health, economic, and environmental consequences could be severe.

Depletion of the ozone layer can adversely affect health and ecosystems in a variety of ways. The most severe ozone depletion so far, over the Antarctic and recently over the Arctic, has been in the early springtime of these regions, when UV radiation is attenuated by a long trajectory through the atmosphere. However, the global increase in ozone-depleting compounds poses the risk of sustained, year-round depletion, the effects of which could be severe.

Column ozone loss will produce variable increases in UV-B transmission depending on the wavelength, with a larger percentage increase at the more damaging shorter wavelengths. As a rule of thumb, a 1 percent decrease in column ozone produces an approximate 2 percent increase in biologically effective UV-B radiation in terms of the DNA action spectrum.

The following list summarizes estimates of the effects of ozone depletion that can be approximately quantified:

Malignant melanoma A 10 percent increase in UV could increase the incidence of malignant melanoma among light-skinned individuals by 3.5–9 percent and could cause an as-yet-unknown increase among dark-skinned individuals.

Nonmelanoma skin cancers A 10 percent increase in UV could increase the incidence of basal-cell and squamous-cell cancer by 10–30 percent.

Cataracts A 10 percent increase in UV could increase the incidence of cataracts by 6–8 percent.

Plants A 25 percent reduction in column ozone could reduce the yield of UV-sensitive varieties of soybeans by 20–25 percent.

Aquatic plant life A 10 percent increase in UV-B could yield a 2.5–5 percent loss of photosynthetically active radiation. A 25 percent reduction in column ozone could yield a 35 percent reduction in primary production near the water surface and a 10 percent reduction in the entire euphotic zone.

Aquatic animal life A 7.5 percent reduction in ozone would reduce the breeding period of shrimp by 50 percent and cause injury to other aquatic microorganisms, with consequent adverse effects on other life forms higher in the food chain.

A number of effects that are at present impossible to quantify may nevertheless be among the most significant sources of damage to health. Among these effects are the following:

damage to the immune system, manifested by a decrease in cellular immune function, leading to increased occurrence of many diseases

increases in diverse infectious diseases, such as leishmaniasis, leprosy, and schistosomiasis

increases in genetic effects due to DNA damage

increases in tropospheric ozone, and attendant effects on health and agriculture

blindness in wildlife and a consequent disruption of ecosystems.

Besides effects on individuals within species, there could be serious effects on entire ecosystems. For instance, there is evidence that declines in populations of many amphibian species, including frogs, may be due in part to UV-B radiation. Reductions in frog populations could, in turn, result in agricultural pest increases and also disrupt natural ecosystems.

A principal concern for ecosystems is the productivity reduction of phytoplankton that accompanies ozone depletion. Phytoplankton are not only food for animals higher in the food chain; they are a principal means for removing carbon from the atmosphere. Therefore, a reduction in phytoplankton productivity could aggravate greenhouse-gas increases. There could be positive feedback between ozone depletion and damage to other biogeochemical systems. Current knowledge is too poor to allow estimation of potential ecosystem damage or disruption.

4. If Third World ODC-consumption increases allowed under present regulations are realized, declines in ozone-depleting chlorine will be delayed approximately 10 years.

Increases in ODC use among the upper and upper-middle classes in the Third World as allowed under current regulations could become a significant source of emissions over the next decade. Our projections indicate that by the year 2000 consumption due to this source alone could grow to about two-thirds of the worldwide ODC consumption before beginning to decline. There are indications that some Third World countries may accelerate their phaseout schedules, but others have declined to agree to the Copenhagen Amendments to the Montreal Protocol because of the inadequate financial assistance.

5. HCFCs are more damaging than has generally been assumed.

The short-term ozone depletion potential of HCFCs (on a time scale of 10–20 years) is generally larger than that indicated by their steady-state ozone-depletion potential (S-S ODP), which is essentially a long-term average figure. For instance, HCFC-22 is three to four times more ozone depleting over 10–20 years than its long-term (200-year average) ODP of 0.05. Steady-state ODPs are the basis for regulating compounds under the U.S. Clean Air Act and are used to categorize production of ozone-depleting compounds in the Montreal Protocol.

6. Production and emissions of HCFCs are increasing.

The production and consumption of HCFCs, and in particular HCFC-22, have been increasing rapidly in the past few years, in part due to substitution for fully halogenated ODCs. HCFCs are increasing in importance as atmospheric chlorine sources.

7. Enough alternative chemicals, processes, and technologies are available so that the phaseout of HCFCs can be accelerated considerably.

A review of the non-ozone-depleting chemicals, processes, and technologies available now or likely to be available soon indicates that HCFCs can be phased out on a much more rapid schedule than that specified in the recently amended Montreal Protocol. Adoption of these alternatives will not entail significant economic disruption, although additional investment will be necessary in some instances, notably when retrofitting of existing equipment is involved.

8. Existing equipment will soon be the dominant source of ODC emissions.

An increasing proportion of the emissions will come from banks (that is, ODCs stored in existing equipment). Current emissions from the service banks of CFC-11, CFC-12, and HCFC-22 (constituting the majority of banks) are of the same order of magnitude as emissions from new production. Emissions of these chemicals from service banks will soon exceed emissions associated with new production. The total amount of

CFCs, halons, and HCFC-22 in service banks in 1992 was about 3 million tonnes—almost three times the world's production of those compounds in that year. Of this about 2.4 million tonnes is CFC-11 and CFC-12; the rest is HCFC-22.

9. Methyl bromide is a substantial contributor to the increase in ozone-depleting atmospheric bromine. More stringent regulation is required.

The contribution of anthropogenic methyl bromide to atmospheric bromine is of the same order of magnitude as the combined contribution of halons. The emissions of methyl bromide associated with commercial activity include its use as a soil, structural, and commodity fumigant and the emissions arising from the exhaust of vehicles using leaded gasoline. Current international regulation calls for a freeze on the fumigant use of methyl bromide in 1995. This exempts post-commodity use and applies only to industrial countries.

10. Anthropogenic biomass burning contributes to the buildup of atmospheric chlorine and bromine.

Burning of biomass in smoldering fires due to human activities, such as the burning of tropical forests and savannas, contributes approximately 4–5 percent of atmospheric chlorine and approximately 20 percent of atmospheric bromine. Though it is a far smaller contributor than ODCs, methyl chloride is an important component of emissions in many Third World countries.

11. Several other unregulated compounds may be contributing to the increase in atmospheric chlorine.

Four chlorinated compounds not yet included in public policy for protecting the ozone layer may be contributing to atmospheric chlorine increases. Their contributions may be of the same order of magnitude as CFC-114 and CFC-115. They are chloroform, methylene chloride, perchloroethylene, and dichloroethane. The emission levels and atmospheric removal rates of these chemicals should be evaluated to determine whether they should be regulated as ODCs.

12. The use of waste heat in absorption air conditioning could eliminate ODCs from many large-scale air conditioning applications and also save energy.

Air conditioning technologies based on use of waste heat from on-site electricity generation have the potential to greatly reduce energy consumption by using waste heat to drive absorption air conditioning systems. This would eliminate CFC use in many large-scale applications immediately. The waste heat can also be used for heating in the winter in place of natural gas or oil that is burned in heating systems on site today, yet this option is not being actively pursued by the EPA, the Department of Energy, or the United Nations Environment Programme. Water/zeolite absorption systems may be attractive for a variety of air conditioning and refrigeration applications. Other underused or underresearched technologies are evaporative cooling (especially combined direct and indirect evaporative cooling) and the use of hydrocarbon refrigerants in systems where they do not pose significant risks of flammability (e.g., small refrigerators). Ammonia can replace HCFC-22 in some chiller applications.

15

Recommendations

1. The schedules for controlling all ODCs must be adhered to globally. A slower phaseout of ODC use in certain applications in the Third World may be necessitated by differences in use patterns. For instance, carbon tetrachloride, which is highly toxic and ozone depleting, is widely used as a solvent and perhaps as a fire suppressant. It may be necessary to use HCFCs or methyl chloroform in certain applications beyond the deadlines for the industrialized countries in order to recover and destroy carbon tetrachloride in existing systems and to make possible the rapid elimination of vastly dispersed uses of carbon tetrachloride (for example, in workshops in small towns). Recovery, recycling, and destruction of CFCs and HCFCs must be instituted, though the timetable may be different than that for the industrialized countries.

2. Consideration of systems that generate electricity on site and use the waste heat for all or part of the air conditioning and heating should be made mandatory for all new buildings and developments that would require more the 300 kilowatts of electrical power for all purposes, including ODC-using air conditioning.

3. Policies to recover CFCs, halons, and HCFC-22 from existing equipment should be implemented fully. This must include recovery from existing mobile air conditioning systems. Recovery from residential refrigeration and air conditioning equipment should be effected after such equipment is discarded. Recovery equipment should be properly tested and certified as to its efficacy.

4. Policies to begin systematic acquisition and destruction of CFCs, halons, and HCFC-22 in existing products should be put in place. A part of the proceeds from taxation of ODCs in industrialized countries should be used to provide economic assistance to convert existing equipment to non-ozone-depleting substitutes (in cases of need) and to provide for the destruction of ODCs with the best available pollution-control technology.

5. The use of methyl bromide as a fumigant should be eliminated in 1996. This deadline may have to be extended should no viable alternatives exist in the Third World.

6. The use of leaded gasoline in the Third World should be eliminated as soon as possible, both because lead is toxic and because the ethylene dibromide in leaded gasoline is emitted in exhaust as methyl bromide.

7. Programs to examine the ozone-depleting potential of chloroform, methylene chloride, perchloroethylene, and dichloroethane should be instituted immediately.

8. A survey of all chlorine- and bromine-containing compounds with lifetimes estimated at more than 3 months and global emissions estimated at more than 200,000 tonnes per year should be undertaken in order to limit and eventually eliminate emissions of these compounds.[1]

9. Policy-development studies as well as scientific studies of the contribution of biomass burning should be initiated as part of the Montreal Protocol process. Negotiations on ways to end the burning of tropical forests should be included in the Montreal Protocol process.

10. Third World subsidiaries of multinational corporations and joint ventures in which multinational corporations have more than 10 percent ownership should be required to follow the same phaseout schedule that applies to the industrialized countries. Laws mandating this will be needed in industrialized countries. This should also be made a part of the Montreal Protocol revision.

11. The multilateral fund of the Montreal Protocol is inadequate and should be increased to cover the actual costs to Third World countries of a rapid phaseout.

12. A substantial, multifaceted, international research project on the various environmental, economic, and health aspects of biomass burning should be initiated.

The adoption of these recommendations would end ozone depletion about 20–30 years earlier than might be expected with strict adherence to the compliance schedule of the Copenhagen Amendments to the Montreal Protocol. In the Saving Our Skins scenario, recommended here, chlorine concentration would peak by approximately 1997 and would then begin to decline immediately. Adherence to this schedule would minimize harm and also considerably reduce the risk of severe ozone depletion over northern latitudes in the next several years.

Appendix
Model Details and Results

Data Sources

The chlorine and bromine concentration model is initialized in the year 1985 with global ODC production data from a variety of sources. These data are compound specific and include the currently used CFCs, halons, carbon tetrachloride (CCl_4), methyl chloroform (CH_3CCl_3), HCFC-22, methyl chloride (CH_3Cl), methyl bromide (CH_3Br), and projections of HCFC-141b and HCFC-123. For most of the compounds, production data are available up to and including the year 1992.[1] Annual production of the remaining compounds is extrapolated to 1992 using trends from the 1980s.

The annual ODC production data are separated into Third World and Industrial World portions in the base year for each compound according to regional estimates made by ICF Incorporated.[2] Starting from 1985, production in the Third World is linearly interpolated to the year-2000 targets described in chapter 12 for each ODC. The remainder for each year between 1985 and 1992 is allotted to the industrialized countries.

Global CFC-11, CFC-12, CFC-113, CFC-114, CFC-115, and HCFC-22 production is initialized for the period 1985–1992 using data compiled by the Alternative Fluorocarbon Environmental Acceptability Study and corrected for underreporting (AFEAS 1993a, 1993b, 1993c, 1992a, 1992b, 1992c). After 1992, CFC production in the industrialized countries is interpolated linearly to the first year in which production and consumption are controlled. HCFC-22 production in the industrialized countries is increased annually to 1996 at the average HCFC production growth rate for 1985–1992. After this, production is linearly interpolated to the first Montreal Protocol control year. Because the estimated 1992 CFC-114 production level for the industrialized countries is below what would have been the 1994 control level specified in the Copenhagen Amendments to the Montreal Protocol, production is frozen at the 1992 level for both 1993 and 1994. After this, production is linearly interpolated to the first control year.

Estimated global halon production is available through 1991.[3] The 1992 production in the industrialized countries is extrapolated, using the trend between the years 1988 and 1991. These four years are used to establish a trend because prior years exhibit an increase in production, whereas these years exhibit a significant decline. Because of the advent of regulation, it is assumed that this decline continued into 1992. After this, production is linearly interpolated to the first control year.

Carbon tetrachloride production data are initialized in 1985 using estimates prepared by Simmonds et al. (1988). For the years 1986–1992, no data could be located on global production. We estimate the production in the industrialized countries during these years by extrapolating the 1985 production using the average global production growth rate for carbon tetrachloride during the period 1980–1985 (approximately 4 percent per year). After this, production is linearly interpolated to the first control year.

Production data for methyl chloroform up to and including 1992 are taken from estimates prepared by Pauline Midgley.[4] Because the estimated 1992 production in the industrialized countries is below the 1993 control level specified in the Montreal Protocol, production of methyl chloroform for 1993 is held constant at the 1992 level. After this, production is linearly interpolated to the first control year.

Methyl chloride emissions were initialized by calculating the production necessary to achieve the 1985 steady-state atmospheric concentration of 600 pptv (WMO 1991, p. 1.15). This amounts to approximately 3582 kilotonnes per year when a 1.5-year lifetime is assumed (ibid., p. 8.8). This level of emissions is held constant until the first control year.

The portion of global methyl bromide emissions derived from methyl bromide's use as a fumigant and from the use of ethylene dibromide in leaded gasoline have been estimated for the years 1985–1990 (UNEP

1992, section II, p. 6; UNEP 1993b). Estimates of methyl bromide emissions from current levels of biomass burning are also available (Maño and Andreae 1994). These data, in combination with the estimated global methyl bromide emissions in 1990, imply a natural component equal to approximately 25 kilotonnes.[5] The fumigant portion of the 1991 and 1992 methyl bromide emissions are extrapolated using the average 1984–1990 fumigant emission trend. The emissions associated with biomass burning and the use of ethylene dibromide are held constant.

Service Banking and Emissions

The delayed emissions from many of the ODC service banks have been explicitly accounted for. The applications for which service banking is represented are CFC-11 residing in closed-cell insulating foam, CFC-12 residing in cooling systems (hermetically sealed ones and non-hermetically-sealed ones), all HCFC-22 residing in medium-term banks (1–10-year release horizon), all halons residing in fire extinguishers, and all CFC-113 residing in cooling systems. Though some CFC-114 and 115 are banked in refrigeration systems, the amounts are small and thus have not been accounted for here.

The method used to calculate the service banks and the subsequent emissions is schematically represented in figure A.1. In a given year i, the production for a given application of a compound $P_i f_a$ for which a service bank exists is first added to the remaining service bank from the previous year B_{i-1}^a. The release from the bank is then calculated by multiplying this increased service bank by a release fraction RF_a. The amount released can thus be expressed as

$$R_i^a = (B_{i-1}^a + P_i f_a) RF_a, \qquad (A.1)$$

where f_a is the fraction of total production allocated to application a. Both this fraction and the release fraction were derived from estimates of annual bank size and sales between the years 1985 and 1992 (or 1990 for the halons) (AFEAS 1993a, 1993c; McColluch 1992). Beyond 1992, the sales fraction and the release fraction are estimated by either averag-

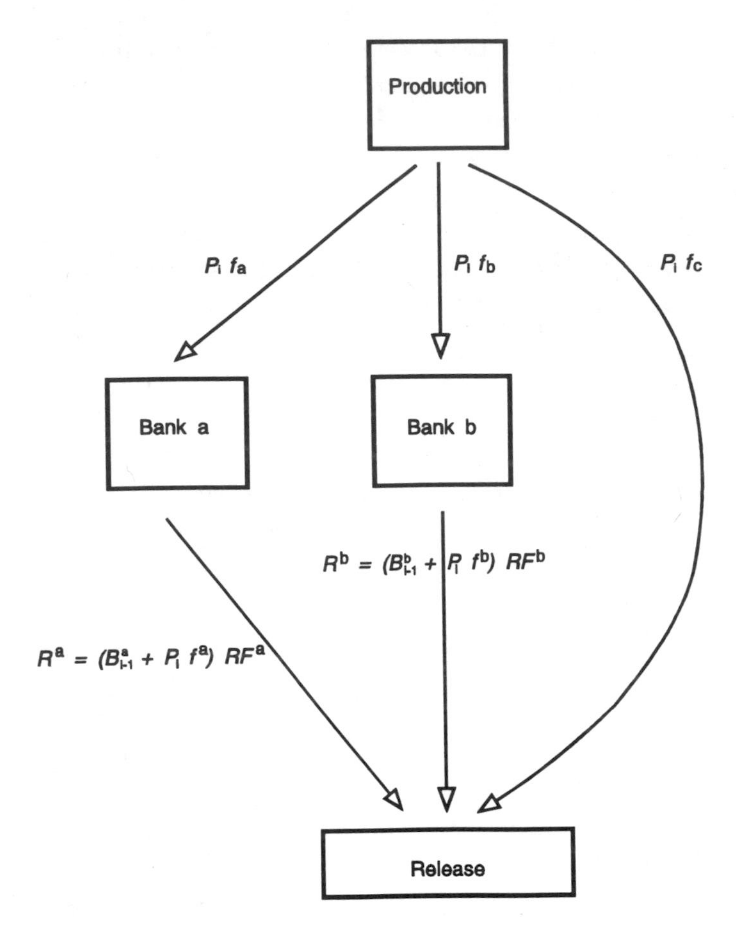

Figure A.1
Method used to calculate service-bank emissions.

ing past data (no discernible trend) or extrapolating according to a calculated trend.

Elimination or control of service banks is considered in the Saving Our Skins scenario. The service banks are divided into a portion that continues to release until exhausted and a portion that experiences a controlled removal or release limitation. For foams and hermetically sealed cooling devices, such as small refrigerators and air conditioners, the majority of the releases are assumed to occur after permanent disposal. Through recovery prior to disposal (or incineration in the case of closed-cell insulating foams), a portion of the bank in these devices is eliminated before release. This effectively reduces the size of the leaking bank by the amounts specified in chapter 12. For non-hermetically-sealed systems, such as large air conditioners and refrigerators, the majority of the releases are avoidable and occur during servicing and maintenance. It is assumed that the bank control enacted here reflects improved servicing and maintenance procedures. In the case of fire extinguishers, we assume reduced testing and training; we also assume retrofitting of many systems to use alternative substances. The recovered halon is ultimately destroyed.

Emissions data (by compound) for the three scenarios are given in tables A.1, A.2, and A.3 for representative years.

Calculating Concentration

The first goal in the exercise of calculating the equivalent chlorine burden in the atmosphere is to ascertain atmospheric ODC concentrations given the emissions outlined in chapter 12. We start with the differential equation describing the mass balance of a particular ODC in the atmosphere,

$$\frac{d[X](t)}{dt} = E_X(t) - \frac{[X](t)}{\tau_X}, \tag{A.2}$$

where $[X](t)$ is the time-dependent, globally averaged mixing ratio of a particular ODC X in units of parts per trillion by volume (pptv), $E_X(t)$ is the time-dependent emissions of a particular ODC in units of pptv per year, and τ_X is the steady-state lifetime of the particular ODC in years. The lifetime representing one e-folding (approximately a 63 percent reduction) in the concentration of a pulsed emission is assumed to remain constant throughout the calculation period. We assume that all the ODCs are well mixed in the atmosphere. This is a reasonable assumption, since mixing times in the atmosphere are of the order of weeks to months and the lifetimes of all the modeled compounds exceed a year.

The unit of concentration, the mixing ratio, is the number of moles or molecules of a particular compound per number of moles or molecules of balance gas. In this case, our balance gas is background air and the concentration of the compound of interest is expressed in units of moles or molecules per trillion of background air.

Table A.1
Emissions, Copenhagen Amendments scenario (kilotonnes/year).

Year	CFC-11	CFC-12	CFC-113	CFC-114	CFC-115	Halon-1211	Halon-1301	HCFC-22	CCl_4	MCF	HCFC-123	HCFC-141b	CH_3Cl	CH_3Br
1985	329.4	424.3	212.5	17.1	10.0	8.9	5.1	152.0	88.7	588.0	0.0	0.0	3581.7	87.4
1990	248.1	342.7	193.0	8.3	11.3	9.5	3.0	219.2	72.4	725.7	0.0	0.0	3581.7	100.0
1992	208.5	312.1	124.5	4.7	10.7	6.9	2.7	247.5	66.7	627.0	0.0	0.0	3581.7	104.4
1994	185.6	236.5	72.6	5.0	3.9	6.0	2.7	286.3	47.2	445.1	102.8	102.8	3581.7	109.5
1996	158.2	180.1	22.9	4.9	1.2	4.9	2.5	322.7	30.7	171.3	205.5	205.5	3581.7	104.7
2000	133.6	128.7	22.7	4.9	1.2	3.4	2.3	329.9	36.1	204.6	187.6	187.6	3581.7	107.5
2005	88.7	64.4	3.5	0.6	0.2	2.1	2.0	307.2	5.2	49.1	160.2	160.2	3581.7	97.3
2010	51.5	20.6	0.6	0.0	0.0	1.2	1.7	243.6	0.0	0.0	107.8	107.8	3581.7	97.3
2020	19.1	3.2	0.3	0.0	0.0	0.4	1.3	83.8	0.0	0.0	27.0	27.0	3426.5	92.3
2050	1.0	0.2	0.1	0.0	0.0	0.0	0.5	0.1	0.0	0.0	0.0	0.0	3116.1	82.3
2090	0.0	0.0	0.0	0.0	0.0	0.0	0.2	0.0	0.0	0.0	0.0	0.0	3116.1	82.3

Table A.2
Emissions, Accelerated Phaseout scenario (kilotonnes/year).

Year	CFC-11	CFC-12	CFC-113	CFC-114	CFC-115	Halon-1211	Halon-1301	HCFC-22	CCl_4	MCF	HCFC-123	HCFC-141b	CH_3Cl	CH_3Br
1985	329.4	424.3	212.5	17.1	10.0	8.9	5.1	152.0	88.7	588.0	0.0	0.0	3581.7	87.4
1990	248.1	342.7	193.0	8.3	11.3	9.5	3.0	219.2	72.4	725.7	0.0	0.0	3581.7	100.0
1992	208.5	312.1	124.5	4.7	10.7	6.9	2.7	247.5	66.7	627.0	0.0	0.0	3581.7	104.4
1994	185.6	236.5	72.6	5.0	3.9	6.0	2.7	286.3	47.2	445.1	102.8	102.8	3581.7	102.2
1996	158.2	180.1	22.9	4.9	1.2	4.9	2.5	322.7	30.7	171.3	164.0	164.0	3581.7	102.2
2000	133.6	128.7	22.7	4.9	1.2	3.4	2.3	329.9	36.1	204.6	153.3	94.4	3581.7	82.8
2005	88.7	64.4	3.5	0.6	0.2	2.1	2.0	303.1	5.2	49.1	130.3	46.1	3581.7	54.6
2010	51.5	20.6	0.6	0.0	0.0	1.2	1.7	210.0	0.0	0.0	71.0	21.4	3581.7	54.6
2020	19.1	3.2	0.3	0.0	0.0	0.4	1.3	49.2	0.0	0.0	10.0	0.0	3426.5	49.6
2050	1.0	0.2	0.1	0.0	0.0	0.0	0.5	0.0	0.0	0.0	0.0	0.0	3116.1	39.6
2090	0.0	0.0	0.0	0.0	0.0	0.0	0.2	0.0	0.0	0.0	0.0	0.0	3116.1	39.6

Table A.3
Emissions, Saving Our Skins scenario (kilotonnes/year).

Year	CFC-11	CFC-12	CFC-113	CFC-114	CFC-115	Halon-1211	Halon-1301	HCFC-22	CCl_4	MCF	HCFC-123	HCFC-141b	CH_3Cl	CH_3Br
1985	329.4	424.3	212.5	17.1	10.0	8.9	5.1	152.0	88.7	588.0	0.0	0.0	3581.7	87.4
1990	248.1	342.7	193.0	8.3	11.3	9.5	3.0	219.2	72.4	725.7	0.0	0.0	3581.7	100.0
1992	208.5	312.1	124.5	4.7	10.7	6.9	2.7	247.5	66.7	627.0	0.0	0.0	3581.7	104.4
1994	185.6	236.5	72.6	5.0	3.9	6.0	2.7	286.3	47.2	445.1	102.8	102.8	3581.7	102.2
1996	74.7	68.6	1.2	0.0	0.0	2.4	1.3	73.8	0.0	51.6	0.0	0.0	3581.7	63.4-
2000	45.6	26.0	0.9	0.0	0.0	1.6	1.1	22.7	0.0	0.0	0.0	0.0	3581.7	54.6
2005	26.2	7.8	0.7	0.0	0.0	0.9	1.0	5.2	0.0	0.0	0.0	0.0	3465.3	50.9
2010	15.6	2.3	0.5	0.0	0.0	0.6	0.8	1.2	0.0	0.0	0.0	0.0	3348.9	47.1
2020	5.8	0.2	0.3	0.0	0.0	0.2	0.6	0.1	0.0	0.0	0.0	0.0	3116.1	39.6
2050	0.3	0.0	0.1	0.0	0.0	0.0	0.2	0.0	0.0	0.0	0.0	0.0	3116.1	39.6
2090	0.0	0.0	0.0	0.0	0.0	0.0	0.1	0.0	0.0	0.0	0.0	0.0	3116.1	39.6

Emissions data are supplied in metric tons (tonnes) of a particular ODC per year. This can be converted to a mixing ratio (pptv) per year as follows:

$$E_X(t, \text{pptv}) = \left[\frac{E_X(t, \text{tonnes})}{M_a}\right]\left(\frac{\text{MW}_a}{\text{MW}_X}\right)(1 \times 10^{12}), \quad \text{(A.3)}$$

where $E_X(t)$ represents the emissions for the modeled ODC X, in tonnes per year, M_a is the mass of the atmosphere (5.14×10^{15} tonnes), and MW_a and MW_X are the molecular weights of air and the modeled ODC, respectively.

Emissions of a few of the ODCs modeled involve the process of banking, as explained previously. All other emissions are considered equivalent to production in a given year.

The continuous solution to equation A.1, assuming constant emissions over the one-year integration period, is

$$[X](t) = E_X(t)\tau_X\{1 - e^{-t/\tau_X}\} + [X](0)e^{-t/\tau_X}. \quad \text{(A.4)}$$

This can be expressed in discrete form as

$$[X](t + 1) = E_X(t + 1)\tau_X\{1 - e^{-1/\tau_X}\} + [X](t)e^{-1/\tau_X}. \quad \text{(A.5)}$$

The concentration model is initialized in the year 1985 from estimated atmospheric concentrations and runs to the year 2090, calculating all time-dependent values on an annual basis (WMO 1991, p. 8.9; UNEP 1992, section II, p. 5).

Tables A.4, A.5, and A.6 list the resulting concentrations by compound for the three scenarios.

As was mentioned in chapter 4, we are comparing the relative ozone-depleting potentials of the selected emission scenarios by examining total chlorine and bromine concentrations in the atmosphere over time. This is a direct function of the atmospheric concentration of each of the ODCs modeled and the number of chlorine and/or bromine atoms present per molecule.

Two different model runs were made. The first simply sums the number of bromine atoms and the number of chlorine atoms. The resulting atmospheric chlorine and bromine concentration is termed the "non-weighted equivalent chlorine concentration." In the second run, we attempt to reflect the greater ozone-depleting efficiency of bromine by

Table A.4
ODC concentrations, Copenhagen Amendments scenario (pptv).

Year	CFC-11	CFC-12	CFC-113	CFC-114	CFC-115	Halon-1211	Halon-1301	HCFC-22	CCl_4	MCF	HCFC-123	HCFC-141b	CH_3Cl	CH_3Br
1985	222.0	382.0	30.0	5.0	4.0	1.5	1.7	80.0	100.0	130.0	0.0	0.0	600.0	11.0
1990	265.8	461.8	64.1	7.4	6.3	2.4	2.3	112.0	103.5	154.8	0.0	0.0	600.0	11.5
1992	273.8	483.3	71.6	7.7	7.1	2.4	2.4	128.2	104.1	156.0	0.0	0.0	600.0	12.0
1994	279.3	497.9	75.5	7.9	7.5	2.4	2.5	147.1	103.5	150.7	4.1	6.9	600.0	12.6
1996	282.7	507.4	76.2	8.2	7.6	2.3	2.7	168.0	101.6	125.5	10.3	21.7	600.0	12.5
2000	285.4	516.7	76.3	8.7	7.7	2.0	2.9	206.7	98.1	89.1	13.1	46.2	600.0	12.7
2005	282.3	516.8	74.9	9.0	7.8	1.6	3.1	239.1	91.2	58.8	11.7	61.9	600.0	12.0
2010	270.2	502.6	71.7	8.8	7.7	1.3	3.2	250.3	82.0	25.9	8.5	63.8	600.0	11.6
2020	236.7	464.7	65.6	8.4	7.6	0.7	3.4	200.2	66.3	5.0	2.3	40.6	576.7	11.0
2050	142.2	359.8	50.1	7.3	7.2	0.1	3.0	36.6	35.0	0.0	0.0	3.1	522.0	9.8
2090	68.9	255.0	34.8	6.1	6.7	0.0	2.1	2.9	15.0	0.0	0.0	0.1	522.0	9.8

Table A.5
ODC concentrations, Accelerated Phaseout scenario (pptv).

Year	CFC-11	CFC-12	CFC-113	CFC-114	CFC-115	Halon-1211	Halon-1301	HCFC-22	CCl_4	MCF	HCFC-123	HCFC-141b	CH_3Cl	CH_3Br
1985	222.0	382.0	30.0	5.0	4.0	1.5	1.7	80.0	100.0	130.0	0.0	0.0	600.0	11.0
1990	265.8	461.8	64.1	7.4	6.3	2.4	2.3	112.0	103.5	154.8	0.0	0.0	600.0	11.5
1992	273.8	483.3	71.6	7.7	7.1	2.4	2.4	128.2	104.1	156.0	0.0	0.0	600.0	12.0
1994	279.3	497.9	75.5	7.9	7.5	2.4	2.5	147.1	103.5	150.7	4.1	6.9	600.0	12.2
1996	282.7	507.4	76.2	8.2	7.6	2.3	2.7	168.0	101.6	125.5	8.7	18.9	600.0	12.2
2000	285.4	516.7	76.3	8.7	7.7	2.0	2.9	206.7	98.1	89.1	10.7	32.1	600.0	11.2
2005	282.3	516.8	74.9	9.0	7.8	1.6	3.1	238.8	91.2	58.8	9.6	30.0	600.0	7.1
2010	270.2	502.6	71.7	8.8	7.7	1.3	3.2	243.7	82.0	25.9	6.0	25.5	600.0	6.5
2020	236.7	464.7	65.6	8.4	7.6	0.7	3.4	179.6	66.3	5.0	1.0	10.6	576.7	6.0
2050	142.2	359.8	50.1	7.3	7.2	0.1	3.0	29.1	35.0	0.0	0.0	0.7	522.0	4.7
2090	68.9	255.0	34.8	6.1	6.7	0.0	2.1	2.3	15.0	0.0	0.0	0.0	522.0	4.7

Table A.6
ODC concentrations, Saving Our Skins scenario (pptv).

Year	CFC-11	CFC-12	CFC-113	CFC-114	CFC-115	Halon-1211	Halon-1301	HCFC-22	CCl_4	MCF	HCFC-123	HCFC-141b	CH_3Cl	CH_3Br
1985	222.0	382.0	30.0	5.0	4.0	1.5	1.7	80.0	100.0	130.0	0.0	0.0	600.0	11.0
1990	265.8	461.8	64.1	7.4	6.3	2.4	2.3	112.0	103.5	154.8	0.0	0.0	600.0	11.5
1992	273.8	483.3	71.6	7.7	7.1	2.4	2.4	128.2	104.1	156.0	0.0	0.0	600.0	12.0
1994	279.3	497.9	75.5	7.9	7.5	2.4	2.5	147.1	103.5	150.7	4.1	6.9	600.0	12.2
1996	277.6	499.0	75.2	7.9	7.5	2.2	2.6	145.0	99.6	118.8	2.2	7.9	600.0	9.8
2000	266.8	489.1	72.7	7.8	7.5	1.7	2.6	120.9	91.5	61.7	0.2	5.4	600.0	7.1
2005	250.0	471.5	69.6	7.6	7.4	1.3	2.7	90.9	82.2	27.2	0.0	3.4	584.5	6.2
2010	232.1	452.5	66.6	7.4	7.3	0.9	2.6	66.8	73.9	12.0	0.0	2.2	565.1	5.7
2020	197.0	415.5	60.9	7.1	7.2	0.4	2.6	35.6	59.8	2.3	0.0	0.9	526.1	4.8
2050	115.7	320.8	46.5	6.2	6.8	0.0	2.1	5.3	31.6	0.0	0.0	0.1	522.0	4.7
2090	56.0	227.3	32.3	5.2	6.3	0.0	1.4	0.4	13.5	0.0	0.0	0.0	522.0	4.7

weighting the bromine atoms in the brominated compounds (halons and methyl bromide) by a factor of 40 (WMO 1991, p. 6.14). The resulting atmospheric concentration of chlorine and bromine is termed the "weighted equivalent chlorine concentration."

The expression for the concentration of equivalent chlorine is given by

$$[\mathrm{Cl}] = \sum_{i=1}^{14} [X_i]\mathrm{Cl\#} + [X_i]\mathrm{WF} \cdot \mathrm{Br\#}, \qquad \text{(A.6)}$$

where [Cl] is the mixing ratio of equivalent chlorine in parts per trillion, WF is the bromine weighting factor (1 or 40), and Cl# and Br# are the number of chlorine and bromine atoms, respectively, per molecule of the ith ODC.

Tables A.7–A.12 list the resulting equivalent chlorine concentrations by compound for the three scenarios for the nonweighted equivalent chlorine concentration run and the weighted equivalent chlorine concentration run.

Table A.7
Nonweighted equivalent chlorine concentrations (ppbv), Copenhagen Amendments scenario.[a]

Year	CFC-11	CFC-12	CFC-113	CFC-114	CFC-115	Halon-1211	Halon-1301	HCFC-22	CCl_4	MCF	HCFC-123	HCFC-141b	CH_3Cl	CH_3Br	Total
1985	0.666	0.764	0.090	0.010	0.004	0.003	0.002	0.080	0.400	0.390	0.000	0.000	0.600	0.011	3.020
1990	0.797	0.924	0.192	0.015	0.006	0.005	0.002	0.112	0.414	0.464	0.000	0.000	0.600	0.012	3.544
1992	0.821	0.967	0.215	0.015	0.007	0.005	0.002	0.128	0.416	0.468	0.000	0.000	0.600	0.012	3.657
1994	0.838	0.996	0.226	0.016	0.007	0.005	0.003	0.147	0.414	0.452	0.004	0.014	0.600	0.013	3.734
1996	0.848	1.015	0.229	0.016	0.008	0.005	0.003	0.168	0.407	0.376	0.010	0.043	0.600	0.012	3.740
2000	0.856	1.033	0.229	0.017	0.008	0.004	0.003	0.207	0.393	0.267	0.013	0.092	0.600	0.013	3.735
2005	0.847	1.034	0.225	0.018	0.008	0.003	0.003	0.239	0.365	0.176	0.012	0.124	0.600	0.012	3.665
2010	0.811	1.005	0.215	0.018	0.008	0.003	0.003	0.250	0.328	0.078	0.009	0.128	0.600	0.012	3.466
2020	0.710	0.929	0.197	0.017	0.008	0.001	0.003	0.200	0.265	0.015	0.002	0.081	0.577	0.011	3.017
2050	0.427	0.720	0.150	0.015	0.007	0.000	0.003	0.037	0.140	0.000	0.000	0.006	0.522	0.010	2.036
2090	0.207	0.510	0.104	0.012	0.007	0.000	0.002	0.003	0.060	0.000	0.000	0.000	0.522	0.010	1.437

a. Transport time has been estimated at approximately 2–5 years. (Assume a midpoint of 3 years.) The timing of all equivalent chlorine values will be delayed by this amount.

Table A.8
Nonweighted equivalent chlorine concentrations (ppbv), Accelerated Phaseout scenario.[a]

Year	CFC-11	CFC-12	CFC-113	CFC-114	CFC-115	Halon-1211	Halon-1301	HCFC-22	CCl_4	MCF	HCFC-123	HCFC-141b	CH_3Cl	CH_3Br	Total
1985	0.666	0.764	0.090	0.010	0.004	0.003	0.002	0.080	0.400	0.390	0.000	0.000	0.600	0.011	3.020
1990	0.797	0.924	0.192	0.015	0.006	0.005	0.002	0.112	0.414	0.464	0.000	0.000	0.600	0.012	3.544
1992	0.821	0.967	0.215	0.015	0.007	0.005	0.002	0.128	0.416	0.468	0.000	0.000	0.600	0.012	3.657
1994	0.838	0.996	0.226	0.016	0.007	0.005	0.003	0.147	0.414	0.452	0.004	0.014	0.600	0.012	3.734
1996	0.848	1.015	0.229	0.016	0.008	0.005	0.003	0.168	0.407	0.376	0.009	0.038	0.600	0.012	3.732
2000	0.856	1.033	0.229	0.017	0.008	0.004	0.003	0.207	0.393	0.267	0.011	0.064	0.600	0.011	3.703
2005	0.847	1.034	0.225	0.018	0.008	0.003	0.003	0.239	0.365	0.176	0.010	0.060	0.600	0.007	3.594
2010	0.811	1.005	0.215	0.018	0.008	0.003	0.003	0.244	0.328	0.078	0.006	0.051	0.600	0.007	3.375
2020	0.710	0.929	0.197	0.017	0.008	0.001	0.003	0.180	0.265	0.015	0.001	0.021	0.577	0.006	2.930
2050	0.427	0.720	0.150	0.015	0.007	0.000	0.003	0.029	0.140	0.000	0.000	0.001	0.522	0.005	2.019
2090	0.207	0.510	0.104	0.012	0.007	0.000	0.002	0.002	0.060	0.000	0.000	0.000	0.522	0.005	1.431

a. Transport time has been estimated at approximately 2–5 years. (Assume a midpoint of 3 years.) The timing of all equivalent chlorine values will be delayed by this amount.

Table A.9
Nonweighted equivalent chlorine concentrations (ppbv), Saving Our Skins scenario.[a]

Year	CFC-11	CFC-12	CFC-113	CFC-114	CFC-115	Halon-1211	Halon-1301	HCFC-22	CCl_4	MCF	HCFC-123	HCFC-141b	CH_3Cl	CH_3Br	Total
1985	0.666	0.764	0.090	0.010	0.004	0.003	0.002	0.080	0.400	0.390	0.000	0.000	0.600	0.011	3.020
1990	0.797	0.924	0.192	0.015	0.006	0.005	0.002	0.112	0.414	0.464	0.000	0.000	0.600	0.012	3.544
1992	0.821	0.967	0.215	0.015	0.007	0.005	0.002	0.128	0.416	0.468	0.000	0.000	0.600	0.012	3.657
1994	0.838	0.996	0.226	0.016	0.007	0.005	0.003	0.147	0.414	0.452	0.004	0.014	0.600	0.012	3.734
1996	0.833	0.998	0.226	0.016	0.008	0.004	0.003	0.145	0.398	0.356	0.002	0.016	0.600	0.010	3.614
2000	0.800	0.978	0.218	0.016	0.007	0.003	0.003	0.121	0.366	0.185	0.000	0.011	0.600	0.007	3.316
2005	0.750	0.943	0.209	0.015	0.007	0.003	0.003	0.091	0.329	0.082	0.000	0.007	0.584	0.006	3.029
2010	0.696	0.905	0.200	0.015	0.007	0.002	0.003	0.067	0.296	0.036	0.000	0.004	0.565	0.006	2.801
2020	0.591	0.831	0.183	0.014	0.007	0.001	0.003	0.036	0.239	0.007	0.000	0.002	0.526	0.005	2.444
2050	0.347	0.642	0.139	0.012	0.007	0.000	0.002	0.005	0.126	0.000	0.000	0.000	0.522	0.005	1.808
2090	0.168	0.455	0.097	0.010	0.006	0.000	0.001	0.000	0.054	0.000	0.000	0.000	0.522	0.005	1.319

a. Transport time has been estimated at approximately 2–5 years. (Assume a midpoint of 3 years.) The timing of all equivalent chlorine values will be delayed by this amount.

Table A.10
Weighted equivalent chlorine concentrations (ppbv). Copenhagen Amendments scenario.[a]

Year	CFC-11	CFC-12	CFC-113	CFC-114	CFC-115	Halon-1211	Halon-1301	HCFC-22	CCl_4	MCF	HCFC-123	HCFC-141b	CH_3Cl	CH_3Br	Total
1985	0.666	0.764	0.090	0.010	0.004	0.062	0.068	0.080	0.400	0.390	0.000	0.000	0.600	0.440	3.574
1990	0.797	0.924	0.192	0.015	0.006	0.096	0.091	0.112	0.414	0.464	0.000	0.000	0.600	0.462	4.174
1992	0.821	0.967	0.215	0.015	0.007	0.098	0.096	0.128	0.416	0.468	0.000	0.000	0.600	0.481	4.313
1994	0.838	0.996	0.226	0.016	0.007	0.098	0.101	0.147	0.414	0.452	0.004	0.014	0.600	0.503	4.416
1996	0.848	1.015	0.229	0.016	0.008	0.095	0.107	0.168	0.407	0.376	0.010	0.043	0.600	0.499	4.420
2000	0.856	1.033	0.229	0.017	0.008	0.084	0.115	0.207	0.393	0.267	0.013	0.092	0.600	0.506	4.421
2005	0.847	1.034	0.225	0.018	0.008	0.068	0.124	0.239	0.365	0.176	0.012	0.124	0.600	0.481	4.319
2010	0.811	1.005	0.215	0.018	0.008	0.051	0.130	0.250	0.328	0.078	0.009	0.128	0.600	0.463	4.093
2020	0.710	0.929	0.197	0.017	0.008	0.027	0.135	0.200	0.265	0.015	0.002	0.081	0.577	0.442	3.605
2050	0.427	0.720	0.150	0.015	0.007	0.003	0.122	0.037	0.140	0.000	0.000	0.006	0.522	0.391	2.538
2090	0.207	0.510	0.104	0.012	0.007	0.000	0.086	0.003	0.060	0.000	0.000	0.000	0.522	0.391	1.901

a. Transport time has been estimated at approximately 2–5 years. (Assume a midpoint of 3 years.) The timing of all equivalent chlorine values will be delayed by this amount.

Table A.11
Weighted equivalent chlorine concentrations (ppbv), Accelerated Phaseout scenario.[a]

Year	CFC-11	CFC-12	CFC-113	CFC-114	CFC-115	Halon-1211	Halon-1301	HCFC-22	CCl_4	MCF	HCFC-123	HCFC-141b	CH_3Cl	CH_3Br	Total
1985	0.666	0.764	0.090	0.010	0.004	0.062	0.068	0.080	0.400	0.390	0.000	0.000	0.600	0.440	3.574
1990	0.797	0.924	0.192	0.015	0.006	0.096	0.091	0.112	0.414	0.464	0.000	0.000	0.600	0.462	4.174
1992	0.821	0.967	0.215	0.015	0.007	0.098	0.096	0.128	0.416	0.468	0.000	0.000	0.600	0.481	4.313
1994	0.838	0.996	0.226	0.016	0.007	0.098	0.101	0.147	0.414	0.452	0.004	0.014	0.600	0.489	4.403
1996	0.848	1.015	0.229	0.016	0.008	0.095	0.107	0.168	0.407	0.376	0.009	0.038	0.600	0.487	4.401
2000	0.856	1.033	0.229	0.017	0.008	0.084	0.115	0.207	0.393	0.267	0.011	0.064	0.600	0.449	4.334
2005	0.847	1.034	0.225	0.018	0.008	0.068	0.124	0.239	0.365	0.176	0.010	0.060	0.600	0.285	4.057
2010	0.811	1.005	0.215	0.018	0.008	0.051	0.130	0.244	0.328	0.078	0.006	0.051	0.600	0.261	3.805
2020	0.710	0.929	0.197	0.017	0.008	0.027	0.135	0.180	0.265	0.015	0.001	0.021	0.577	0.239	3.320
2050	0.427	0.720	0.150	0.015	0.007	0.003	0.122	0.029	0.140	0.000	0.000	0.001	0.522	0.188	2.323
2090	0.207	0.510	0.104	0.012	0.007	0.000	0.086	0.002	0.060	0.000	0.000	0.000	0.522	0.188	1.698

a. Transport time has been estimated at approximately 2–5 years. (Assume a midpoint of 3 years.) The timing of all equivalent chlorine values will be delayed by this amount.

Table A.12
Weighted equivalent chlorine concentrations (ppbv), Saving Our Skins scenaro.[a]

Year	CFC-11	CFC-12	CFC-113	CFC-114	CFC-115	Halon-1211	Halon-1301	HCFC-22	CCl_4	MCF	HCFC-123	HCFC-141b	CH_3Cl	CH_3Br	Total
1985	0.666	0.764	0.090	0.010	0.004	0.062	0.068	0.080	0.400	0.390	0.000	0.000	0.600	0.440	3.574
1990	0.797	0.924	0.192	0.015	0.006	0.096	0.091	0.112	0.414	0.464	0.000	0.000	0.600	0.462	4.174
1992	0.821	0.967	0.215	0.015	0.007	0.098	0.096	0.128	0.416	0.468	0.000	0.000	0.600	0.481	4.313
1994	0.838	0.996	0.226	0.016	0.007	0.098	0.101	0.147	0.414	0.452	0.004	0.014	0.600	0.489	4.403
1996	0.833	0.998	0.226	0.016	0.008	0.090	0.104	0.145	0.398	0.356	0.002	0.016	0.600	0.392	4.183
2000	0.800	0.978	0.218	0.016	0.007	0.071	0.105	0.121	0.366	0.185	0.000	0.011	0.600	0.286	3.765
2005	0.750	0.943	0.209	0.015	0.007	0.052	0.106	0.091	0.329	0.082	0.000	0.007	0.584	0.249	3.424
2010	0.696	0.905	0.200	0.015	0.007	0.037	0.106	0.067	0.296	0.036	0.000	0.004	0.565	0.229	3.163
2020	0.591	0.831	0.183	0.014	0.007	0.018	0.103	0.036	0.239	0.007	0.000	0.002	0.526	0.194	2.749
2050	0.347	0.642	0.139	0.012	0.007	0.002	0.084	0.005	0.126	0.000	0.000	0.000	0.522	0.188	2.075
2090	0.168	0.455	0.097	0.010	0.006	0.000	0.056	0.000	0.054	0.000	0.000	0.000	0.522	0.188	1.557

a. Transport time has been estimated at approximately 2–5 years. (Assume a midpoint of 3 years.) The timing of all equivalent chlorine values will be delayed by this amount.

Notes

Chapter 1

1. The Dobson unit is named after G. M. B. Dobson, a pioneer in the study of atmospheric ozone. For an excellent account of his work see Dobson 1968.

2. The energy of a photon, $h\nu$, is the product of the frequency and Planck's constant. The product of frequency and wavelength (λ) is a constant equal to the speed of light.

3. Reactive forms of chlorine (Cl and ClO) are more precisely referred to as **radicals**, because of their electronic configuration. For simplicity, we will not use this term.

4. The term **aerosol** is used to refer to any suspended solid or liquid particles (or a mixture) in the atmosphere.

5. A **dimer** is a molecule in which a sequence of atoms is repeated twice. A **trimer** is a molecule in which three repetitions occur. A **polymer** is a molecule in which a sequence is repeated more than three times.

6. Sulfate aerosols are typically droplets of sulfuric acid and water.

7. On May 7, 1993, the TOMS instrument aboard Nimbus 7 ceased functioning.

8. Concerning area-weighted weekly column ozone between 65°S and 65°N.

9. The spatial extent of the Antarctic ozone hole is usually defined as the area over which ozone levels are approximately 200 Dobson units or less.

10. See, e.g., Madronich 1992.

11. Approximately 25% of the sulfur compounds in the stratosphere are estimated to be anthropogenic.

12. The extent to which subsonic aircraft (which fly primarily in the troposphere) affect stratospheric ozone is unclear. Recent two-dimensional modeling indicates that the impact is negligible (WMO 1991, pp. 9.7–9.8).

13. This has diminished somewhat as a result of the more stringent regulation now in place.

Chapter 2

1. The spectrum for UV-B is sometimes defined as 280–320 nm (see, e. g., Smith 1992). The upper limit for UV-C wavelengths would be correspondingly changed to 280 nm in this case.

2. Because the absorption cutoff is not sharp, the wavelength range comprising UV-B radiation as defined in the literature is sometimes defined as 280–320 nm and sometimes as 290–320 nm. We will use the latter definition as a convention.

3. The discussion in this section is based on chapters 242 and 522 of Wyngaarden et al. 1992 unless otherwise stated.

4. The discussion in this section is based on Morison 1989 unless otherwise stated.

5. Basal-cell and squamous-cell skin cancers are also called "nonmelanoma" cancers.

6. This calculation is based on the assumption that all the increased UV actually reaches the surface, and therefore ignores physical and meteorological phenomena decreasing the amount of UV reaching the Earth under certain conditions.

Chapter 3

1. Only a minute quantity of astatine exists in nature. It is a short-lived radioactive element first discovered when it was artificially made by bombarding bismuth with alpha particles. It is not relevant to our discussion of halocarbons.

2. The most common at the time were ammonia, methyl chloride, and sulfur dioxide.

3. Earlier, Midgely had developed lead additives for gasoline. An excellent discussion of his inventions can be found in Cagin and Dray 1993.

4. Although some halocarbons are produced naturally and contribute to what we have referred to as the natural ozone balance, all references to halocarbons in the remainder of this document refer to human-derived compounds unless specifically stated otherwise.

5. Manufacturers use various brand names, such as Suva (DuPont) and Klea (ICI), when referring to these compounds.

6. Many compounds have been proposed as substitutes for ozone-depleting halocarbons. Figure 3.1 lists only the most promising candidates being considered or used as of late 1993.

7. It is estimated that approximately 60% of the chloroform emissions ($\sim$1000 kilotonnes per year) are due to natural processes (WMO 1991, p. 1.15).

8. For example, R-502 (51.2% CFC-115, 48.8% HCFC-22) is a blend currently used in large refrigeration systems. Many mixtures of HCFCs and HFCs have

been proposed as alternatives to currently used ODCs. For example, two blends proposed as replacements for HCFC-22 are HFC-32/HFC-125 (60%/40%) and HFC-32/HFC-134a (30%/70%).

9. These are the most recent data for which comprehensive regional or national estimates have been performed. Many of the regional use patterns have not changed since 1985.

10. Source: UNEP 1991a, p. ES-1. There are conflicting estimates of the amount of ODCs used as aerosol propellants. For example, AFEAS 1993a estimates that approximately 71,000 tonnes of CFC-11 and CFC-12 were produced for propellant purposes in 1990.

Chapter 4

1. The total atmospheric concentration of chlorine and bromine referred to here includes all the chlorine and bromine associated with human-derived as well as natural halocarbons in the atmosphere. Because not all chlorine and bromine released into the atmosphere will participate in stratospheric ozone depletion, this quantity should be considered an upper limit to the amount of chlorine and bromine potentially involved in ozone loss.

2. To be more specific: 2 ppbv of chlorine and 0.4 ppbv of chlorine equivalent bromine.

3. The criterion we propose here is specific to ozone-layer protection. Its underlying rationale is that it provides for some margin of safety against humans' triggering additional phenomena like the Antarctic ozone hole.

4. The relative steady-state ozone loss can differ from the relative steady-state chlorine contribution at a particular time in that the altitude dependence of chlorine release is explicitly accounted for and translated into ozone loss.

5. The T-ODP values at a time horizon of 500 years can be considered representative of the S-S ODP.

Chapter 5

1. Mixtures are of two principal types. Azeotropic mixtures behave as though the mixture was a single substance so far as boiling and condensation are concerned. In nonazeotropic mixtures, the component chemicals boil and condense at different temperatures. In addition, there are some mixtures which are not completely azeotropic but nearly so.

2. IEER estimate based on UNEP 1991d, AFEAS 1993b, and AFEAS 1993c. This is the most recent published sector-specific consumption data.

3. IEER estimate based on UNEP 1991d, AFEAS 1993a, AFEAS 1993b, and AFEAS 1993c.

4. Based on AFEAS 1993a.

5. Based on AFEAS 1993a and AFEAS 1993c.

6. Worldwide production of ODCs in 1990 is estimated at 1.9 million tonnes. This excludes human-derived CH_3Cl and CH_3Br emissions. See appendix for data sources.

7. Derived from data given in Fischer et al. 1991, p. 8.2, table 8.2.

8. Jerry Stofflet (General Motors Corporation), telephone communications, May 1988. Note that a "ton" of air conditioning is defined as 12,000 Btu per hour cooling capacity and is not to be confused with the measure of weight.

9. Based on data presented in AFEAS 1993a.

10. Based on data presented in AFEAS 1993c.

11. IEER estimate derived from data presented in UNEP 1991d.

12. Derived from data presented in UNEP 1991d.

13. R-500 is a 73.8%/26.2% mixture of CFC-12 and HFC-152a.

14. Several examples of such retrofitting are presented in EPA 1993b.

15. Jeffrey Levy, Significant New Alternatives Program, U.S. EPA, personal communication, April 1994.

16. IKON is the trade name for these compounds.

17. As of early 1994, CF_3I had passed acute-toxicity tests (*GECR* 1994).

18. A 50-45-5 mixture would replace HCFC-22; a 25-70-5 mixture would replace R-502.

19. The coefficient of performance (COP) for the HCFC-22 replacement mixture was estimated to be a 15% improvement over HCFC-22. The R-502 replacement exhibited a gain roughly of 25%.

20. See "Greenfreeze," a pamphlet issued by Greenpeace (probably in 1993).

21. "Cool Storage Roofs ETAP Project," Consultant Report, California Energy Commission, Document P500-92-014, 1992, appendix C, p. 1.

Chapter 6

1. IEER estimate based on data presented in UNEP 1991b. AFEAS (1993a) estimates a significantly larger amount of 1990 CFC-11 consumption in this application: approximately 200,000 tonnes. The source of this discrepancy is unknown.

2. IEER estimate based on data presented in UNEP 1991b and AFEAS 1993c. On AFEAS's (1993a) estimate see note 1.

3. IEER estimate based on data presented in UNEP 1991b.

4. Telephone communication, Ski Fuseo, National Fiber, October 1992.

5. Telephone communication, Roger Caffier, consultant, Degussa Corporation,

Ridgefield Park, N.J., October 1992; telephone communication, Ram Srikanth, staff engineer, Admiral, Galesburg, Ill., November 1992.

6. Telephone communication, Ram Srikanth, staff engineer, Admiral. Galesburg, Ill., November 1992.

7. Telephone communication, Tom Potter, Solar Energy Research Institute, Golden, Colorado, October 1992.

8. Telephone communication, Stan Rusek, senior engineer, Owens-Corning Corp., Granville, Ohio, October 1992.

9. R values refer to the insulating ability of material per unit thickness. The unit used here is hr-ft2/Btu-inch.

10. The CFC-11 foam was partially blown with water. An additional test yielding approximately the same result is reported in York 1993.

11. The process using IPC is called "LBL2 technology."

12. After time, the insulating ability of CO2-blown foams decreases substantially.

Chapter 7

1. These values are corrected for estimated underreporting. We estimate that in 1989 approximately 15% of total CFC-113 use was in nonsolvent (e.g. foam, refrigerant, and aerosol) applications.

2. IEER estimate based on Fischer et al. 1991, p. 11.4, table 11.2, AFEAS 1993b, and AFEAS 1992b.

3. Written communication, Pauline Midgley, December 8, 1993.

4. Percent estimated from data supplied by Pauline Midgley (ibid.).

5. Written communication, Pauline Midgley, December 8, 1993.

6. Originally called "Turner's Crazy Flux" after Ray Turner, a Hughes Aircraft employee who discovered that lemon juice made an excellent water-soluble soldering bond (IEEE Spectrum 1993).

7. CBE 1992, citing manufacturers' claims.

Chapter 8

1. Personal communication, Pauline Midgley, December 7, 1993.

2. David Rosage, facilities engineer, Goddard Networks Division, NASA (notes taken by Arjun Makhijani at International CFC and Halon Alternatives Conference, Baltimore, December 3, 1991).

3. Telephone communication, Ted Moore, Center for Environmental Technologies, University of New Mexico, October 1992.

Chapter 9

1. This does not include the use of methyl chloroform or HCFC-22 as propellants. It is estimated that the amounts of HCFCs and methyl chloroform used in aerosol products were approximately 20,000 and 34,000 tonnes, respectively, in 1990.

2. AFEAS (1993a) estimates that the 1992 consumption of CFC-11 and CFC-12 as propellants was approximately 30,800 tonnes. However, the 1990 AFEAS estimate was about 71,000 tonnes, whereas the UNEP estimate was 115,000 tonnes (UNEP 1991a). The source of this discrepancy is unknown.

3. Telephone communication, Ron Davis, technical director, CCL Custom Manufacturing, Danville, Ill., October 1992.

4. It is estimated that methyl chloroform accounts for 10% of all U. S. auxiliary blowing agent use in the production of flexible foams (UNEP 1991b, p. III-42).

Chapter 10

1. Chapter 1 contains a discussion of volcanic eruptions and their contribution to atmospheric chlorine.

2. Derived from WMO 1989, vol. I, p. 246. We include only chlorine from organic compounds that play significant roles in stratospheric ozone depletion.

3. Rasmussen et al. 1980; personal verbal communication, Wei M. Hao, Max-Planck-Institut, Mainz, March 6, 1990; Crutzen et al. 1979.

4. For more details on the choice of this range see Makhijani and Makhijani 1990. Andreae (1991, p. 10, table 1.4) gives a range of 20–300 ppmv, with a "best guess" of 50 ppmv.

5. World Congress on Climate and Development, Hamburg, 1988.

6. We use a rather conservative value of 80% carbon dioxide, the rest being carbon monoxide. Since we are making an order-of-magnitude estimate, the exact ratio of carbon dioxide to carbon monoxide does not make a significant difference in our calculation relative to the other uncertainties.

7. These estimates were based on global biomass emissions of between 2.5 million and 4.5 million kilotonnes of carbon.

8. Telephone communication, Mary O'Brien, University of Montana, October 1992.

Chapter 11

1. The appendixes to Benedick 1991 contain the full text of the Vienna Convention, the Montreal Protocol, and the amendments and adjustments agreed to in London in 1990.

2. Signature does not, by itself, entail a legal obligation. It represents an intention to ratify or accede to the treaty. Ratification or accession (ratification without signing) makes the treaty legally binding for that country and is achieved by domestic approval. At that point, a country is considered a "party" to the protocol. When the protocol enters into force, it does so only for those countries party to it.

3. Written communication, The Depository, United Nations Office of Legal Affairs, New York.

4. In meeting these reductions, each compound is weighted by its ozone-depleting potential.

5. As of early 1994, 51 parties had met this criteria with an additional 35 pending.

6. An English translation of the text of the Swedish program can be found in appendix II of Makhijani et al. 1988.

7. "Adjustments" modify control measures for substances already included in the protocol. "Amendments," on the other hand, refer to new substances or the alteration of provisions other than control measures already in place. Because adjustments do not require ratification, they enter into force immediately for all parties to the Montreal Protocol. Amendments require ratification.

8. In order to classify as an Article 5 country, consumption of carbon tetrachloride and methyl chloroform must not exceed 0.2 kg/person on the date at which the protocol enters into force for that country or during the additional 10 years.

9. IEER estimate.

10. Written communication, The Depository, United Nations Office of Legal Affairs, New York.

11. Industry delegates outnumbered delegates not from industry or government three to one. DuPont alone had more representatives at the negotiations than all but six countries (Rowlands 1993).

12. HCFC-22 is identified explicitly on the right-hand side of this equation because it was the only HCFC produced in significant quantities in 1989.

13. UNEP press release, March 22, 1994.

14. Base-line amounts have not been specified as of early 1994.

15. We list the most important here. For a complete list of Title VI rules, call the Stratospheric Ozone Information Hotline: 1 800 296-1996.

Chapter 12

1. By "Third World countries" we mean the countries defined in Article 5 of the Montreal Protocol.

2. Data and references are supplied in the appendix.

3. Production controls included in international regulations and those introduced in this work are usually not specified for consecutive years. In these instances, production in the intervening years is linearly interpolated. ODC-production data from the last few years indicate that this is a reasonable assumption.

4. These compounds are listed under Title III of the amended U.S. Clean Air Act (US CAA 1990).

5. The provisions of the Copenhagen Amendments are detailed in chapter 11.

6. Written communication, ICF, 1991.

7. For example, per-capita use in the United States in 1985 was approximately 3 kg (including methyl chloroform and HCFC-22).

8. HCFC-22 captures approximately 10% of the total estimated consumption in the year 2000. HCFC-123 and HCF-141b capture 5% each.

9. Production quantities are not available. Recent estimates indicate that the combined production of HCFC-141b and HCFC-141b from two producers, Elf Atochem and Solvay, were roughly 50,000 tonnes in 1993 (EIN 1993).

10. HCFC-142b may be produced in greater quantities than HCFC-123 in the future. Were we to simulate HCFC use with HCFC-141b and 142b, the impact of HCFCs on our simulated atmospheric chlorine concentration would be slightly larger because HCFC-142b has a longer lifetime than HCFC-123 (and one chlorine atom per molecule, rather than two for HCFC-123).

11. The interim reduction schedule and cap specification is described in chapter 11.

12. A growth rate of 8.5% is used. This is the current average ODC growth rate in Third World (UNEP/ExCom 1992, p. ii).

13. This rate reflects growth between 1985 and 1992.

14. Our estimate for 1991 is based on an assumed growth from 1990 levels of 6.3%. This average rate for the growth of the fumigant component between 1984 and 1990 is extended to the first control year.

15. Quarantine and preshipment (commodity) fumigation constitute approximately 20% of the total methyl bromide emissions associated with fumigant use. See chapter 10 for more details.

16. The 1992 Copenhagen revisions of the Montreal Protocol did not specify a control schedule for methyl bromide in Third World countries. This is to be determined in 1995.

17. FR 1993b; *GECR* 1993a; personal communication, Brad Hurley (*GECR* editor), February 9, 1994.

18. ODCs currently residing in air conditioners, refrigerators, and foams are referred to as **service banks**.

19. The weighting factor for methyl bromide ranges from 30 to 120. The factor of 40 represents a consensus of the WMO's experts.

20. As was discussed in chapter 4, this method results in an upper limit on the amounts of chlorine and bromine available to deplete ozone.

21. Transport time has been estimated at approximately 2–5 years. We assume a midpoint of 3 years. The timings of all equivalent chlorine values will be delayed by this amount.

22. In 1991 the atmospheric concentration of carbon monoxide began to decline and the growth rate of methane slowed noticeably. It is not clear whether this is connected to changes in OH (Novelli et al. 1994).

23. This works as follows: Consider a large leaky bucket with water flowing into it. The water will build up to a point where the outflow from the leaks equals the inflow. Now, if we plug up some of the leaks, the water will build up to a higher level even though the rate of inflow has not changed.

Chapter 13

1. CFCs are an important subset of a larger group of ozone-depleting compounds—see chapter 3.

2. IEER estimate based on use patterns at the time.

3. A number of chemicals that were later confirmed as ozone-depleting compounds were not considered at the time. The most important of these are methyl chloroform, carbon tetrachloride, HCFC-22, and methyl bromide.

4. Telephone communication, Kathleen H. Forte, DuPont External Affairs Division, May 26, 1988.

5. Ibid.

6. We have not been able get an exact date from DuPont representatives.

7. Telephone communication, Kathleen H. Forte, DuPont External Affairs Division, May 26, 1988.

8. Ibid. This quotation was repeated to Ms. Forte on June 6, 1988, so as to confirm that it accurately represented DuPont's position.

9. Personal facsimile communication, Cecelia Hellstadius, Embassy of Sweden, Washington, April 5, 1994.

10. Absorption air conditioning uses the alternate absorption and desorption of ammonia in water to produce the same cooling effect as vapor compression and condensation in the more familiar, electrically driven, ODC-using air conditioning systems. Lithium bromide may be used instead of ammonia. Since absorption and desorption do not require compression but only a heat source, the system can be run using waste heat from electricity generation, or some other source of energy (e.g. natural gas).

11. Notes of Arjun Makhijani from Workshop on Refrigeration and Air Conditioning, International CFC and Halon Alternatives Conference, UNEP Overview Presentations, Baltimore, December 3, 1991.

12. Personal facsimile communication, Cecelia Hellstadius, Embassy of Sweden, Washington, April 5, 1994.

13. EPA Environmental Protection Agency 1993b, Eastman Chemical Co. case history.

Chapter 15

1. For a compound containing two chlorine atoms and having a molecular weight of approximately 100 g/mole, the resulting contribution to equivalent atmospheric chlorine would be of the same order of magnitude as our long-term goal of 0.2 pptv (see chapter 4).

Appendix

1. This applies to all the CFCs, to HCFC-22, and to CH_3CCl_3.

2. Written communication, ICF, 1991. This separation was not performed for methyl bromide or methyl chloride.

3. McColluch 1992; written communication, Pauline Midgley, December 7, 1993.

4. Personal communication, Pauline Midgley, December 8, 1993.

5. Khalil (1993) estimates global emissions to be approximately 100 kilotonnes.

References

Abbatt, J. P. D., and M. J. Molina. 1992. "The Heterogeneous Reaction of $HOCl + HCl \rightarrow Cl_2 + H_2O$ on Ice and Nitric Acid Trihydrate: Reaction Probabilities and Stratospheric Implications." *Geophysical Research Letters* 19, no. 5: 461–464.

ACDA (Arms Control and Disarmament Agency). 1978. An Assessment of Frequently Neglected Effects in Nuclear Attacks. Civil Defense Study Report 5.

AFEAS (Alternative Fluorocarbons Environmental Acceptability Study). 1992a. Chlorofluorocarbons (CFCs) 11, 12: Cumulative Production through 1979 and Annual Production and Sales for the Years 1980–1990.

AFEAS (Alternative Fluorocarbons Environmental Acceptability Study). 1992b. Chlorofluorocarbons (CFCs) 113, 114, and 115: Cumulative Production through 1979 and Annual Production and Sales for the Years 1980–1990.

AFEAS (Alternative Fluorocarbons Environmental Acceptability Study). 1992c. 1970–1991 Production and Sales of HCFC-22.

AFEAS (Alternative Fluorocarbons Environmental Acceptability Study). 1993a. Chlorofluorocarbons (CFCs) 11 and 12 Annual Production for the Years 1931–1975 and Annual Production and Sales for the Years 1976–1992.

AFEAS (Alternative Fluorocarbons Environmental Acceptability Study). 1993b. Chlorofluorocarbons (CFCs) 113, 114 and 115 Cumulative Production through 1979 and Annual Production and Sales for the Years 1980–1992.

AFEAS (Alternative Fluorocarbons Environmental Acceptability Study). 1993c. Hydrochlorofluorocarbon-22 (HCFC-22) Annual Production and Sales for the Years 1970–1992.

Alliance for Responsible CFC Policy. 1991. *Ozone Protection Policies: A Briefing Book.*

Anderson, J. G., W. H. Brune, S. A. Lloyd, D. W. Toohey, S. P. Sander, W. L. Starr, M. Loewenstein, and J. R. Podolske. 1989. "Kinetics of O_3 Destruction by

ClO and BrO Within the Antarctic Vortex: An Analysis Based on in Situ ER-2 Data." *Journal of Geophysical Research* 94, no. D9: 11,480–11,520.

Anderson, J. G., D. W. Toohey, and W. H. Brune. 1991. "Free Radicals Within the Antarctic Vortex: The Role of CFCs in Antarctic Ozone Loss." *Science* 251: 39–46.

Andreae, M. O. 1991. "Biomass Burning: Its History, Use, and Distribution and Its Impact on Environmental Quality and Global Climate." In *Global Biomass Burning*, ed. J. S. Levine. MIT Press.

Angell, J. K., and J. Korshover. 1964. "Quasi-Biennial Variations in Temperature, Total Ozone, and Tropopause Height." *Journal of Atmospheric Science* 21, p. 479.

Angell, J. K., and J. Korshover. 1967. "Biennial Variation in Springtime Temperature and Total Ozone in Extratropical Latitudes." *Monthly Weather Review* 95, p. 757.

Armstrong, B. K. 1986. "Sunlight and Malignant Melanoma in Western Australia." See EPA 1986.

Atkinson, R. J., W. A. Matthews, P. A. Newman, and R. A. Plumb. 1989. "Evidence of Mid-altitude Impact of Antarctic Ozone Depletion." *Nature* 340: 290–294.

Atkinson, W. 1993. "Mobile Air Conditiong: Mobile Air Conditioning Service under the Clean Air Act." In Proceedings of the 1993 International CFC and Halon Alternatives Conference, Washington.

Atwood, T. 1991. "HFC 32 ... Part of the 'New Math' of Refrigerants." In Proceedings of the International CFC and Halon Alternatives Conference, Baltimore.

Austin, J., N. Butchart, and K. P. Shine. 1992. "Possibility of an Arctic Ozone Hole in a Doubled-CO_2 Climate." *Nature* 360: 221–225.

Avallone, L. M., D. W. Toohey, M. H. Proffitt, J. J. Margitan, K. R. Chan, and J. G. Anderson. 1993. "In Situ Measurements of ClO at Mid-Latitudes: Is There an Effect from Mt. Pinatubo?" *Geophysical Research Letters* 20, no. 22: 2519–2522.

Baker, J. 1991. Talk given at Refrigeration and Air Conditioning Workshop, International CFC and Halon Alternatives Conference, Baltimore.

Benedick, R. E. 1991. *Ozone Diplomacy: New Directions in Safeguarding the Planet.* Harvard University Press.

Benson, D. K., and T. F. Potter. 1992. The Effect of New Priorities and New Materials on Residential Refrigerator Design. National Renewable Energy Laboratory, Golden, Colorado.

Berchowitz, D. M. n.d. Free-Piston Rankine Compression and Stirling Cycle Machines for Domestic Refrigeration. Sunpower Co., Athens, Ohio.

Bhaduri, S. C. 1993. "Performance of HCFC-22 Based Refrigerant-Absorption Mixtures." In Proceedings of the 1993 International CFC and Halon Conference, Washington.

Billimoria, S. B., & Co., for Ministry of Environment and Forests, Government of India, 1990. Interim Report on the Supply and Use of Ozone Depleting Substances in India, Part B, Sectoral Analysis.

Blaustein, A., P. Hoffman, D. G. Hokit, J. Kiesecker, S. Walls, and J. Hays. 1994. "UV Repair and Resistance to Solar UV-B in Amphibian Eggs: A Link to Population Decline." *Proceedings of the National Academy of Sciences* 91: 1791–1795.

Blessing, J. 1991. "A Unit for the Recovery of CFC-11 from Rigid Polyurethane Foams." In Proceedings of the International CFC and Halon Alternatives Conference, Baltimore.

Blevins, E., J. Sharpe, A. Fine, and B. Kopko. n.d. Zero Ozone Depleting Blowing Agents for Use in Polyurethane Based Foam Insulation.

Blumthaler, M., and W. Amback. 1990. "Indication of Increasing Solar Ultraviolet-B Radiation Flux in Alpine Regions." *Science* 248: 206–208.

Bojkov, R. D., C. S. Zerefos, D. S. Balis, I. C. Ziomas, and A. F. Bais. 1993. "Record Low Total Ozone During Northern Winters of 1992 and 1993." *Geophysical Research Letters* 20, no. 13: 1351–1354.

Brasseur, G., and C. Granier. 1992. "Mount Pinatubo Aerosols, Chlorofluorocarbons, and Ozone Depletion." *Science* 257: 1239–1242.

Browell, E. V., C. F. Butler, M. A. Fenn, W. B. Grant, S. Ismail, M. R. Schoeberl, O. B. Toon, M. Loewenstein, and J. R. Podolske. 1993. "Ozone and Aerosol Changes During the 1991–1992 Airborne Arctic Stratospheric Expedition." *Science* 261: 1155–1157.

Bruhl, C., and P. J. Crutzen. 1989. "On the Disproportionate Role of Tropospheric Ozone as a Filter against Solar UV-B Radiation." *Geophysical Research Letters* 16, no. 7: 703–706.

Bruhl, C., and P. J. Crutzen. 1990. "Ozone and Climate Changes in the Light of the Montreal Protocol: A Model Study." *Ambio* 19, no. 6–7: 293–301.

Brune, W. H., D. W. Toohey, J. G. Anderson, and K. R. Chan. 1990. "In Situ Observations of ClO in the Arctic Stratosphere: ER-2 Aircraft Results from 59 N to 80 N Latitude." *Geophysical Research Letters* 17, no. 4, March supplement: 505–508

Brune, W. H., J. G. Anderson, D. W. Toohey, D. W. Fahey, S. R. Kawa, R. Jones, D. S. McKenna, and L. R. Poole. 1991. "The Potential for Ozone Depletion in the Arctic Polar Stratosphere." *Science* 252: 1260–1266.

Bruno, K. 1991. "Poison Petrol, Leaded Gas Exports to the Third World." *Multinational Monitor*, July-August: 24–27.

Cagin, S., and P. Dray. 1993. *Between Earth and Sky: How CFCs Changed Our World and Endangered the Ozone Layer*. Pantheon.

Calkins, J., and M. Blakefield. 1986. "An Estimate of the Role of Current Levels of Solar Ultraviolet Radiation in Aquatic Ecosystems." See EPA 1986.

Catchpole, D.1991. "Halon Phase-out: An Alaskan Oil & Gas Industry Perspective." In Proceedings of the International CFC and Halon Alternatives Conference, Baltimore.

CATs (Californians for Alternatives to Toxics). 1992. Ozone Depletion Caused by Methyl Bromide.

CBE (Citizens for a Better Environment). 1992. California's Worst Corporate Ozone Destroyers.

C&EN. 1992a. "New Propellant System Devised for Aerosol Packaging." *Chemical and Engineering News,* January 13: 21–22.

C&EN. 1992b. "Ozone Depletion: Arctic Hole Feared; Sulfate Aerosol Blamed." *Chemical and Engineering News*, February 10: 4–5.

Chandra, S., and R. S. Stolarski. 1991. "Recent Trends in Stratospheric Total Ozone: Implications of Dynamical and El Chichon Perturbations." *Geophysical Research Letters* 18, no. 12: 2277–2280.

Charman, W. N. 1990. "Ocular Hazards Arising from Depletion of the Natural Atmospheric Ozone Layer: AA Review." *Ophthalmic and Physiological Optics* 10 (October): 333–341.

Chiou, J. P. 1986. Application of Solar Powered Ventilator in Automobiles. SHAH technical paper 860585.

Cleve, U. 1989. "Application of Carbon Based Adsorbers for Washing of Flue Gases." In Proceedings of the 1989 Incineration Conference, Knoxville.

CMR. 1989. "Chemical Profile: Methyl Chloride." *Chemical Marketing Reporter*, March 28.

Cooper, K. D., L. Oberhelman, T. Hamilton, O. Baadsgaard, M. Terhune, G. LeVee, T. Anderson, and H. Koren. 1993. "UV Exposure Reduces Immunization Rates and Promotes Tolerance to Epicutaneous Antigens in Humans: Relationship to Dose, CD1a-DR+ Epidermal Macrophage Induction, and Langerhans Cell Depletion." In *Proceedings of the National Academy of Sciences* 89: 8497–8501.

Corr, S., E. J. Goodwin, R. D. Gregson, A. Halse, A. A. Lindley, S. H. Colmery, T. W. Dekleva, and R. W. Yost. 1991. "Retrofitting Mobile Air-conditioning Systems with HFC-134a." In Proceedings of the International CFC and Halon Alternatives Conference, Baltimore.

Cowan, R. S. 1983. *More Work for Mother*. Basic Books.

Creyf, H. 1991. "Substitution of CFCs in Flexible and Rigid Polyurethane Foams: State of the Art, A Europur Company Viewpoint." Presented at The Global Business Outlook for CFC Alternatives, London.

Crompton, G. K. 1991. "Dry Powder Inhalers: Advantages and Limitations." *Journal of Aerosol Medicine* 4, no. 3: 151–154.

Crutzen, P. J., et al. 1979. "Biomass Burning as a Source of Atmospheric Trace Gases CO, H_2, N_2O, CH_3Cl, and COS." *Nature* 282: 253–256.

Crutzen, P. J., and F. Arnold. 1986. "Nitric Acid Cloud Formation in the Cold Antarctic Stratosphere: A Major Cause for the Springtime 'Ozone Hole.'" *Nature* 324: 651–655.

Crutzen, P. J., R. Muller, C. Bruhl, and T. Peter. 1992. "On the Potential Importance of the Gas Phase Reaction $CH_3O_2 + ClO \rightarrow ClOO + CH_3O$ and the Heterogeneous Reaction $HOCl + HCl \rightarrow H_2O + Cl_2$ in 'Ozone Hole' Chemistry." *Geophysical Research Letters* 19, no. 11: 1113–1116.

Cunnold, D. M., P. J. Fraser, R. F. Weiss, R. G. Prinn, P. G. Simmonds, B. R. Miller, F. N. Alyea, and A. J. Crawford. 1994. "Global Trends and Annual Releases of CCl_3F and CCl_2F_2 Estimated from ALE/GAGE and Other Measurements from July 1978 to June 1991." *Journal of Geophysical Research* 99, no. D1: 1107–1126.

Dekleva, T. W., B. Durr, P. D. Guy, L. Petersson, and T. Widgren. 1993. "Retrofit of a Swedish District Heat Pump from R-500 to KLEA-134a." In Proceedings of the 1993 International CFC and Halon Alternatives Conference, Washington.

Delmas, R. J. 1992. "Environmental Information from Ice Cores." *Reviews of Geophysics* 30, no. 1: 1–21.

Deshler, T., B. J. Johnson, and W. R. Rozier. 1994. "Changes in the Character of Polar Stratospheric Clouds over Antarctica in 1992 Due to Pinatubo Volcanic Aerosol." *Geophysical Research Letters* 21, no. 4: 273–276.

de Vos, R., and I. D. Rosbotham. 1993. "Polyurethane Foam Based Vacuum Panel Technology." In Proceedings of the 1993 International CFC and Halon Alternatives Conference, Washington.

Dobson, G. M. B. 1968. "Fory Years' Research on Atmospheric Ozone at Oxford: A History." *Applied Optics* 7, no. 3: 387–405.

DOC (U.S. Department of Commerce). 1992. *Statistical Abstract of the United States 1992*. Government Printing Office.

DoD (U.S. Department of Defense). 1991. Recommendations for Eliminating the Use of Ozone Depleting Compounds in the Defense Sector. Chlorofluorocarbon Advisory Committee report.

Dohlinger, M. 1993. "Comparative Energy Efficiency of Hydrocarbon Refrigerants." Presented at Ozone-Safe Cooling 1993 Conference, Washington.

Donate, F., and J. Papajesk. 1991. "A New Semi-Aqueous Alternative to 1,1,1-Trichloroethane in Metal Cleaning." In Proceedings of the International CFC and Halon Alternatives Conference, Baltimore.

DuPont. 1980. Fluorocarbon/Ozone Update—Alternatives to Fully Halogenated Chlorofluorocarbons: The DuPont Development Program.

DuPont. 1987. Fluorocarbon/Ozone Update—Alternatives to Fully Halogenated Chlorofluorocarbons: The DuPont Development Program.

Dwyer, F., and K. Thrun. 1991. "CFC Foam Blowing Agent Substitutes in Rigid Polyurethane Insulation." In Proceedings of the International CFC and Halon Alternatives Conference, Baltimore.

EIN (Environmental Information Networks, Inc.). 1991. CFC Executive News Summary, November.

EIN (Environmental Information Networks, Inc.). 1992a. "Toyota to Install Ozone-Friendly Air Conditioners by 1994." March 3.

EIN (Environmental Information Networks, Inc.). 1992b. "Ford Taurus, First CFC-Free Car from U.S., Rolls Off Line in Atlanta." March 4.

EIN (Environmental Information Networks, Inc.). 1992c. "Editorial Cites Two Potential Substitute Refrigeration Technologies." March 11.

Eisenstark, A. 1989. "Bacterial Genes Involved in Response to Near-Ultraviolet Radiation." *Advances in Genetics* 26: 99–147.

Elkins, J. W., T. M. Thompson, T. H. Swanson, J. H. Butler, B. D. Hall, S. O. Cummings, D. A. Fisher, and A. G. Raffo. 1993. "Decrease in the Growth Rates of Atmospheric Chlorofluorocarbons 11 and 12." *Nature* 364: 780–783.

Elkins, J. W., T. M. Thompson, J. H. Butler, R. C. Myers, A. D. Clarke, T. H. Swanson, D. J. Endres, A. M. Yoshinaga, R. C. Schnell, M. Winey, B. G. Mendonca, M. V. Losleben, N. B. A. Trivett, D. E. J. Worthy, V. Hudec, V. Chorney, P. J. Fraser, and L. W. Porter. 1994. Pages 426–427 of Trends '93: A Compendium of Data on Global Change, ed. T. Bowden et al. (report ORBL/CDIAC-65, Carbon Dioxide Information Analysis Center, Oak Ridge National Laboratory).

In Trends '93: A Compendium of Data on Global Change, ed. T. A. Boden et al. Report ORBL/CDIAC-65, Carbon Dioxide Information Analysis Center, Oak Ridge National Laboratory.

Elmets, C. A., M. LeVine, and D. R. Bickers. 1985. "Action Spectrum Studies for Induction of Immunological Unresponsiveness to Dinitrofluorobenzene Following *In Vivo* Low Dose Ultraviolet Radiation." *Photochemistry and Photobiology* 42, no. 4: 391–397.

Elwood, J. M., and T. G. Hislop. 1982. Solar Radiation in the Etiology of Cutaneous Malignant Melanoma in Caucasians. National Cancer Institute monograph 62.

Emmet, E. A. 1986. "Health Effects of Ultraviolet Radiation." See EPA 1986.

Energy Concepts Co. 1993. "Summary of Research and Development Activities." Reprinted in Proceedings of the Ozone-Safe Cooling Conference, Washington.

EPA (U.S. Environmental Protection Agency). 1986. *Effects of Changes in Stratospheric Ozone and Global Climate.* Proceedings of the International Conference on Health and Environmental Effects of Ozone Modification and Climate Change.

EPA (U.S. Environmental Protection Agency). 1988. Future Concentrations of Stratospheric Chlorine and Bromine. Report EPA 400/1-88/005.

EPA (U.S. Environmental Protection Agency). 1993a. Stratospheric Ozone Protection Final Rule Summary, Complying with the Refrigerant Recycling Rule.

EPA (U.S. Environmental Protection Agency). 1993b. Moving to Alternative Refrigerants.

EPA Office of Air and Radiation. 1987. Assessing the Risks of Trace Gases That Can Modify the Stratosphere.

EPA Office of Air and Radiation. 1987. Regulatory Impact Analysis: Protection of Stratospheric Ozone.

Ertinger, R. 1991. "HCFC-22: Requirements for Unitary Products." In Proceedings of the International CFC and Halon Alternatives Conference, Baltimore.

Ewan, C., E. A. Bryant, G. D. Calvert, J. Marthick, and D. Condon-Paoloni. 1991. "Potential Effects of Greenhouse Effect and Ozone Layer Depletion in Australia." *Medical Journal of Australia* 154: 554–559.

Evans, W. F. J., and C. T. McElroy. 1985. "The Conversion of N_2O_5 to HNO_3 at High Latitudes in Winter." *Geophysical Research Letters* 12, no. 12: 825–828.

EW. 1993. "Denmark Freezes Ozone Protection Rules under EC Pressure." *Environment Watch: Western Europe*, October 15: 12–13.

Fahey, D. W., K. K. Kelly, S. R. Kawa, A. F. Tuck, M. Loewenstein, K. R. Chan, and L. E. Heidt. 1990. "Observations of Denitrification and Dehydration in the Winter Polar Stratospheres." *Nature* 344: 321–324.

Federal Register. 1992. 40 CFR Part 82, "Protection of Stratospheric Ozone." *Federal Register* 57, no. 135: 31242–31268.

Federal Register. 1993a. 40 CFR Part 82, "Protection of Stratospheric Ozone." *Federal Register* 58, no. 27: 8136–8169.

Federal Register.1993b. 40 CFR Part 82, "Protection of Stratospheric Ozone." *Federal Register* 58, no. 236: 65018–65082.

Fischer, S. K., P. J. Hughes, P. D Fairchild, C. L. Kusik, J. T. Dieckmann, E. M. McMahon, and N. Hobday. 1991. Energy and Global Warming Impacts of CFC Alternative Technologies. Alternative Fluorocarbons Environmental Acceptability Study and U.S. Department of Energy.

Fisher, D. A., C. H. Hales, D. L. Filkin, M. K. W. Ko, N. D. Sze, P. S. Connell, D. J. Wuebbles, I. S. A. Isaksen, and F. Stordal. 1990. "Model Calculations of the Relative Effects of CFCs and Their Replacements on Stratospheric Ozone." *Nature* 344: 508–512.

Frederick, J. E., H. E. Snell, and E. K. Haywood. 1989. "Solar Ultraviolet Radiation at the Earth's Surface." *Photochemistry and Photobiology* 50, no. 8: 443–450.

Frederick, J. E., and A. D. Alberts. 1991. "Prolonged Enhancement in Surface Ultraviolet Radiation During the Antarctic Spring of 1990." *Geophysical Research Letters* 18, no. 10: 1869–1871.

Frederick, J. E., P. F. Soulen, S. B. Diaz, I. Smolskaia, C. R. Booth, T. Lucas, and D. Neuschuler. 1993. "Solar Ultraviolet Irradiance Observed from Southern Argentina: September 1990 to March 1991." *Journal of Geophysical Research*

Fried, A., B. E. Henry, and J. G. Calvert. 1994. "The Reaction Probability of N2O5 with Sulfuric Acid Aerosols at Stratospheric Temperatures and Compositions." *Journal of Geophysical Research* 99, no. D2: 3517–3532.

Friends of the Earth Ltd. 1991. Cold Comfort for the Ozone Layer: Local Authority Recovery and Recycling of CFCs from Domestic Refrigeration Equipment.

Frutin, B. 1991. "Polygas—An Alternative Propellant." In Proceedings of the International CFC and Halon Alternatives Conference, Baltimore.

Gallagher, P., and V. Krukonis. 1991. "Precision Parts Cleaning with Supercritical Carbon Dioxide." In Proceedings of the International CFC and Halon Alternatives Conference, Baltimore.

Garrett, S. L., and T. Hofler. 1992. "Thermoacoustic Refrigeration." *ASHRAE Journal* 34, no.12: 28–36.

Gillum, W., and A. Jackson. 1991. "Replacement of Chlorinated Solvents for In-line Pre-plate Metal Cleaning with Environmentally Sound Alternatives." In Proceedings of the International CFC and Halon Alternatives Conference, Baltimore.

Glass, A. G., and R. Hoover. 1989. "The Emerging Epidemic of Melanoma and Squamous Cell Skin Cancer." *Journal of the American Medical Association* 292, no. 15: 2097–2100.

Gleason, J. F., P. K. Bhartia, J. R. Herman, R. McPeters, P. Newman, R. S. Stolarski, L. Flynn, G. Labow, D. Larko, C. Seftor, C. Wellemeyer, W. D. Komhyr, A. J. Miller, and W. Planet. 1993. "Record Low Global Ozone in 1992." *Science* 260: 523–525.

GECR. 1991. "From a Million Kilograms to Nearly Zero in Three Years: How Northern Telecom Cut the Cord on CFCs." *Global Environmental Change Report* III, no. 24: 1–3.

GECR. 1992a. "Hughes Aircraft to Cut CFC Use with Citrus-Based Flux." *Global Environmental Change Report* IV, no. 2, p. 7.

GECR. 1992b. "Canada Will Phase Out CFC by 1996." *Global Environmental Change Report* IV, no. 6, p. 4.

GECR. 1992c. "First CFC-Free Supermarket Opens in the US." *Global Environmental Change Report* IV, no. 6, p. 7.

GECR. 1992d. "EPA Identifies HFC-32 as Possible Substitute for R-22." *Global Environmental Change Report* IV, no. 6, p. 7.

GECR. 1992e. "Sweden's CFC Product Import Bans May Conflict with EC and GATT." *Global Environmental Change Report* IV, no. 18, p. 4.

GECR. 1992f. "AT&T Develops New Methyl Chloroform Substitute." *Global Environmental Change Report* IV, no. 23, p. 7.

GECR. 1993. "Phaseout Deadlines in OECD Countries for Ozone-Depleting Compounds." Special Issue.

GECR. 1994. "FICs Could Be Next CFC and Halon Replacements." *Global Environmental Change Report* V, no. 19, p. 1.

Godwin, G. S., and M. Menzer 1993. "Results of Compressor Calorimeter Tests in ARI's R-22 Alternative Refrigerants Evaluation Program "AREP." In Proceedings of the 1993 International CFC and Halon Alternatives Conference, Washington.

Graedel, T. E. 1978. *Chemical Compounds in the Atmosphere*. Academic Press.

Gramlen, P. H., B. C. Lane, P. M. Midgley, and J. M. Steed. 1986. "The Production and Release to the Atmosphere of CCl_3F and CCl_2F_2 (chlorofluorocarbons CFC 11 and CFC 12)." *Atmospheric Environment* 20: 1077–1085.

Grant, W. B. 1988. "Global Stratospheric Ozone and UBV Radiation." *Science* 242, p. 1111.

Grant, W. B., J. Fishman, E. V. Browell, V. G. Brackett, D. Nganga, A. Minga, B. Cros, R. E. Veiga, C. F. Butler, M. A. Fenn, and G. D. Nowicki. 1992. "Observations of Reduced Concentrations in the Tropical Stratosphere after the Eruption of Mt. Pinatubo." *Geophysical Research Letters* 19, no. 11: 1109–1112.

Greenpeace. 1992a. "Green Freeze Press Release." London, October 3.

Greenpeace. 1992b. Making the Right Choice.

Griffith, B. T., D. Arasteh, and S. Selkowitz. 1991. "Gas-Filled Panel High-Performance Thermal Insulation." Lawrence Berkeley Laboratory Report LBL-30160.

Hader, D.-P. 1986. "Effects of Enhanced UV-B Radiation on the Survival of Micro-Organisms." See EPA 1986.

Hader, D.-P., M. Hader, and S.-M. Liu. 1990. "Effects of Solar Radiation on Photoorientation, Motility and Pigmentation in a Freshwater Peridinium." *Bio-Systems* 23: 335–343.

Hamill, P., and O. B. Toon. 1991. "Polar Stratospheric Clouds and the Ozone Hole." *Physics Today*: 34–42, (December).

Hammitt, J. K., F. Camm, P. S. Connell, W. E. Mooz, K. A. Wolf, D. J. Wuebbles, and A. Bamezai. 1987. "Future Emission Scenarios for Chemicals That May Deplete Stratospheric Ozone." *Nature* 330: 711–716.

Hansen, J., A. Lacis, and M. Prather. 1989. "Greenhouse Effect of Chlorofluorocarbons and Other Trace Gases." *Journal of Geophysical Research* 94, no. D13: 16,4417–16,421.

Hanson, D. R., and A. R. Ravishankara. 1991. "Reaction Probabilities of $ClONO_2$ and N_2O_5 on 40 to 75% Sulfuric Acid Solutions." *Journal of Geophysical Research* 96, no. D9: 17307–17314.

Hanson, D. R., and A. R. Ravishankara. 1992."Investigation of the Reactive and Nonreactive Processes Involving $ClONO_2$ and HCl on Water and Nitric Acid Doped Ice." *Journal Physical Chemistry* 96: 2682–2691.

Harper, D. B. 1985. "Halomethanes from Halide Ion—A Highly Efficient Fungal Conversion of Environmental Significance." *Nature* 315: 55–57.

Hartmann, D. L., L. E. Heidt, M. Loewenstein, J. R. Podolske, J. Vedder, W. L. Starr, and S. E. Strahan. 1989. "Transport Into the South Polar Vortex in Early Spring." *Journal of Geophysical Research* 94, no. D14: 16779–16795.

Hayes, M. E. 1991. Progress and Developments in Semi-Aqueous Cleaning. Petrofirm, Inc., Fernandina, Florida.

Heath, D. F., A. J. Krueger, H. A. Roeder, and B. H. Henderson. 1975. "The Solar Backscatter Ultraviolet and Total Ozone Mapping Spectrometer (SBUV/TOMS) for Nimbus G." *Optical Engineering* 14: 323–331.

Heimann, R. P. 1993. "Desiccant-Based Open Cycle Cooling Opportunities in a Post-CFC World." In Proceedings of 1993 International CFC and Halon Alternatives Conference, Washington.

Henderson-Sellers, A. 1987. "Effects of Change in Land Use on Climate in the Humid Tropics." In *The Geophysiology of Amazonia*, ed. R. Dickinson. Wiley.

Herndl, G., G. Müller-Niklas, and J. Frick. 1993. "Major Role of Ultraviolet-B in Controlling Bacterioplankton Growth in the Surface Layer of the Ocean." *Nature* 361: 717–719.

Hirota, I., T. Hirooka, and M. Shiotani. 1983. "Upper Stratospheric Circulations in the Two Hemispheres Observed by Satellites." *Quarterly Journal of the Royal Meteorological Society* 109, no. 461: 443–454.

Hofmann, D. J., J. W. Harder, S. R. Rolf, and J. M. Rosen. 1987. "Balloonborne Observations of the Temporal Development and Vertical Structure of the Antarctic Ozone Hole in 1986." *Nature* 326: 59–62.

Hofmann, D. J., and S. Solomon. 1989. "Ozone Destruction through Heterogeneous Chemistry Following the Eruption of El Chichon." *Journal of Geophysical Research* 94, no. D4: 5029–5041.

Hofmann, D. J., T. L. Deshler, P. Aimedieu, W. A. Matthews, P. V. Johnston, Y. Kondo, W. R. Sheldon, G. J. Byrne, and J. R. Benbrook. 1989. "Stratospheric Clouds and Ozone Depletion in the Arctic During January 1989." *Nature* 340: 117–121.

Hofmann, D.J. 1990. "Increase in the Stratospheric Background Sulfuric Acid Aerosol Mass in the Past 10 Years." *Science* 248: 996–1000.

Hofmann, D.J., and T. L. Deshler. 1991. "Evidence from Balloon Measurements for Chemical Depletion of Stratospheric Ozone in the Arctic Winter of 1989–90." *Nature* 349: 300–304.

Hofmann, D. J. 1991. "Aircraft Sulfur Emissions." *Nature* 349, p. 659.

Hofmann, D. J., S. J. Oltmans, J. M. Harris, S. Solomon, T. Deshler, and B. J. Johnson. 1992. "Observations and Possible Causes of New Ozone Depletion in Antarctica in 1991." *Nature* 359: 283–287.

Hofmann, D. J., and S. J. Oltmans. 1993. "Anomalous Antarctic Ozone During 1992: Evidence for Pinatubo Volcanic Aerosol Effects." *Journal of Geophysical Research* 98, no. D10: 18555–18561.

Hofmann, D., S. Oltmans, W. D. Komhyr, J. M. Harris, J. A. Lathrop, A. O. Langford, T. Deshler, B. J. Johnson, A. Torres, and W. A. Matthews. 1994. "Ozone Loss in the Lower Stratosphere over the United States in 1992–1993: Evidence for Heterogeneous Chemistry on the Pinatubo Aerosol." *Geophysical Research Letters* 21, no. 1: 65–68.

Hollis, D. E., and A. Scheibner. 1988. "Ultrastructural Changes in Epidermal Langerhans Cells and Melanocytes in Response to Ultraviolet Radiation, in Australians of Aboriginal and Celtic Descent." *British Journal of Epidemiology* 119: 21–31.

Hübler, G., D. W. Fahey, K. K. Kelly, D. D. Montzka, M. A. Carroll, A. F. Tuck, L. E. Heidt, W. H. Pollock, G. L. Gregory, and J. F. Vedder. 1990. "Redistribution of Reactive Odd Nitrogen in the Lower Arctic Stratosphere." *Geophysical Research Letters* 17, no. 4, March supplement: 453–456.

Huselton, C. A., and H. Z. Hill. 1990. "Melanin Photosensitises Ultraviolet Light (UVC) DNA Damage in Pigmented Cells." *Environmental and Molecular Mutagenisis* 16, no. 1: 37–43.

IAC (International Alternatives Conference). 1988. J. E. Poole, "Alternatives to CFC 12 in Mobile Air Conditioning." In Proceedings of the Conference and Trade Fair: Substitutes and Alternatives to CFCs and Halons, Washington.

IAC (International Alternatives Conference). 1991. In Proceedings of the International CFC and Halon Alternatives Conference, Baltimore.

IEEE Spectrum. 1993. "Cleaning Up." February: 20–26.

Irvine, California. 1989. "An Ordinance of the City Council of the City of Irvine Governing the Manufacture, Distribution, Sale and Recycling of Products Which Utilize Ozone-Depleting Compounds" (ordinance 89-21).

Isaksen, I. S. A. 1987. "Is the Oxidizing Capacity of the Atmosphere Changing?" In *The Changing Atmosphere*, ed. F. S Rowland and I. S. A Isaksen. Wiley.

Johnston, H. 1971. "Reduction of Stratospheric Ozone by Nitrogen Oxide Catalysts from Supersonic Transport Exhaust." *Science* 173: 517–522.

Johnston, H. S., D. E. Kinnison, and D. J. Wuebbles. 1989. "Nitrogen Oxides From High-Altitude Aircraft: An Update of Potential Effects on Ozone." *Journal of Geophysical Research* 94, no. D13: 16,351–16,363.

Jones, R. R. 1987. "Ozone Depletion and Cancer Risk." *Lancet* 2, no. 8556: 443–446.

Jones, R. L., D. S. McKenna, L. R. Poole, and S. Solomon. 1990. "On the Influence of Polar Stratospheric Cloud Formation on Chemical Composition During the 1988/89 Arctic Winter." *Geophysical Research Letters* 17, no. 4, March supplement: 545–548.

JPR. 1991. "Methyl Bromide is More Bad News for Stratospheric Ozone." *Journal of Pesticide Reform* 11, no. 4, p. 30.

Kazachki, G. S., and R. V. Hendricks. 1993a. "Calorimeter Tests of HFC-236ea as a CFC-114 Alternative and HFC-245ca as a CFC-11 Alternative." In Proceedings of the 1993 International CFC and Halon Alternatives Conference, Washington.

Kazachki, G. S., and C. L. Gage. 1993b. "Thermodynamic Evaluation and Compressor Characteristics of HFC-236ea and HFC-245ca as CFC-114 and CFC-11 Replacements in Chillers." In Proceedings of the 1993 International CFC and Halon Alternatives Conference, Washington.

Kelfkens, G., F. R. de Gruijl, and J. C. van der Leun. 1990. "Ozone Depletion and Increase in Annual Carcinogenic Ultraviolet Dose." *Photochemistry and Photobiology* 52, no. 4: 819–823.

Kelly, J. R. 1986. "How Might Enhanced Levels of Solar UV-B Radiation Affect Marine Ecosystems?" See EPA 1986.

Kerr, J. B., D. Wardle, and D. W. Tarasick. 1993. "Record Low Ozone Values over Canada in Early 1993." *Geophysical Research Letters* 20, no. 18: 1979–1982.

Kerr, J. B., and C. T. McElroy. 1993. "Evidence for Large Upward Trends of Ultraviolet-B Radiation Linked to Ozone Depletion." *Science* 262: 1032–1034.

Khalil, M. A. K., R. A. Rasmussen, and R. Gunawardena. 1993. "Atmospheric Methyl Bromide: Trends and Global Mass Balance." *Journal of Geophysical Research* 98, no. D2: 2887–2896.

Kimball, R. 1993. "Chemist Hopeful His Discovery Can Rescue the Ozone." *Albuquerque Journal*, October 29.

Klausen, K. H., and E. Larsen. 1991. "PUR Foam for Refrigeration with Good Insulation Properties and Zero ODP." Tectrade Kemi A/s, Naerum, Denmark.

Ko, M. K. W., N. D. Sze, J. M. Rodriguez, D. K. Weistenstein, C. W. Heisey, R. P. Wayne, P. Biggs, C. E. Canosa-Mas, H. W. Sidebottom, and J. Treacy. 1994a. "CF_3 Chemistry: Potential Implications for Stratospheric Ozone." *Geophysical Research Letters* 21, no. 2 (January 15): 101–104.

Kolar, D. 1993. "Durability of Refrigerant Recovery Equipment." In Proceedings of the 1993 International CFC and Halon Conference, Washington.

Komhyr, W. D., R. D. Grass, R. D. Evans, R. K. Leonard, D. M. Quincy, D. Hofmann, and G. L. Koenig. 1994. "Unprecedented 1993 Ozone Decrease over the United States from Dobson Spectrophotometer Observations." *Geophysical Research Letters* 21, no. 3 (February 1): 201–204.

Kripke, M. 1991. Effects of Ultraviolet Radiation on Immune Responses of Mice and Men. Testimony before U.S. Senate Committee on Commerce, Science and Transportation, November 15.

Lacis, A. A., D. J. Wuebbles, and J. A. Logan. 1990. "Radiative Forcing of Climate by Changes in the Vertical Distribution of Ozone." *Journal of Geophysical Research* 95, no. D7: 9971–9981.

Lankford, L., and J. Nimitz. 1993. "A New Class of High-Performance Environmentally Sound Refrigerants." In Proceedings of the 1993 International CFC and Halon Alternatives Conference, Washington.

Lashof, D. A., and D. R. Ahuja. 1990. "Relative Contributions of Greenhouse Gas Emissions to Global Warming." *Nature* 344: 529–531.

LeJeune, F. J. 1986. "Epidemiology and Etiology of Malignant Melanoma." *Biomedicine and Pharmacotherapy* 40, no. 3: 91–99.

Levine, J. S. (ed.). 1991. *Global Biomass Burning: Atmospheric, Climatic and Biospheric Implications*. MIT Press.

Liu, S. C., S. A. McKeen, and S. Madronic. 1991. "Effect of Anthropogenic Aerosols on Biologically Active Ultra Violet Radiation." *Geophysical Research Letters* 18, no. 12: 2265–2268.

Lobert, J. M., D. H. Scharffe, W.-M. Hao, T. A. Kuhlbusch, R. Seuwen, P. Warneck, and P. J. Crutzen. 1991. "Experimental Evaluation of Biomass Burning Emissions: Nitrogen and Carbon Containing Compounds." In *Global Biomass Burning*, ed. J. S. Levine. MIT Press.

Logan, J. A. 1985. "Tropospheric Ozone: Seasonal Behavior, Trends, and Anthropogenic Influence." *Journal of Geophysical Research* 90, no. D6: 10,463–10,482.

MacKenzie, D. 1990. "Cheaper Alternatives for CFCs." *New Scientist*, June 30, 1990: 39–40.

Madronich, S. 1992. "Implications of Recent Total Atmospheric Ozone Measurements for Biologically Active Ultraviolet Radiation Reaching the Earth's Surface. *Geophysical Research Letters* 19 (January 3): 37–40.

Mahlman, J. D. 1992. "A Looming Arctic Ozone Hole?." *Nature* 360: 209–210.

Makhijani, A., A. Makhijani, and A. Bickel. 1988. Saving Our Skins: Technical Potential and Policies for the Elimination of Ozone-depleting Chlorine Compounds. Environmental Policy Institute and Institute for Energy and Environmental Research.

Makhijani, A., and A. Makhijani. 1990. Biomass Burning and Ozone Depletion: An Assessment of the Problem and Its Implications for the Protection of the Ozone Layer. Institute for Energy and Environmental Research.

Malaspina, M., K. Schafer, and R. Wiles. 1992. "What Works Report #1: Air Pollution Solutions." Environmental Exchange, Washington.

Mankin, W. G., and M. T. Coffey. 1984. "Increased Stratospheric Hydrogen Chloride in the El Chichon Cloud." *Science* 226: 170–172.

Maño, S., and M. O. Andreae. 1994. "Emission of Methyl Bromide from Biomass Burning." *Science* 263: 1255–1257.

Manzer, L. E. 1990. "The CFC-Ozone Issue: Progress on the Development of Alternatives to CFCs." *Science* 249: 31–35.

Markandya, A., and University College London. 1992. The Montreal Protocol, Funding the Incremental Costs for Article 5 Countries: A Review and Update, Prepared on Behalf of UNEP for discussion at the Open-ended Working Group of the Parties to the Montreal Protocol, Geneva.

Mertens, J. 1991. "1,1,1-Trichloroethane Uses in Coatings Application." In Proceedings of the International CFC and Halon Alternatives Conference, Baltimore.

Mather, J. H., and W. H. Brune. 1990. "Heterogeneous Chemistry on Liquid Sulfate Aerosols: A Comparison of In Situ Measurements with Zero-Dimensional Model Calculations." *Geophysical Research Letters* 17, no. 9: 1283–1286.

MBr Workshop. 1993. Executive Summary, International Workshops on Alternatives to Methyl Bromide for Soil Fumigation, Rotterdam and Rome.

McCormick, M. P., H. M. Steele, P. Hamill, W. P. Chu, and T. J. Swissler. 1982. "Polar Stratospheric Cloud Sightings by SAM II." *Journal of the Atmospheric Sciences* 39: 1387–1397.

McCulloch, A. 1992. "Global Production and Emissions of Bromochlorodifluoromethane and Bromotrifluoromethane (Halons 1211 and 1301)." *Atmospheric Environment* 26A, no. 7: 1325–1329.

McElroy, D. L., and M. P. Scofield. 1990. Chlorofluorocarbon (CFC) Technologies Review of Foamed-Board Insulation. Report ORNL/TM-11291, Oak Ridge National Laboratory.

McElroy, M. B., R. J. Salawitch, S. C. Wofsy, and J. A. Logan. 1986. "Reductions of Antarctic Ozone Due to Synergistic Interactions of Chlorine and Bromine." *Nature* 321: 759–762.

McKenzie, R. L., and J. M. Elwood. 1990. "Intensity of Solar Ultraviolet Radiation and Its Implications for Skin Cancer." *New Zealand Medical Journal* 103, no. 887: 152–154.

Mei, V. C., F. C. Chen, and R. A. Sullivan. 1991. "Analysis of Non-CFC Automotive Air-conditioning." In Proceedings of the International CFC and Halon Alternatives Conference, Baltimore.

Meyer, A. 1993. "C- and FHC-free Refrigerators by Foron Hausgeraete Co.—A Breakthrough in Household Refrigeration." Foron Corporation, Niederschmiedeberg, Germany.

Microsoft Corp. 1994. Encarta.

Midgley, P. M. 1989. "The Production and Release to the Atmosphere of 1,1,1-trichlroethane (Methyl Chloroform)." *Atmospheric Environment* 23, no. 12: 2663–2665.

Midgley, T., and A. L. Henne. 1930. "Organic Fluorides as Refrigerants." *Industrial and Engineering Chemistry* 22, no. 5: 542–545.

Miller, A. S., and I. S. Mintzer. 1986. *The Sky Is The Limit: Strategies for Protecting the Ozone Layer*, World Resources Institute, Washington.

Molina, M. J., and F. S. Rowland. 1974. "Stratospheric Sink for Chlorofluoromethanes: Chlorine Atom-Catalysed Destruction of Ozone." *Nature* 239: 810–812.

Molina, L. T., and M. J. Molina. 1987. "Production of Cl_2O_2 from the Self-Reaction of the ClO Radical." *Journal of Physical Chemistry* 91, no. 2: 433–436.

Molina, M. J., T. L. Tso, L. T. Molina, and F. C. Y. Wang. 1987. "Antarctic Stratospheric Chemistry of Chlorine Nitrate, Hydrogen Chloride, and Ice: Release of Active Chlorine." *Science* 238: 1253–1257.

Molina, M.., R. Zhang, P. J. Wooldridge, J. R. McMahon, J. E. Kim, H. Y. Chang, and K. D. Beyer. 1994. "Physical Chemistry of the $H_2SO_4/HNO_3/H_2O$ System: Implications for Polar Stratospheric Clouds." *Science* 261: 1418–1423.

Morison, W. L. 1989. "Effects of Ultraviolet Radiation on the Immune System in Humans." *Photochemistry and Photobiology* 50, no. 4: 515–524.

Mossel, J. 1988. "Uses of Halons and Opportunities for Emission Reductions; Size and Structure of the Market." In Proceedings of the Conference and Trade Fair: Substitutes and Alternatives to CFCs and Halons, Washington.

Murphy, J. A., M. Bolmer, M. Elsheikh, J. D. Roux, C. Meynard, and O. Volkert. 1993. "HFC-365 as a Zero ODP Blowing Agent for Foams." In Proceedings of the 1993 International CFC and Halon Alternatives Conference, Washington.

NASA (National Aeronautics and Space Administration). 1986. *Present State of Knowledge of the Upper Atmosphere: An Assessment Report.*

NASA (National Aeronautics and Space Administration). 1992. Press release, End of Mission Statement, Second Airborne Arctic Stratospheric Expedition (AASE-II).

NOAA (National Oceanic and Atmospheric Administration). 1993. "1993 Antarctic Ozone Hole Reaches Record Lows." Press release, October 18, 1993.

New Scientist. 1992a. "Fire-fighters Find Gas That's Easy On Ozone." *New Scientist*, April 25, p. 19.

New Scientist. 1992b. "Barbequed CFCs Are Kind to the Atmosphere." *New Scientist*, August 22, p. 13.

New Scientist. 1992c. "Hard on Insects, Easy on Ozone." *New Scientist*, October 17, p. 10.

Newman, P., R. Stolarski, M. Schoeberl, L. R. Lait, and A. Krueger. 1990. "Total Ozone During the 88–89 Northern Hemisphere Winter." *Geophysical Research Letters* 17, no. 4, March supplement: 317–320.

Newman, P., R. Stolarski, M. Schoeberl, R. McPeters, and A. Krueger. 1991. "The 1990 Antarctic Hole as Observed by TOMS." *Geophysical Research Letters* 18, no. 4: 661–664.

Newman, P., L. R. Lait, M. Schoeberl, E. R. Nash, K. Kelly, D. W. Fahey, R. Nagatani, D. Toohey, L. Avallone, and J. Anderson. 1993. "Stratospheric Meteorological Conditions in the Arctic Polar Vortex, 1991 to 1992." *Science* 261: 1143–1146.

Nigrey, P. J., and J. S. Arzigian. 1991. "Non-CFC Cleaning of Printed Wiring Boards." Sandia National Laboratories.

Nimitz, J., and L. Lankford. 1993a. "Fluoroiodocarbons as Halon Replacements." In Proceedings of the 1993 International CFC and Halon Alternatives Conference, Washington.

Nimitz, J., and L. Lankford. 1993b. "New Class of Nonflammable Environmentally Safe Solvents." Abstract prepared for Fourth Annual International Workshop on Solvent Substitution, Phoenix.

Nimitz, J., and L. Lankford. 1994. "New Foam Blowing Agents Containing Fluoroiodocarbons." Abstract prepared for SPI Conference, Boston.

Novelli, P. C., K. A. Masarie, P. P. Tans, and P. M. Lang. 1994. "Recent Changes in Atmospheric Carbon Monoxide." *Science* 263: 1587–1590.

NRC (National Research Council). 1975. *Long-Term Worldwide Effects of Multiple Nuclear Weapons Detonations*. National Technical Information Services order no. PB-279-976.

NRC (National Research Council). 1976. *Halocarbons: Effects on Stratospheric Ozone*. National Academy Press.

NRC (National Research Council). 1982. *Causes and Effects of Stratospheric Ozone Reduction: An Update.*

NRC (National Research Council). 1983. *Causes and Effects of Stratospheric Ozone Reduction: An Update.*

NRDC (Natural Resources Defense Council). 1990. "Wanted: For Destruction of the Ozone Layer Public Enemy no. 1,1,1."

Owens-Corning. 1992. *Fiber Glass Insulation Concept Promises High Energy Efficiency for Appliances.* Press release, Owens-Corning World Headquarters, Toledo, Ohio, October 21.

OzonAction. 1993a. "Contributions to the Multilateral Fund." *OzonAction* no. 8. United Nations Environment Programme, September.

OzonAction. 1993b. "5th Meeting of the Parties." *OzonAction* no. 9. United Nations Environment Programme, December.

Ozone Depletion Network Online Today. 1993. "HCFC Investors Satisfied with EU's New HCFC Phaseout Deadline." Environmental Information Networks, Inc., December 15.

Ozone Depletion Network Online Today. 1994. "AT&T Develops Ozone-Friendly Cleaning Process Using Dry Ice." Environmental Information Networks, Inc., March 31.

Ozone Secretariat. 1993. *Handbook for the Montreal Protocol on Substances that Deplete the Ozone Layer*, third edition. Nairobi.

PAN (Pesticide Action Network). 1992. "Action Needed to Ban Ozone-Depleting Pesticide." North America Updates Service, San Francisco, June 18.

Pearman, G. I., D. Etheridge, F. de Silva, and P. J. Fraser. 1986. "Evidence of Changing Concentrations of Atmospheric CO_2, N_2O and CH_4 from Air Bubbles in Antarctic Ice." *Nature* 320: 248–250.

Pedersen, N. M. 1991. "Destruction of CFC by Incineration of Refrigerators and Freezers." In Proceedings of the International CFC and Halon Alternatives Conference, Baltimore.

Pitt, D. E. 1994. "U.N. Envoys Fear New Cod Wars as Fish Dwindle." *New York Times*, March 20.

Planet, W. G., J. H. Lienesch, A. J. Miller, R. Nagatani, R. D. McPeters, E. Hilsenrath, R. P. Cebula, M. T. DeLand, C. G. Wellemeyer, and K. Horvath. 1994. "Northern Hemisphere Total Ozone Values from 1989–1993 Determined with the NOAA-11 Solar Backscatter Ultraviolet (SBUV/2) Instrument." *Geophysical Research Letters* 21, no. 3: 205–208.

Potter, T. F., and D. K. Benson. 1991. Non-CFC Vacuum Alternatives for the Energy-Efficient Insulation of Household Refrigerators: Design and Use. Report SERI/TP-253-4124, Solar Energy Research Institute, Golden, Colorado.

Prather, M., M. M. Garcia, R. Suozzo, and D. Rindl. 1990. "Global Impact of the Antarctic Ozone Hole: Dynamical Dilution With a Three-Dimensional Chemical Transport Model." *Journal of Geophysical Research* 95, no. D4: 3449–3471.

Prather, M., and A. H. Jaffe. 1990. "Global Impact of the Antarctic Ozone Hole: Chemical Propagation." *Journal of Geophysical Research* 95, no. D4: 3473–3492.

Prather, M. J., and R. T. Watson. 1990. "Stratospheric Ozone Depletion and Future Levels of Atmospheric Chlorine and Bromine." *Nature* 344: 729–734.

Prather, M. J., M. M. Garcia, A. R. Douglass, C. H. Jackman, M. W. Ko, and N. D. Sze. 1990. "The Space Shuttle's Impact on the Stratosphere." *Journal of Geophysical Research* 95, no. D11: 18,583–18,590.

Prather, M. J. 1992a. "More Rapid Polar Ozone Depletion through the Reaction of HOCl with HCl on Polar Stratospheric Clouds." *Nature* 355: 534–537.

Prather, M. J. 1992b. "Catastrophic Loss of Stratospheric Ozone in Dense Volcanic Clouds." *Journal of Geophysical Research* 97, no. D9: 10187–10191.

Prinn, R. G., D. Cunnold, R. Rasmussen, P. Simmonds, F. Alyea, A. Crawford, P. Fraser, and R. Rosen. 1990. "Atmospheric Emissions and Trends of Nitrous Oxide Deduced from 10 Years of ALE-GAGE Data." *Journal of Geophysical Research* 95, no. D11: 18,369–18,385.

Prinn, R., D. Cunnold, P. Simmonds, F. Alyea, R. Boldi, A. Crawford, P. Fraser, D. Gutzler, D. Hartley, R. Rosen, and R. Rasmussen. 1992. "Global Average Concentration and Trend for Hydroxyl Radicals Deduced from ALE/GAGE Trichloroethane (Methyl Chloroform) Data for 1978–1990." *Journal of Geophysical Research* 97, no. D2: 2445–2461.

Proffitt, M. H., D. W. Fahey, K. K. Kelly, and A. F. Tuck. 1989. "High-latitude Ozone Loss Outside the Antarctic Ozone Hole." *Nature* 342: 233–237.

Proffitt, M. H., J. J. Margitan, K. K. Kelly, M. Loewenstein, J. R. Podolske, and K. R. Chan. 1990. "Ozone Loss in the Arctic Polar Vortex Inferred from High-Altitude Aircraft Measurements." *Nature* 347: 31–36.

Proffitt, M. H., K. Aikin, J. J. Margitan, M. Loewenstein, J. R. Podolske, A. Weaver, K. R. Chan, H. Fast, J. W. Elkins. 1993. "Ozone Loss Inside the Northern Polar Vortex During the 1991–1992 Winter." *Science* 261: 1150–1154.

Purushothama, S. 1991. "Halons—The Problems of Developing Countries (with Specific Reference to India)." In Proceedings of the International CFC and Halon Alternatives Conference, Baltimore.

Ramanathan, V. 1975. "Greenhouse Effect Due to Chlorofluorocarbons: Climate Implications." *Science* 190: 50–52.

Ramaswamy, V., M. D. Schwarzkopf, and K. P. Shine. 1992. "Radiative Forcing of Climate from Halocarbon-induced Global Stratospheric Ozone Loss." *Nature* 355: 810–812.

Rasmussen, R. A., L. E. Rasmussen, M. A. K. Khalil, and R. W. Dalluge. 1980. "Concentration Distribution of Methyl Chloride in the Atmosphere." *Journal of Geophysical Research* 85, no. C12: 7350–7356.

Ravishankara, A. R., A. A. Turnipseed, N. R. Jensen, S. Barone, M. Mills, C. J. Howard, and S. Solomon. 1994. "Do Hydrofluorocarbons Destroy Stratospheric Ozone?" *Science* 263, 71–75.

Reeves, C. E., and S. A. Penkett. 1994. "An Estimate of the Anthropogenic Contribution to Atmospheric Methyl Bromide." *Geophysical Research Letters* 20, no. 15: 1563–1566.

Reilly, W. K. 1991. Statement, October 22.

Revelle, R. 1987. "Comments on 'Causes of Deforestation in the Brazilian Amazon'." In Robert Dickinson, ed., *The Geophysiology of Amazonia*. Wiley.

Rice, W., K. L. Whitfield, and D. S. C. Chau. 1993. "Hydraulic Refrigeration system (HRS) Using N-Butane as the Refrigerant." In Proceedings of the Ozone-Safe Cooling Conference, Washington.

Riffe, R., and T. Dekleva. 1991. "Effects of Lubricants on the Energy Efficiency of Home Appliances." In Proceedings of the International CFC and Halon Alternatives Conference, Baltimore.

Roan, S. L. 1990. *Ozone Crisis: The 15 Year Evolution of a Sudden Global Emergency*. Wiley.

Rodriguez, J. M., K. W. Ko, and N. D. Sze. 1988. "Antarctic Chlorine Chemistry: Possible Global Implications." *Geophysical Research Letters* 15: 257–260.

Rodriguez, J. M., M. K. W. Ko, and M. D. Sze. 1991. "Role of Heterogeneous Conversion of N_2O_5 on Sulfate Aerosols in Global Ozone Losses." *Nature* 352: 134–137.

Rodriguez, J. M. 1993. "Probing Stratospheric Ozone." *Science* 261: 1128–1129.

Rodriguez, J. M., M. K. W. Ko, N. D. Sze, C. W. Heisey, G. K. Yue, and M. P. McCormick. 1994. "Ozone Response to Enhanced Heterogeneous Processing After the Eruption of Mt. Pinatubo." *Geophysical Research Letters* 21, no. 3: 209–212.

Rosencranz, A., and R. Milligan. 1990. "CFC Abatement: The Needs of Developing Countries." *Ambio* 19, no. 6–7: 312–316.

Rowland, F. S. 1988. Testimony before U.S. Senate Committee on Environment and Public Works Subcommittee on Hazardous Wastes and Toxic Substances, March 30.

Rowland, F. S. 1990. "Stratospheric Ozone Depletion by Chlorofluorocarbons." *AMBIO* 19, no. 6–7: 281–292.

Rowlands, I. H. 1993. "The Fourth Meeting of the Parties to the Montreal Protocol: Report and Reflection." *Environment* 35, no. 6: 25–34.

Rowse, J. 1992. "Insulation: Cellulose Is Better than Fiberglass." *Boston Globe*, October 14.

Roy, C. H., H. P. Gies, and G. Elliot. 1990. "Ozone Depletion." *Nature* 347: 235–236.

Ryan, M. 1993. "Is Dry Cleaning All Wet?" *World Watch*, May-June: 7–9.

Salawitch, R. J., S. C. Wofsy, E. W. Gottlieb, L. R. Lait, P. A. Newman, M. R. Schoeberl, M. Lowenstein, J. R. Podolske, S. E. Strahan, M. H. Proffitt, C. R. Webster, R. D. May, D. W. Fahey, D. Baumgardner, J. E. Dye, J. C. Wilson, K. K. Kelly, J. W. Elkins, K. R. Chan, and J. G. Anderson. 1993. "Chemical Loss of Ozone in the Arctic Polar Vortex in the Winter of 1991–1992." *Science* 261: 1146–1149.

Sami, S. M. 1993. "A Performance Comparative Study of a New HFC Blend for Existing CFCs and HCFCs." View-graphs prepared for International CFC and Halon Alternatives Conference, Washington.

Schoeberl, M. R., and D. L. Hartmann. 1991. "The Dynamics of the Stratospheric Polar Vortex and Its Relation to Springtime Ozone Depletions." *Science* 251: 46–52.

Schoeberl, M. R., L. R. Lait, P. A. Newman, and J. E. Rosenfield. 1992. "The Structure of the Polar Vortex." *Journal of Geophysical Research* 97, no. D8: 7859–7882.

Schoeberl, M. R., R. S. Stolarski, A. R. Douglass, P. A. Newman, L. R. Lait, J. W. Waters, L. Froidevaux, and W. G. Ready. 1993. "MLS ClO Observations and Arctic Polar Vortex Temperatures." *Geophysical Research Letters* 20, no. 24: 2861–2864.

Schwartz, J., P. Maier-Laxhuber, and R. Wõrz. 1993. "Cooling and Air Conditioning with Water/Zeolite." In Proceedings of the Ozone-Safe Cooling Conference, Washington.

Science. 1992. "Pinatubo Fails to Deepen the Hole." *Science* 258, p. 395.

Science. 1994. "Antarctic Ozone Hole Fails to Recover." *Science* 266, p. 217.

SCOPE (Scientific Committee on Problems of the Environment). 1992. *Effects of Increased Ultraviolet Radiation on Biological Systems*, Proceedings of a Workshop held in February 1992 in Budapest; published by SCOPE Secretariat, Paris.

SCOPE (Scientific Committee on Problems of the Environment). 1993. *Effects of Increased Ultraviolet Radiation on Global Ecosystems*, Proceedings of a Workshop held in October 1992 in Tramariglio (Sassari), Sardinia; published by SCOPE Secretariat, Paris.

Scotto, J., G. Cotton, F. Urbach, D. Berger, and T. Fears. 1988. "Biologically Effective Ultraviolet Radiation: Surface Measurements in the United States, 1974 to 1985." *Science* 239: 762–764.

Seckmeyer, G., and R. L. McKenzie. 1992. "Increased Ultraviolet Radiation in New Zealand (45° S) Relative to Germany (48° N)." *Nature* 359: 135–137.

SEPA (Swedish Environmental Protection Agency). 1991. *Experience from CFC Phaseout: Summary*, Swedish Environmental Protection Agency Information Department, Solna, Sweden.

Shabecoff, P. 1988. "DuPont to Halt Chemicals that Peril Ozone." *New York Times*, March 25.

Simmonds, P. G., D. M. Cunnold, F. N. Alyea, C. A. Cardelino, A. J. Crawford, R. G. Prinn, P. J Fraser, R. A. Rasmussen, and R. D. Rosen. 1988. "Carbon Tetrachloride Lifetimes and Emissions Determined from Daily Global Measurements During 1978–1985." *Journal of Atmospheric Chemistry* 7: 35–58.

Singh, H. B., L. J. Salas, H. Shigeishi, and E. Scribner. 1979. "Atmospheric Halocarbons, Hydrocarbons, and Sulfur Hexafluoride: Global Distributions, Sources, and Sinks." *Science* 203: 899–903.

Singh, H. B., L. J. Salas, and R. E. Stiles. 1983. "Methyl Halides in and over the Eastern Pacific (40° N–32° S)." *Journal of Geophysical Research* 88: 3684–3490.

Singh, H. B., and M. Kanakidou. 1994. "An Investigation of the Atmospheric Sources and Sinks of Methyl Bromide." *Geophysical Research Letters* 20, no. 2: 133–136.

Skole, D., and C. Tucker. 1993. "Tropical Deforestation and Habitat Fragmentation in the Amazon: Satellite Data from 1978 to 1988." *Science* 260: 1905–1910.

Skolnick, A. 1991. "Melanoma Epidemic Yields Grim Statistics." Medical News and Perspectives, *Journal of the American Medical Association* 265, no. 24: 3217–3218.

Smith, K. C. 1977. *The Science of Photobiology*. Plenum.

Smith, R. C., B. B. Prezelin, K. S. Baker, R. R. Bidigare, N. P. Boucher, T. Coley, D. Karentz, S. MacIntyre, H. A. Matlick, D. Menzies, M. Ondrusek, Z. Wan, and K. J. Waters. 1992. "Ozone Depletion: Ultraviolet Radiation and Phytoplankton Biology in Antarctic Waters." *Science* 255, no. 5047: 952–959.

Smith, D. N., K. Ratanphruks, M. W. Tufts, and A. S. Ng. 1993. "HFC-236ea: A Potential Alternative to CFC-114." In Proceedings of the 1993 International CFC and Halon Alternatives Conference, Washington.

Smithart, E. L. 1991. "HCFC-23's Use in Chillers." In Proceedings of the International CFC and Halon Alternatives Conference, Baltimore.

Snyder General Corp. 1992. Keeping the Ozone Whole. Minneapolis.

Solomon, S., R. R. Garcia, F. S. Rowland, and D. J. Wuebbles. 1986. "On the Depletion of Antarctic Ozone." *Nature* 321: 755–758.

Solomon, S., and D. L. Albritton. 1992. "Time-Dependent Ozone Depletion Potentials for Short- and Long-Term Forecasts." *Nature* 357: 33–37.

Solomon, S., R. W. Sanders, R. R. Garcia, and J. G. Keys. 1993. "Increased Chlorine Dioxide over Antarctica Caused by Volcanic Aerosols from Mount Pinatubo." *Nature* 363: 245–248.

SORG (Stratospheric Ozone Research Group). 1991. *Stratospheric Ozone 1991*. HMSO.

SORG (Stratospheric Ozone Research Group). 1993. *Stratospheric Ozone 1993*. HMSO.

Stolarski, R. S., and R. J. Cicerone. 1974. "Stratospheric Chlorine: A Possible Sink for Ozone." *Canadian Journal of Chemistry* 52: 1610–1615.

Stolarski, R. S., P. Bloomfield, R. D. McPeters, and J. R. Herman. 1991. "Total Ozone Trends Deduced from Nimbus 7 TOMS Data." *Geophysical Research Letters* 18, no. 6: 1015–1018.

Stolarski, R., R. Bojkov, L. Bishop, C. Zerefos, J. Staehelin, and J. Zawodny. 1992. "Measured Trends in Stratospheric Ozone." *Science* 256: 342–349.

Streilein, J. 1991. "Immunological Factors in Skin Cancer." *New England Journal of Medicine* 325, no. 12: 884–886.

Sze, N. D., M. K. W. Ko, D. K. Weisenstein, and J. M. Rodriguez. 1989. "Antarctic Ozone Hole: Possible Implications for Ozone Trends in the Southern Hemisphere." *Journal of Geophysical Research* 94, no. D9: 11,521–11,528.

Tabazadeh, R., and R. P. Turco. 1993. "Stratospheric Chlorine Injection by Volcanic Eruptions: HCl Scavenging and Implications for Ozone." *Science* 260: 1082–1086.

Teramura, A. H. 1986. "The Potential Consequences of Ozone Depletion Upon Global Agriculture." See EPA 1986.

Teramura, A. H. 1991. Testimony before the Senate Committee on Commerce, Science and Transportation, November 15, 1991.

Thompson, A. M. 1992. "The Oxidizing Capacity of the Earth's Atmosphere: Probable Past and Future Changes." *Science* 256: 1157–1165.

Thornton, A. 1993. "Environmentally Acceptable Sterilization Using HFC-227ea." In Proceedings of the 1993 International CFC and Halon Alternatives Conference, Washington.

Tolbert, M. A., M. J. Rossi, R. Malhotra, and D. M. Golden. 1987. "Reaction of Chlorine Nitrate with Hydrogen Chloride and Water at Antarctic Stratospheric Temperatures." *Science* 238: 1258–1260.

Tolbert, M. A., M. J. Rossi, and D. M. Golden. 1988. "Antarctic Ozone Depletion Chemistry: Reactions of N_2O_5 with H_2O and HCl on Ice Surfaces." *Science* 240: 1018–1021.

Toohey, D. W., L. M. Avallone, L. R. Lait, P. A. Newman, M. R. Schoeberl, D. W. Fahey, E. L. Woodbridge, and J. G. Anderson. 1993. "The Seasonal Evolution of Reactive Chlorine in the Northern Hemisphere Stratosphere." *Science* 261: 1134–1136.

Toon, O. B., P. Hamill, R. P. Turco, and J. Pinto. 1986. "Condensation of HNO_3 and HCl in the Winter Polar Stratospheres." *Geophysical Research Letters* 13, no. 12, November supplement: 1284–1287.

Toon, O. B., R. P. Turco, J. Jordan, J. Goodman, and G. Ferry. 1989. "Physical Processes in Polar Stratospheric Ice Clouds." *Journal of Geophysical Research* 94, no. D9: 11,359–11,380.

Toon, O., E. Browell, B. Gary, L. Lait, J. Livingston, P. Newman, R. Pueschel, P. Russell, M. Schoeberl, G. Toon, W. Traub, F. P. J. Valero, H. Selkirk, and J. Jordan. 1993. "Heterogeneous Reaction Probabilities, Solubilities, and the Physical State of Cold Volcanic Aerosols." *Science* 261: 1136–1140.

Treadwell, D. 1991. "Application of Propane (R-290) to a Single Packaged Unitary Air-Conditioning Product." In Proceedings of the International CFC and Halon Alternatives Conference, Baltimore.

Tuck, A. F. 1989. "Synoptic and Chemical Evolution of the Antarctic Vortex in Late Winter and Early Spring 1987." *Journal of Geographic Research* 94, 11,687–11,737.

Turco, R. P., and R. S. Golitsyn. 1988. "A Status Report: Global Effects of Nuclear War." *Environment* 30, no. 5.

Umweltbundesamt (German Federal Environment Agency). 1989. "Responsibility Means Doing Without."

UNEP (United Nations Environment Programme). 1989a. *Technical Progress on Protecting the Ozone Layer: Halon Fire Extinguishing Agents Technical Options Report*, June 30.

UNEP (United Nations Environment Programme). 1989b. *Economic Panel Report*, June 30.

UNEP (United Nations Environment Programme). 1991a. *Montreal Protocol 1991 Assessment: Report of the Aerosol Products, Sterilants, Miscellaneous Uses and Carbon Tetrachloride Technical Options Committee*, December.

UNEP (United Nations Environment Programme). 1991b. *Montreal Protocol 1991 Assessment: Report of the Flexible and Rigid Foams Technical Options Committee*, December.

UNEP (United Nations Environment Programme). 1991c. *Montreal Protocol 1991 Assessment: Report of the Halons Technical Options Committee*, December.

UNEP (United Nations Environment Programme). 1991d. *Montreal Protocol 1991 Assessment: Report of the Refrigeration, Air Conditioning and Heat Pumps Technical Options Committee*, December.

UNEP (United Nations Environment Programme). 1991e. *Montreal Protocol 1991 Assessment: Report of the Solvents, Coatings, and Adhesives Technical Options Committee*, December.

UNEP (United Nations Environment Programme). 1991f. *Montreal Protocol 1991 Assessment: Report of the Technology and Economic Assessment Panel*, December.

UNEP (United Nations Environment Programme). 1991g. *Environmental Effects of Ozone Depletion: 1991 Update*, November.

UNEP (United Nations Environment Programme). 1992. *Synthesis Report of the Methyl Bromide Interim Scientific Assessment and Methyl Bromide Interim Technology and Economic Assessment*, June.

UNEP (United Nations Environment Programme). 1993a. *Action On Ozone*, September.

UNEP (United Nations Environment Programme). 1993b. *1993 Report of the Technology and Economic Assessment Panel*, July.

UNEP/ExCom (Executive Committee of the Interim Multilateral Fund for the Implementation of the Montreal Protocol). 1992. *Meeting the Needs of Article 5 Parties for Controlled Substances During the Grace and Phase-out Periods (Draft Report)*, UNEP/OzL.Pro/ExCom/8/25, September 21.

U.S. CAA (U.S. Clean Air Act). 1990. *Congressional Record* 136, no. 149, part IV, October 26.

U.S. International Trade Commission. 1989. *Synthetic Organic Chemicals: Production and Sales 1988.*

U.S. International Trade Commission. 1991. "Preliminary Report on U.S. Production of Selected Synthetic Organic Chemicals" (June 28 and November 27).

Vineyard, E. A., and L. J. Swatkowski. 1993. "Energy Efficiency of HFC-134a Versus HFC-152a." In Proceedings of the 1993 International CFC and Halon Alternatives Conference, Washington.

von Eynatten, Solray Fluor, und Derivate GmbH. 1993. "HFC-134a/HFC-23: A Very Promising Alternative to HCFC-22." In Proceedings of the 1993 International CFC and Halon Alternatives Conference, Washington.

Wald, M. 1992. "Car Need Freon? Better Fill Up Soon." *New York Times*, February 1.

Wallace, J. M., and P. V. Hobbs. 1977. *Atmospheric Science: An Introductory Survey*. Academic Press.

Warneck, P. 1988. *Chemistry of the Natural Atmosphere*. Academic Press.

Waters, J. W., L. Froidevaux, W. G. Read, G. L. Manney, L. S. Elson, D. A. Flower, R. F. Jarnot, and R. S. Harwoo. 1993d. "Stratospheric ClO and Ozone from the Microwave Limb Sounder on the Upper Atmosphere Research Satellite." *Nature* 362: 597–602.

Wayne, R. P. 1991. *Chemistry of Atmospheres*, Oxford: Oxford University Press.

Weaver, A., M. Loewenstein, J. R. Podolske, S. E. Strahan, M. H. Proffitt, K. Aikin, J. J. Margitan, H. H. Jonnson, C. A. Brock, J. C. Wilson, and O. B. Toon. 1993. "Effects of Pinatubo Aerosol on Stratospheric Ozone at Mid-latitudes." *Geophysical Research Letters* 20, no. 22: 2515–2518.

Webster, C. R., R. D. May, D. W. Toohey, L. M. Avallone, J. G. Anderson, P. Newman, L. Lait, M. R. Schoeberl, J. W. Elkins, and K. R. Chan. 1993. "Chlorine Chemistry on Polar Stratospheric Cloud Particles in the Arctic Winter." *Science* 261: 1130–1134.

Weiler, S. 1991. "Effects of Ozone-related Increases in UV-B Radiation on Marine Phytoplankton." Testimony before U.S. Senate Committee on Commerce, Science and Transportation, November 15.

Weisenstein, D. K., M. K. Ko, and N. D. Sze. 1992. "The Chlorine Budget of the Present-Day Atmosphere: A Modeling Study." *Journal of Geophysical Research* 97, no. D2: 2547-2559.

Welsh, G. 1991. "The Use of 100% Carbon Dioxide as a Blowing Agent for Polystyrene Foam Sheet." In Proceedings of the International CFC and Halon Alternatives Conference, Baltimore.

Wenning, U. G. 1993. "Hydrocarbons as PU Blowing Agents in Domestic Appliances." In Proceedings of the 1993 International CFC and Halon Alternatives Conference, Washington.

Wilson, J. C., H. H. Jonsson, C.A. Brock, D. W. Toohey, L. M. Avallone, D. Baumgardner, J. E. Dye, L. R. Poole, D. C. Woods, R. J. DeCoursey, M. Osborn, M. C. Pitts, K. K. Kelly, K. R. Chan, G. V. Ferry, M. Loewenstein, J. R. Podolske, and A. Weaver. 1993. "In Situ Observations of Aerosol and Chlorine Monoxide After the 1991 Eruption of Mount Pinatubo: Effect of Reactions on Sulfate Aerosol." *Science* 261: 1140–1143.

WMO (World Meteorological Organization). 1985. *Atmospheric Ozone 1985: Assessment of our Understanding of the Processes Controlling its Present Distribution and Change*, Geneva, Switzerland: WMO.

WMO (World Meteorological Organization). 1988. *Report of the International Ozone Trends Panel*, volumes I and II.

WMO (World Meteorological Organization). 1989. *Scientific Assessment of Stratospheric Ozone: 1989*, volumes I and II, Geneva, Switzerland: WMO.

WMO (World Meteorological Organization). 1990. Intergovernmental Panel on Climate Change, *Scientific Assessment of Climate Change*, Geneva, Switzerland: WMO, June.

WMO (World Meteorological Organization). 1991. *Scientific Assessment of Ozone Depletion: 1991*. WMO.

WMO (World Meteorological Organization). 1994. *Scientific Assessment of Ozone Depletion: 1994 Executive Summary*. WMO.

Wofsy, S. C., M. B. McElroy, and N. D. Sze. 1975. "Freon Consumption: Implications for Atmospheric Ozone." *Science* 187: 535–537.

Wofsy, S. C., M. J. Molina, R. J. Salawitch, L. E. Fox, and M. B. McElroy. 1988. "Interactions Between HCl, NO_X and H_2O Ice in the Antarctic Stratosphere: Implications for Ozone." *Journal of Geophysical Research* 93, no. D3: 2442–2450.

Wofsy, S. C., R. J. Salawitch, J. H. Yatteau, and M. B. McElroy. 1990. "Condensation of HNO3 on Falling Ice Particles: Mechanism for Denitrification of the Polar Stratosphere." *Geophysical Research Letters* 17, no. 4, March supplement: 449–452.

Wolff, E. W., and R. Mulvaney. 1991. "Reactions on Sulphuric Acid Aerosol and on Polar Stratospheric Clouds in the Antarctic Stratosphere." *Geophysical Research Letters* 18, no. 6: 1007–1010, (June).

World Bank. 1993. "China Acts to Protect Ozone Layer." Press Release, World Bank, November 30.

Worrest, R. 1991. Testimony before U.S. Senate Committee on Commerce, Science and Transportation, November 15, 1991.

WRI (World Resources Institute). 1992. *World Resources 1992–93: A Guide to the Global Environment*. Oxford University Press.

Wuebbles, D. J. 1981. "The Relative Efficiency of a Number of Halocarbons for Destroying Stratospheric Ozone." Report UCID-18924, Lawrence Livermore National Laboratory.

Wuebbles, D. J. 1983. "Chlorocarbon Emission Scenarios: Potential Impact on Stratospheric Ozone." *Journal of Geophysical Research* 88, no. C2: 1433–1443.

Wyngaarden, James B., Lloyd H. Smith, and J. Claude Bennett, editors. 1992. *Cecil Textbook of Medicine*, 19th edition. Saunders.

Yannuzzi, L. A., Y. L. Fisher, J. S. Slakter, and A. Krueger. 1989. "Solar Retinopathy: A Photobiologic and Geophysical Analysis." *Retina* 9, no. 1: 28–43.

Yoon, C. K. 1994. "Thinning Implicated in Decline of Frogs and Toads." *New York Times*, March 1.

York, R. 1993. "Zero ODP Blowing Agents for Polyurethane Foams." In Proceedings of the 1993 International CFC and Halon Alternatives Conference, Washington.

Zurer, P. 1992. "Industry, Consumers Prepare for Compliance with Pending CFC Ban." *Chemical and Engineering News*, June 22: 7–13.

Zurer, P. 1993. "Ozone Depletion's Recurring Surprises Challenge Atmospheric Scientists." *Chemical and Engineering News*, May 24, pp. 8–18.

Index